Atmospheric Boundary Layer Flows

Atmospheric Boundary Layer Flows

Their Structure and Measurement

J. C. KAIMAL
J. J. FINNIGAN

New York Oxford
OXFORD UNIVERSITY PRESS
1994

Oxford University Press

Oxford New York Toronto
Delhi Bombay Calcutta Madras Karachi
Kuala Lumpur Singapore Hong Kong Tokyo
Nairobi Dar es Salaam Cape Town
Melbourne Auckland Madrid

and associated companies in
Berlin Ibadan

Published by Oxford University Press, Inc.,
200 Madison Avenue, New York, New York 10016

Library of Congress Cataloging-in-Publication Data
Kaimal, J. C.
Atmospheric boundary layer flows : their structure and measurement
J.C. Kaimal, J.J. Finnigan.
p. cm. Includes bibliographical references and index.
ISBN 0-19-506239-6
1. Boundary layer (Meteorology)
I. Finnigan, J. J. (John J.) II. Title.
QC880.4.B65K35 1994 551.5—dc20
92-20792

1 3 5 7 9 8 6 4 2

Printed in the United States of America
on acid-free paper

PREFACE

The boundary layer is the lowest 1–2 km of the atmosphere, the region most directly influenced by the exchange of momentum, heat, and water vapor at the earth's surface. Turbulent motions on time scales of an hour or less dominate the flow in this region, transporting atmospheric properties both horizontally and vertically through its depth. The mean properties of the flow in this layer—the wind speed, temperature, and humidity—experience their sharpest gradients in the first 50–100 m, appropriately called the surface layer. Turbulent exchange in this shallow layer controls the exchange of heat, mass, and momentum at the surface and thereby the state of the whole boundary layer. It is hardly surprising we should have a lively curiosity about this region.

What is surprising is that systematic scientific investigation of the surface layer and its overlying boundary layer is confined to the last 30 years. Almost all of that inquiry, furthermore, has been devoted to the simplest situation, the boundary layer over flat, open land. During this period, micrometeorologists in the main concentrated on atmospheric behavior above simple surfaces such as grass or bare soil, while agricultural meteorologists grappled with the complexities of turbulent exchange within plant canopies. With some isolated exceptions, the extension to more complex situations such as flat but heterogeneous surfaces and hills and valleys has been pursued only during the last 15 years by a small but growing group of researchers.

The reasons for the slow accumulation of understanding in this area are twofold: difficulties in measurement and difficulties in handling the mathematical description. The root cause of both is turbulence—the chaotic, essentially unpredictable variations in the atmospheric properties. Recent advances in sensing technology and computing power have greatly enhanced our ability to tackle both problems.

The earliest attempts to make quantitative measurements of atmospheric turbulence can be traced back (as can so much in modern fluid dynamics) to G. I. Taylor. In 1917, Taylor deduced such fundamental features of turbulent flow as anisotropy and the eddy flux of momentum from observations of the gyrations of balloons and universally jointed wind vanes. Other isolated essays into turbulence measurement appeared sporadically. In 1930, for example, we have a description by F. J. Scrase of his measurements over Salisbury Plain in England with a small propeller-type air meter, a swinging plate anemometer, and a bivane attached to a recording pen. Fascinating as these early attempts were, useful measurements of turbulence in the atmosphere had to await the 1960s.

At that time availabili y of sonic anemometers and other fast-response sensors was matched by modern digital computers that made it possible to handle the vast amounts of data such instruments generated. Early descriptions of boundary layer turbulence had relied entirely on data from slow-response instruments such as cup anemometers and simple thermometers that could furnish only the average values of wind speed and temperature. Even major scientific field expeditions such as the 1967 Australian Wangara experiment made no turbulence measurements and had to use empirical formulas borrowed from aerodynamics to account for the influence of turbulence on their results. The 1968 Kansas experiment was the first to deploy fast-response sensors on a large scale and to process the data in real time with a mobile computer system.

The mathematical difficulties were just as intractable as the practical ones. The governing equations of atmospheric flow are the Navier-Stokes equations—nonlinear, second-order, partial differential equations—attributes that spell difficulty when it comes to obtaining practically useful solutions. Furthermore, while the Navier-Stokes equations describe the instantaneous state of the wind, what we actually measure, or work with, are the averaged properties of the turbulent wind. We can obtain the equations governing the behavior of these averaged properties using the Navier-Stokes equations as a starting point, but exact solutions for these equations are just as difficult to obtain as those of the Navier-Stokes equations themselves. Since solutions of some kind are what we need to make predictions about the way the boundary layer will evolve or to deduce some physically important property such as momentum flux from easily measured variables, we abandon mathematical rigor and adopt an engineering approach. In pursuing this approach, the digital computer has been an indispensable tool. The equations and formulas you will encounter throughout this book reflect this blend of mathematical analysis, physical insight often based on experience, and unabashed empiricism. This is the state of the art.

In Chapter 1, we describe the surface layer within the context of a boundary layer over flat ground. The emphasis is on similarity relations, formulas that describe universal relationships between measurable quantities, but we shall try to draw, as we progress, a physical picture of the properties that give rise to these formulas. We have chosen those forms that best illustrate the process we are describing. Our objective is an uncluttered narrative that conveys a sense of the order and symmetry, idealized perhaps, that emerges in atmospheric turbulence measurements.

In Chapter 2, we discuss surface layer structure again in the context of a boundary layer over flat ground but from the point of view of spectral behavior. Here the functional forms complement those of the time-averaged properties in

Chapter 1. Together, they set the stage for our discussions of flow over more complicated terrain. Chapter 3 draws us into the first level of complexity, flat terrain with uniform vegetation. We investigate the interaction between the wind and the plant canopy and how the canopy modifies the universal relations described in the first two chapters.

In Chapter 4, we abandon horizontal homogeneity and the one-dimensional simplicity that it brought to surface layer formulations to discuss changes in surface conditions and the advective boundary layers they generate. Chapter 5 takes us from the relatively well understood flat earth to the complications of hills and valleys. New coordinate systems and new methods of analysis are needed to describe the flow over such a landscape. Chapters 6 and 7 deal with the practical aspects of boundary layer investigation. A knowledge of instrumentation options and analysis strategies is crucial to the planning of any experiment, but even users of processed data benefit from the knowledge of how the data were obtained and treated.

This book brings under one cover a description of the basic flow structures that are observed in the atmospheric boundary layer and the techniques available for studying them. The topics reflect our interest and perceptions, although we have striven for objectivity. As is necessary in a basic text, many interesting topics have received a cursory treatment; internal gravity waves and mesoscale features such as sea breezes and valley winds are examples that spring immediately to mind. We hope you will be sufficiently motivated by the material included here to pursue such fascinating topics through more specialized sources.

ACKNOWLEDGMENTS

Many friends and colleagues helped in the preparation of this book. Akiva Yaglom, Donald Lenschow, Thomas Horst, Leif Kristensen, Christopher Fairall, Michael Raupach, Peter Coppin, Frank Bradley, Dick Lang, and Yaping Shao lead the long list of reviewers who patiently read various chapters and offered comments we found invaluable. Our special thanks to Leif Kristensen and Christopher Fairall for reading the final manuscript in its entirety and identifying the inconsistencies and errors we missed. The painstaking efforts of Loys Balsley and Irene Page in preparing the manuscript are most gratefully acknowledged, as are those of Alison Aragon, Ron Youngs, and Greg Heath in preparing the figures. Our thanks also to Steven Clifford for the initial idea for the book and his continued support of the project and to Joyce Berry, our editor, for her patience and confidence in its outcome.

Hamilton, N.Y. J. C. K.
Canberra (Australia) J. J. F.

CONTENTS

Atmospheric Boundary Layer Flows

1

FLOW OVER FLAT UNIFORM TERRAIN

We start with the simplest of boundary layers, that over an infinite flat surface. Here we can assume the flow to be horizontally homogeneous. Its statistical properties are independent of horizontal position; they vary only with height and time. This assumption of *horizontal homogeneity* is essential in a first approach to understanding a process already complicated by such factors as the earth's rotation, diurnal and spatial variations in surface heating, changing weather conditions, and the coexistence of convective and shear-generated turbulence. It allows us to ignore partial derivatives of mean quantities along the horizontal axes (the advection terms) in the governing equations. Only ocean surfaces come close to the idealized infinite surface. Over land we settle for surfaces that are locally homogeneous, flat plains with short uniform vegetation, where the advection terms are small enough to be negligible.

If, in addition to horizontal homogeneity, we can assume *stationarity*, that the statistical properties of the flow do not change with time, the time derivatives in the governing equations vanish as well. This condition cannot be realized in its strict sense because of the long-term variabilities in the atmosphere. But for most applications we can treat the process as a sequence of steady states. The major simplification it permits is the introduction of time averages that represent the properties of the process and not those of the averaging time.

These two conditions clear the way for us to apply fluid dynamical theories and empirical laws developed from wind tunnel studies to the atmosphere's boundary layer. We can see why micrometeorologists in the 1950s and 1960s scoured the countryside for flat uniform sites. The experiments over the plains of Nebraska, Kansas, and Minnesota (USA), Kerang and Hay (Australia), and Tsimliansk (USSR) gave us the first inklings of universal behavior in boundary layer turbulence.

1.1 Boundary layer structure and depth

Our concept of the atmospheric boundary layer (ABL) and its vertical extent has changed significantly over the last few decades. Sutton (1953) separated the boundary layer into two regions:

1. A *surface layer* region 50–100 m deep of approximately constant (in the vertical) shearing stress, where the flow is insensitive to the earth's rotation and the wind structure is determined primarily by surface friction and the vertical gradient of temperature
2. A region above that layer extending to a height of 500–1000 m, where the shearing stress is variable and the wind structure is influenced by surface friction, temperature gradient, and the earth's rotation

The two-layer concept roughly parallels that of the inner and outer regions in laboratory shear flows, although the true extent of the similarity in the scaling laws in each of those regions to laboratory flow was not known at that time. (Velocity fluctuations scale with distance from the surface in the inner layer and with the thickness of the whole boundary layer in the outer layer.) Above these two layers is the free atmosphere, where the flow is in near-geostrophic balance and no longer influenced by surface friction.

Viewed as the height at which the wind first attains geostrophic balance, the ABL depth z_h can be expressed as (Sutton, 1953)

$$z_h = \pi \left(\frac{2K_m}{f} \right)^{1/2}, \tag{1.1}$$

where K_m is the exchange coefficient for momentum (discussed in later sections) and f is the Coriolis parameter representing the effect of the earth's rotation. f is equal to $2\pi\Omega \sin\phi, \Omega$ being the earth's rotation rate and ϕ the latitude. This definition was based on the assumption that K_m is constant with height, an unfeasible assumption, and was soon abandoned.

Theoretical considerations (Tennekes, 1982) pointed to a boundary layer depth that is proportional to u_*/f, where u_* is the friction velocity (also discussed later) representing the wind stress at the surface

$$z_h = C \left(\frac{u_*}{f} \right), \tag{1.2}$$

where C is an empirical constant. A value of $C = 0.25$ yields boundary layer heights close to observed daytime heights. In reality, the frictional and Coriolis effects are often overwhelmed by external factors such as subsidence and lapse rate evolution, so u_*/f cannot be depended on to provide reliable estimates of z_h.

Both of the above definitions assume a neutrally stable boundary layer, one in which air parcels displaced up and down adiabatically maintain exactly the same density as the surrounding air and thus experience no net buoyancy forces. The temperature stratification that produces this state, referred to in meteorology as the adiabatic lapse rate ($\simeq 1°C/100$ m drop in temperature with height), is often transitory. The more persistent states are the daytime convectively mixed boundary layer, where the temperature drops more rapidly with height than the adiabatic rate in the lower regions of the layer and displaced parcels accelerate vertically away from their original positions, and the nighttime stable boundary layer, where the temperature drops less rapidly with height and displaced parcels return to their original positions. The upper limits for these two states define the depths of the daytime and nighttime boundary layers. They are represented by

$$z_h = \begin{cases} z_i \ (daytime) \\ h \ \ (nighttime) \end{cases}, \tag{1.3}$$

where z_i is the height of the base of the inversion layer capping the daytime boundary layer, and h is the height of the top of the nighttime ground-based turbulent layer identified in different studies as the top of the surface inversion, as the level of the wind speed maximum that develops sometimes within and sometimes above the inversion, or simply as the top of the strongest near-surface echo layer in sodar (see Chapter 6) records.

In the convective boundary layer (CBL), the capping inversion acts as a lid damping out vertical motions. The steepest gradients in the mean wind speed, wind direction, and temperature occur in the first 10% or so of the CBL; in the upper 90% of the CBL strong convective mixing smoothes out almost all vertical variations in the mean profiles (Fig. 1.1). At height z_i, the mean profiles begin

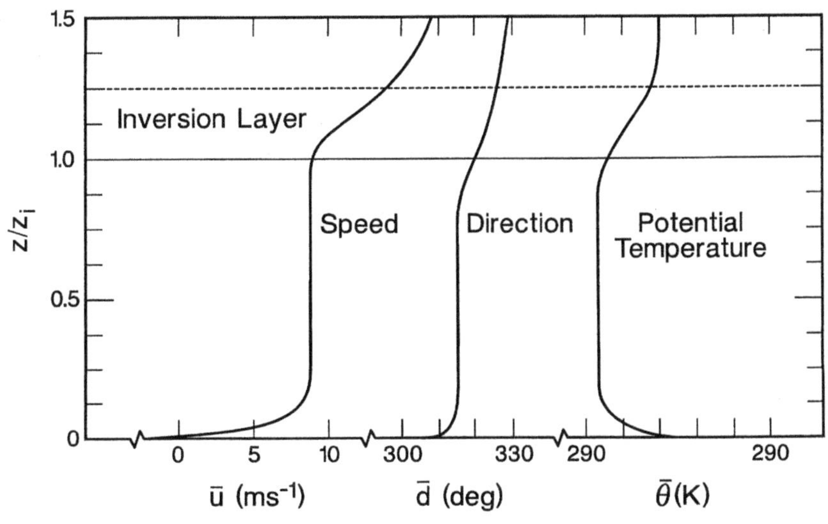

Fig. 1.1. Mean vertical profiles of wind speed, wind direction, and potential temperature in the convective boundary layer.

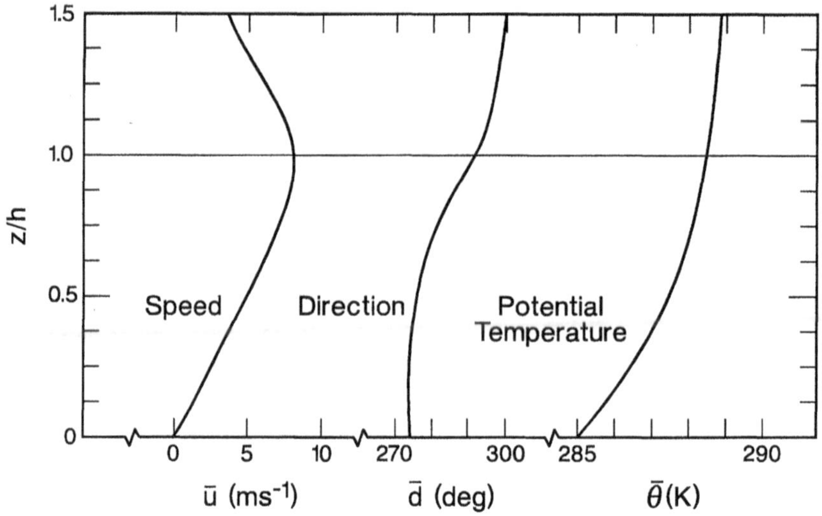

FIG. 1.2. Mean vertical profiles of wind speed, wind direction, and potential temperature in the stable boundary layer.

to depart from their near-constant CBL values to approach the free atmosphere values above.

The top of the stable boundary layer (SBL) is not as sharply defined as the top of the CBL. Turbulence levels decrease gradually with height, damped out by a combination of static stability and diminished wind shear. Although there is general agreement that h should be the height where turbulence drops to negligible levels, say 5% of surface values, its correspondence to the height of the surface inversion and/or that of the wind maximum depends very much on the history of the SBL evolution. All these significant features are represented in the wind and temperature profiles of Fig. 1.2 without the stipulation that the inversion top and the wind maximum simultaneously coincide with h . The wind maximum may be above or below the inversion top, but the sodar echoes usually stop at the lower of the two. (There is no temperature turbulence at the wind maximum and therefore no echoes from that height.) This height is probably the best estimate of h we can obtain in the absence of direct turbulence measurements.

In both Figs. 1.1 and 1.2 we have plotted temperature in the form of potential temperature θ, the temperature an air parcel with absolute temperature T and pressure p would have if brought adiabatically to the pressure at the 1000-mb (millibar) level. To the first order we can write $\theta = T + (g/c_p)\Delta z$, where g is the acceleration due to gravity, c_p is the specific heat at constant pressure, and Δz is the height difference from the 1000-mb level.

We then have

$$\frac{\partial \theta}{\partial z} = \left(\frac{\partial T}{\partial z} + \frac{g}{c_p} \right) . \tag{1.4}$$

g/c_p is the adiabatic lapse rate mentioned earlier. At 1000 mb, T and θ are, by definition, equal. In neutrally stratified air, $\partial\theta/\partial z = 0$; unstable and stable layers are easily recognized in this representation by their respectively negative ($\partial\theta/\partial z < 0$) and positive ($\partial\theta/\partial z > 0$) slopes.

When significant moisture is present, such as over water surfaces or in plant canopies, the contribution of moisture to the air density is included by defining a potential virtual temperature

$$\theta_v \equiv \left(\frac{\theta}{T}\right) T_v, \tag{1.5}$$

where $T_v = T(1 + 0.61q)$. T_v, the virtual temperature, is the temperature at which dry air has the same density as moist air at the same pressure and q is the specific humidity. Over land the difference between θ and θ_v is small and often ignored. We follow the same practice here, recognizing that wherever buoyancy effects are involved, θ_v would replace θ.

1.2 States of the ABL

Although the overland ABL is evolving continuously in response to the heating and cooling of the earth's surface, it does have distinct states (Figs. 1.1 and 1.2) that can be described in fairly simple terms. Following sunrise, a convective layer develops near the ground (Fig. 1.3) as the sun's rays heat the surface and, indirectly, the air in contact with it. The convective layer grows through the morning, reaching

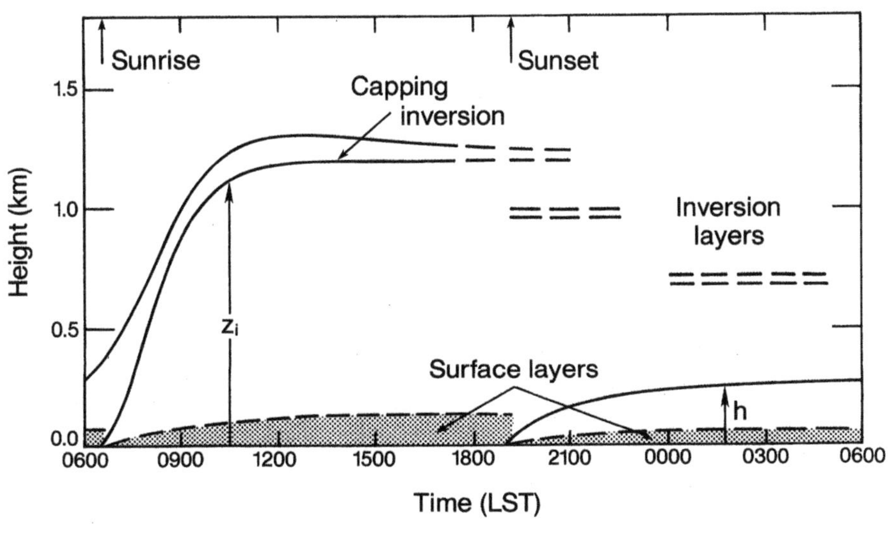

Fig. 1.3. Evolution of the convective and stable boundary layers in response to surface heating and cooling. The time indicated is Local Standard Time (LST).

heights of 1–2 km by midafternoon. The surface inversion that prevailed before sunrise evolves as the capping layer, rising with the convective layer as it grows upward. Often, the capping inversion will stay at about the same height through much of the afternoon, its strength and persistence determined largely by prior lapse rate history. A strong morning inversion followed by subsidence through the day invariably produces a well-defined capping inversion. Close to the ground, an unstable surface layer approximating Sutton's (1953) definition is clearly in evidence during the day, as seen in Fig. 1.3.

Within the CBL, convection is carried out by small plumes that merge into larger plumes (thermals) that transport the heat all the way to the capping inversion base, as depicted in Fig. 1.4. The figure also shows the entrainment process by which air from above the inversion base is drawn into the CBL in the regions of sinking motion. Occasionally, the more energetic thermals penetrate the capping inversion, but for the most part, they simply distort the upper interface of the CBL, making it appear highly convoluted. Horizontal roll vortices and dust devils also appear when conditions favor their development, adding to the modes available for convective mixing in the CBL.

With the approach of sunset, the capping inversion weakens and becomes patchy as one or more shallow inversion layers form below it (Fig. 1.5). At this time there is a rapid collapse of turbulent motions in the boundary layer as the buoyant plumes that maintain them lose their energy source near the surface where the ground is cooling quickly from radiative heat loss to space. The air immediately above the surface cools and mixes progressively upward through the action of turbulence generated by wind shear. The inversion that begins to form at the surface grows steadily to a depth of 100–200 m by midnight. A shallow, stable

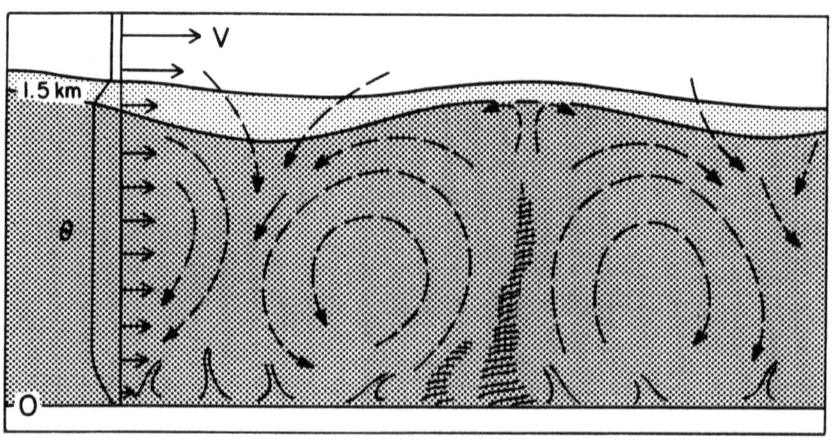

Fig. 1.4. Schematic of convective boundary layer circulation and entrainment of air through the capping inversion (from Wyngaard, 1990).

surface layer can be discerned here as well (Fig. 1.3), where the flow remains sensitive to the presence of the boundary below. Flow within the SBL is characterized by strong wind shear, small eddies, and occasional wave activity (Fig. 1.5); stratified inversion layers may appear and disappear above the SBL. The wind maximum (or low-level jet, as it is often called) may develop within the inversion or above it, as mentioned earlier. Gravity-driven drainage winds usually have their maxima within the inversion layer, but jets that form in response to the inertial oscillations induced by the earth's rotation develop considerably above the surface inversion and move downward as the night progresses.

The SBL seldom, if ever, attains the same state of equilibrium reached by the CBL. The mean wind and temperature profiles continue to evolve through the night and even small slopes in the terrain generate drainage winds. Internal gravity waves may grow and propagate. Usually by sunrise, these disturbances settle down and the stage is set for the development of a new CBL.

This sequence of CBL and SBL evolution is typical for land surfaces in the midlatitudes. In the tropics, the base of the trade wind inversion ($\sim$ 1.5 km) serves as the upper boundary for all vertical transfers, including (almost always) cumulus convection. When clouds are present, the subcloud layer (which extends about 100 m below the cloud base) acts much like the capping inversion in Fig. 1.1. The base of this subcloud layer ($\sim$ 600 m), where θ_v begins to increase to its cloud base values, is taken as the boundary layer top and used as the scaling height for the convection processes below (LeMone, 1978). Over oceans within the tropics, temperature gradients tend to be very close to the adiabatic value, and convection is supported mainly by latent heat from moisture flux at the surface. Nevertheless, the marine CBLs show surprising similarities to CBLs over land (LeMone, 1978).

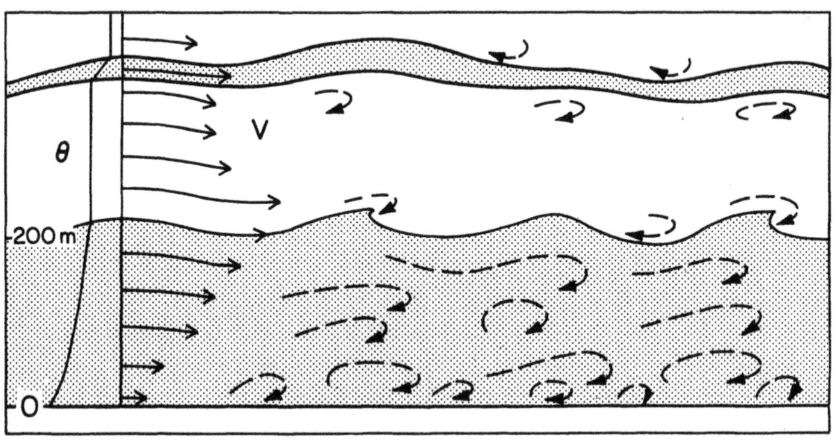

FIG. 1.5. Schematic of stable boundary layer flow showing eddy structure, waves, and elevated inversion layer (from Wyngaard, 1990).

1.3 Surface layer

1.3.1 *Profiles and fluxes*

The idealizations embodied in our assumptions of horizontal homogeneity and stationarity are more easily realized in the surface layer than elsewhere in the ABL. In this layer, we also have strong vertical gradients controlling the transfer of momentum, mass, and heat through it. They raise the possibility that simple laws, analogous to those that govern molecular diffusion in laminar flows, might apply here. The relationship between the vertical fluxes of these properties and their time-averaged vertical gradients was perceived by investigators in the 1950s and 1960s as the central problem in micrometeorology. Our perception is broader today, but the vast amount of data collected and analyzed over the years did yield relationships we now take for granted and use as a baseline for testing surface layer behavior over more complicated terrain.

Carrying the analogy of molecular diffusion to turbulent transport in the atmosphere, we can express the fluxes of momentum (τ), heat (H), and moisture (E) in terms of gradients in the vertical direction z:

$$\tau = K_m \rho \frac{\partial \overline{u}}{\partial z}, \tag{1.6}$$

$$H = -K_h \rho c_p \frac{\partial \overline{\theta}}{\partial z}, \tag{1.7}$$

$$E = -K_q \rho \frac{\partial \overline{q}}{\partial z}, \tag{1.8}$$

where K_m, K_h, and K_q are the turbulent exchange coefficients for momentum, heat, and moisture; $\overline{u}$, $\overline{\theta}$, and $\overline{q}$ are the mean streamwise wind component, mean potential temperature, and mean specific humidity; and ρ is the density of air. (Note that τ, in contrast to H and E, is defined positive downward.) The exchange coefficients are the counterparts of viscosity, conductivity, and diffusivity in laminar flow, but their magnitudes are typically three orders of magnitude larger. The extension of that concept to turbulent exchange in the atmosphere is known as K-theory.

The downward flux of momentum has dimensions of stress (force/area). It is a vector quantity with components τ_x and τ_y. In the surface layer over uniform terrain, the decrease in stress with height is small enough to be negligible. Constancy of stress implies constancy also in the wind direction with height, and this allows us to treat the momentum transport as a one-dimensional problem in (1.6). Because the flow is horizontally homogeneous, we can assume that $\overline{u}$, like the scalars $\overline{\theta}$ and $\overline{q}$, varies only in the vertical. Aligning the x direction with the wind direction, we have $\tau_x = \tau_0$ and $\tau_y = 0$ in the surface layer.

We now define a reference velocity u_*, known as the friction velocity, to represent the effect of wind stress τ_0 on the ground,

$$\tau_0 = \rho u_*^2.$$

This velocity, which varies with the nature of the surface and the magnitude of the wind, has emerged as an important scaling parameter in surface layer studies.

1.3.2 *Logarithmic wind profile*

The introduction of u_* leads to a simple formulation for wind profiles in a neutrally stratified atmosphere. We can represent K_m, which has dimensions of length $\times$ velocity, as the product of the two surface layer scaling parameters with those dimensions, height z and u_*:

$$K_m = k\, u_*\, z, \tag{1.9}$$

where k is the constant of proportionality. Substituting for τ and K_m in (1.6), we get

$$\frac{\partial \overline{u}}{\partial z} = \frac{u_*}{kz}. \tag{1.10}$$

Integration of (1.10) yields the classical logarithmic wind profile

$$\overline{u}(z) = \frac{u_*}{k} \ln\left(\frac{z}{z_0}\right). \tag{1.11}$$

Here z_0, the constant of integration (known as the roughness length), is the height at which $\overline{u}$, extrapolated downward, vanishes. Laboratory measurements indicate z_0 to be approximately 1/30 the height of the roughness elements, but over natural flat terrain the factor is often observed to be larger (Yaglom, 1979). The constant k is known as the von Karman constant, with reported values ranging from 0.35 to 0.43. For this constant we adopt a value of 0.4, the one most generally accepted in boundary layer work today.

The logarithmic wind profile is strictly valid only for the neutral atmosphere. As the atmosphere becomes more stable or unstable, the profile departs from being logarithmic (Fig. 1.6). For most applications very close to the ground (i.e., $z < 10$ m), however, the wind profile can be assumed to be almost always logarithmic. This allows us to use the relationships in (1.6), (1.7), and (1.8) to compute the fluxes from gradient measurement of properties in the layer.

Thus, if $\overline{u}_1$ and $\overline{u}_2$ are the mean winds measured at two heights z_1 and z_2, we have

$$\overline{u}_2 - \overline{u}_1 = \frac{u_*}{k} \ln\left(\frac{z_2}{z_1}\right). \tag{1.12}$$

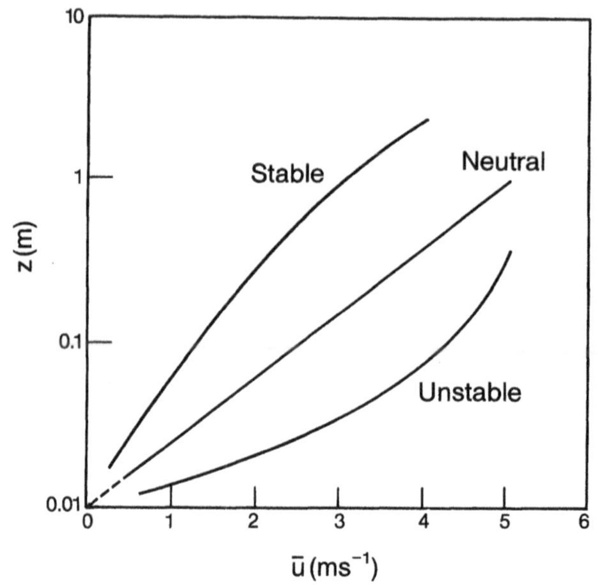

FIG. 1.6. Wind profile in stable, neutral, and unstable air.

We can now write

$$K_m = kz_r\,u_* = \frac{k^2 z_r}{\ln(z_2/z_1)}(\overline{u}_2 - \overline{u}_1), \tag{1.13}$$

substituting for u_* from (1.12). If the reference height z_r is the geometric mean of z_1 and z_2, that is, $(z_1 z_2)^{1/2}$, then at $z = z_r$, the mean wind speed is equal to $(\overline{u}_1 + \overline{u}_2)/2$.

Now, if we assume that close to the ground $K_m = K_h = K_q$ (Panofsky and Dutton, 1984), all we need for computing the vertical fluxes are the mean values of θ and q measured at the same heights at which $\overline{u}_1$ and $\overline{u}_2$ are observed. If we can also assume that their profiles are approximately logarithmic, we can, using (1.10) and (1.12), express the gradient of the mean of any such variable ζ as

$$\frac{\partial \overline{\zeta}}{\partial z} \simeq \frac{\overline{\zeta}_2 - \overline{\zeta}_1}{z_r \ln(z_2/z_1)}. \tag{1.14}$$

The fluxes of momentum, heat, and moisture assume the form

$$\frac{\tau}{\rho} = \left(\frac{k}{\ln(z_2/z_1)}\right)^2 (\overline{u}_2 - \overline{u}_1)^2, \tag{1.15}$$

$$\frac{H}{\rho c_p} = -\left(\frac{k}{\ln(z_2/z_1)}\right)^2 (\overline{u}_2 - \overline{u}_1)(\overline{\theta}_2 - \overline{\theta}_1), \tag{1.16}$$

$$\frac{E}{\rho} = -\left(\frac{k}{\ln(z_2/z_1)}\right)^2 (\overline{u}_2 - \overline{u}_1)(\overline{q}_2 - \overline{q}_1). \tag{1.17}$$

Estimation of fluxes from profiles is subject to several constraints. We require the fluxes to be constant with height, which implies uniform flat terrain and observing heights well within the surface layer. The logarithmic law applies strictly to neutral surface layers; the estimates become less reliable as the stability departs from neutral and the profiles deviate more and more from logarithmic.

1.3.3 *Eddy correlation and fluxes*

Earlier in this chapter we defined the fluxes of momentum, heat, and moisture in terms of the mean vertical gradients of those properties. Because these fluxes are almost entirely the result of turbulent mixing, we should be able to define them more directly in terms of the turbulent (or eddy) components of velocities and of the properties being transferred. Mean flux across any plane implies correlation between the wind component normal to that plane and the entity in question. In the covariance between the two, we have a direct measure of the flux across the plane.

We start with the definition of velocity components u, v, and w along right-handed coordinate axes x, y, and z, respectively. These velocities and scalars such as θ and q can each be separated into a mean component denoted by an overbar and an eddy component denoted by a prime:

$$\begin{aligned} u &= \overline{u} + u' \\ v &= \overline{v} + v' \\ w &= \overline{w} + w' \\ \theta &= \overline{\theta} + \theta' \\ q &= \overline{q} + q'. \end{aligned}$$

By definition, we have $\overline{u'} = \overline{v'} = \overline{w'} = \overline{\theta'} = \overline{q'} = 0$. If the x axis is defined in the direction of the mean flow, $\overline{v} = \overline{w} = 0$; only $\overline{u} \neq 0$. Over a flat, level, homogeneous surface, we take x and y to be horizontal and z to be vertical and positive upwards.

The fluxes can now be written as

$$\tau = -\rho\overline{u'w'} \tag{1.18}$$

$$H = \rho c_p \overline{w'\theta'} \tag{1.19}$$

$$E = \rho\overline{w'q'}. \tag{1.20}$$

The covariance terms on the right are unambiguous measures of the fluxes, requiring no assumptions about the mixing properties of turbulence. To realize them, however, we need fast-response sensors that can faithfully follow the fluctuations in all the variables and digital sampling at rates (10–20 Hz depending on height), much faster than needed for mean profile measurements. (The mean products of the velocity fluctuations are referred to, in general, as Reynolds stresses, and $-\rho\overline{u'w'}$, in particular, as the Reynolds shear stress.)

1.3.4 *Measures of stability*

We have seen how the ABL responds to changes in stability brought about by the heating and cooling of the ground. As we search for universal behavior in the boundary layer, we need to establish a proper measure of stability in the surface layer. The most widely used indicator of stability in laboratory and early atmospheric work is the gradient Richardson number Ri, a nondimensional parameter representing the relative importance of buoyancy and shear in producing turbulence:

$$Ri = \frac{(g/\overline{\theta})(\partial\overline{\theta}/\partial z)}{(\partial\overline{u}/\partial z)^2}, \tag{1.21}$$

where g is the acceleration due to gravity (the term $g/\overline{\theta}$ is referred to as the buoyancy parameter[1]). The advantage in using the Richardson number is that it contains gradients of mean quantities that are easy to measure. Ri is positive for stable stratification, negative for unstable stratification, and zero for neutral stratification. Above a critical value Ri_c (= 0.25 for inviscid flow), the flow undergoes a transition from turbulent flow to laminar. Between values 0 and Ri_c, turbulence is almost entirely mechanical in origin, generated by wind shear; at $Ri < 0$, we have a mixture of both mechanical and convective turbulence. The disadvantage in using Ri is that it is an unknown function of height and therefore not an appropriate stability parameter for characterizing surface layer structure.

Two other forms of the Richardson number, R_f the flux Richardson number and R_b the bulk Richardson number, are sometimes used. R_f is the ratio of the rate of production of turbulence by buoyancy to that by shear:

$$R_f = \frac{(g/\overline{\theta})\overline{w'\theta'}}{\overline{u'w'}(\partial\,\overline{u}/\partial z)} \simeq \frac{K_h}{K_m}\,Ri. \tag{1.22}$$

In (1.22) the numerator represents the production (or destruction) of turbulence by thermal stratification, and the denominator represents the production by turbulent stress working against the mean strain. We will be discussing these concepts

[1] Strictly speaking, the buoyancy parameter $g/\overline{\theta}$ should be $g/T_r(z)$, where $T_r(z)$ is a hydrostatic adiabatic reference state defined by $\partial T_r/\partial z = -g/c_p$ and $T_r(z_r) = \overline{T}(z_r)$ (z_r being the reference height). For practical purposes, in the ABL we can write $g/T_r(z_r) \simeq g/\overline{T}(z) \simeq g/\overline{\theta}(z)$.

later in the context of the turbulent kinetic energy budget in the surface layer. Although R_f properly characterizes the effect of flow stratification on turbulence, it is awkward to use in practice, being a mixture of eddy correlations and mean gradients, and is therefore seldom used.

R_b is a useful indicator of stability close to the ground, particularly under conditions of low wind shear when Ri becomes undependable:

$$R_b = \frac{(g/\overline{\theta})(\overline{\theta}_z - \overline{\theta}_0)/2}{(\overline{u}_z/z)^2}, \tag{1.23}$$

where $\overline{\theta}_z$ and $\overline{\theta}_0$ denote mean potential temperatures at height z and the surface, respectively, and $\overline{u}_z$ denotes the mean wind at height z. In light winds, the denominator in this expression behaves more predictably than in (1.21). Its application is limited because it is only a crude approximation of Ri and requires nomograms (Panofsky and Dutton, 1984) relating it to Ri and $\ln(z/z_0)$ to render it useful.

The stability parameter now recognized by boundary layer meteorologists as appropriate for the surface layer is the ratio of height z to the scaling length L (the Obukhov length), which can be expressed as

$$\frac{z}{L} = -\frac{(g/\overline{\theta})(\overline{w'\theta'})_0}{u_*^3/kz} \tag{1.24}$$

where $(\overline{w'\theta'})_0$ denotes temperature flux at the surface. The ratio z/L is essentially the same as R_f with substitutions in the denominator of u_*/kz for $\partial u/\partial z$ and u_*^2 for $-(\overline{u'w'})$. The negative sign is introduced so z/L has the same sign as Ri and R_f, negative when the atmosphere is unstable and positive when it is stable. This quantity is more useful than Ri because L can be assumed constant through the surface layer. It also implies that within the surface layer [where this parameter forms the basis for the similarity hypothesis proposed by Monin and Obukhov (1954)], the effects of varying height and stability (as represented by L) are interchangeable.

1.3.5 *Monin-Obukhov similarity*

Empirical evidence from field experiments conducted over flat terrain points to a surface layer where the structure of turbulence is determined by a few key parameters as proposed by Monin and Obukhov (1954). These are the height z, the buoyancy parameter $g/\overline{\theta}$, the kinematic surface stress τ_0/ρ, and the surface temperature flux $H_0/\rho c_p$. According to the Monin-Obukhov (M-O) hypothesis, various atmospheric parameters and statistics, such as gradients, variances, and covariances, when normalized by appropriate powers of the scaling velocity u_* and the scaling temperature T_* (as defined below), become universal functions of z/L:

$$u_* = \left[-(\overline{u'w'})_0\right]^{1/2} \tag{1.25a}$$

$$T_* = \frac{-(\overline{w'\theta'})_0}{u_*}. \tag{1.25b}$$

Although defined strictly in terms of fluxes at the surface, u_* and T_* are evaluated, in practice, from measurements of the fluxes at some convenient height within the surface layer where their vertical variations can be assumed negligible with height, a reasonable assumption for $z \lesssim |L|$ (Haugen et al., 1971). The important nondimensional forms to emerge in the surface layer are

$$\phi_m = (kz/u_*)(\partial\overline{u}/\partial z) \quad \text{wind shear,} \tag{1.26}$$

$$\phi_h = (kz/T_*)(\partial\overline{\theta}/\partial z) \quad \text{thermal stratification,} \tag{1.27}$$

$$\phi_w = \sigma_w/u_* \quad \text{variability in } w, \tag{1.28}$$

$$\phi_\theta = \sigma_\theta/|T_*| \quad \text{variability in } \theta, \tag{1.29}$$

$$\phi_\epsilon = kz\epsilon/u_*^3 \quad \text{dissipation of turbulent kinetic energy,} \tag{1.30}$$

where σ_w and σ_θ are the standard deviations of w and θ, and ϵ is the rate of dissipation of turbulent kinetic energy. We introduce ϵ here because of its relevance to discussions of the turbulent kinetic energy budget later in this chapter. Its relationship to velocity spectra will be discussed in Chapter 2.

All the above functions follow M-O scaling with surprisingly small scatter, as evident in the plots of the Kansas data (Businger et al., 1971; Wyngaard and Coté, 1971). The following formulations are essentially the Kansas results, reexamined and refined through comparison with other observations (Dyer, 1974; Högström, 1988):

$$\phi_m = \begin{cases} (1 + 16\,|z/L|)^{-1/4}, & -2 \le z/L \le 0 \\ (1 + 5\,z/L), & 0 \le z/L \le 1 \end{cases} \tag{1.31}$$

$$\phi_h = \begin{cases} (1 + 16\,|z/L|)^{-1/2}, & -2 \le z/L \le 0 \\ (1 + 5\,z/L), & 0 \le z/L \le 1 \end{cases} \tag{1.32}$$

$$\phi_w = \begin{cases} 1.25(1 + 3\,|z/L|)^{1/3}, & -2 \le z/L \le 0 \\ 1.25(1 + 0.2\,z/L), & 0 \le z/L \le 1 \end{cases} \tag{1.33}$$

$$\phi_\theta = \begin{cases} 2(1 + 9.5\,|z/L|)^{-1/3}, & -2 \le z/L \le 0 \\ 2(1 + 0.5\,z/L)^{-1}, & 0 \le z/L \le 1 \end{cases} \tag{1.34}$$

$$\phi_\epsilon = \begin{cases} (1 + 0.5\,|z/L|^{2/3})^{3/2}, & -2 \le z/L \le 0 \\ (1 + 5\,z/L), & 0 \le z/L \le 1. \end{cases} \tag{1.35}$$

The forms of these functions, plotted in Fig. 1.7, cannot be predicted from dimensional analysis. They have to be determined empirically from field experiments. Conspicuously absent from this list are forms for the standard deviations

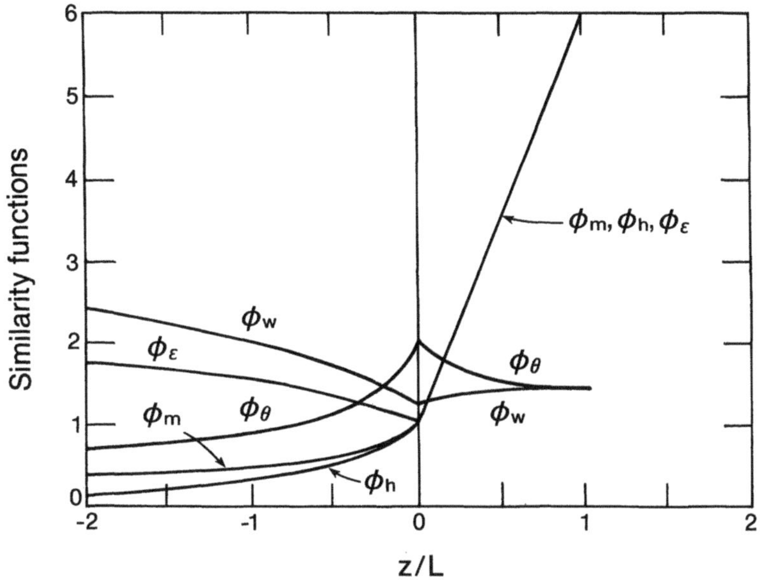

FIG. 1.7. Similarity functions in the surface layer.

of u and v. They do not follow M-O similarity for reasons that will become clear as we examine the behavior of their spectra in Chapter 2. We can, however, expect scalar properties such as q and their fluxes to follow M-O similarity and to have functional forms resembling those of ϕ_h and ϕ_θ when normalized by the appropriate scaling parameter, in this case q_*, defined as $-(\overline{w'q'})_0/u_*$.

In our idealizations of these wind profiles, we have forced $\phi_m = \phi_h = \phi_\epsilon = 1$ at $z/L = 0$. Not all experiments support the assumption of unity for neutral ϕ_m and ϕ_h (Businger et al., 1971) but many do (Dyer, 1974; Dyer and Bradley, 1982). The forms we have chosen for ϕ_m and ϕ_h are known as the Businger-Dyer relations (Businger, 1988; Panofsky and Dutton, 1984). We have ϕ_ϵ also approaching unity at neutral stability because of the expected balance between shear production and viscous dissipation of turbulence in the absence of any buoyant production and transport:

$$\begin{aligned} \epsilon &= -(\overline{u'w'})\,\frac{\partial \overline{u}}{\partial z} \\ &= u_*^2\left(\frac{u_*}{kz}\right). \end{aligned} \tag{1.36}$$

Having chosen a form for ϕ_m, we can derive an expression for the diabatic (nonneutral) wind profile. (The mean profiles $\overline{u}/u_*, \overline{\theta}/T_*$, etc., are not functions of z/L; they have to be derived from the gradients.) Integration of ϕ_m yields a modified logarithmic profile (Panofsky and Dutton, 1984):

$$\overline{u}(z) = \frac{u_*}{k}\,[\ln(z/z_0) - \psi_m], \tag{1.37}$$

where the diabatic term ψ_m is the integral of $(1 - \phi_m)/(z/L)$ over limits z_0/L to z/L. ψ_m is thus a function of z/L. The expression for ψ_m is simple for the stable surface layer ($-5\,z/L$, for our choice of ϕ_m) but more cumbersome for the unstable layer. ψ_m values corresponding to ϕ_m in unstable air (1.31) are presented in Table 1.1.

The basic definitions for ϕ_m and ϕ_h in (1.26) and (1.27) lead to the following identities:

$$K_h/K_m \equiv \phi_m/\phi_h, \tag{1.38a}$$

$$Ri \equiv (z/L)(\phi_h/\phi_m^2), \tag{1.38b}$$

$$R_f \equiv (z/L)/\phi_m. \tag{1.38c}$$

Table 1.1. ϕ_m and ψ_m in Unstable Air

z/L	ϕ_m	ψ_m
0.00	1.000	0.000
−0.10	0.788	0.284
−0.20	0.699	0.461
−0.30	0.644	0.595
−0.40	0.606	0.702
−0.50	0.577	0.793
−0.60	0.554	0.873
−0.70	0.535	0.943
−0.80	0.519	1.006
−0.90	0.505	1.063
−1.00	0.493	1.116
−1.10	0.482	1.165
−1.20	0.472	1.211
−1.30	0.463	1.253
−1.40	0.455	1.293
−1.50	0.447	1.331
−1.60	0.440	1.367
−1.70	0.434	1.401
−1.80	0.428	1.434
−1.90	0.422	1.465
−2.00	0.417	1.495

From (1.31) and (1.32) we have

$$\phi_h = \begin{cases} \phi_m^2, & -2 \le z/L \le 0 \\ \phi_m, & 0 \le z/L \le 1, \end{cases} \tag{1.39}$$

which leads to simple expressions for K_h/K_m and Ri.

$$K_h/K_m = \begin{cases} (1 + 16\,|z/L|)^{1/4}, & -2 \le z/L \le 0 \\ 1, & 0 \le z/L \le 1 \end{cases} \tag{1.40}$$

$$Ri = \begin{cases} z/L, & -2 \le z/L \le 0 \\ (z/L)(1 + 5\,z/L)^{-1}, & 0 \le z/L \le 1 \end{cases} \tag{1.41}$$

$$R_f = \begin{cases} (z/L)(1 + 16\,|z/L|)^{1/4}, & -2 \le z/L \le 0 \\ (z/L)(1 + 5\,z/L)^{-1}, & 0 \le z/L \le 1. \end{cases} \tag{1.42}$$

The equivalence of Ri and z/L in unstable air is a consequence of the Businger-Dyer formulation. (Ri is derived purely from the mean profiles and z/L from the eddy fluxes, so their close functional dependence has been viewed as a test of M-O similarity in the surface layer.) Plots of K_h/K_m, Ri, and R_f as functions of z/L are presented in Fig. 1.8.

The correlation coefficients for uw, $w\theta$, and $u\theta$ (stable only) also follow M-O similarity. Figure 1.9, based on a replot of the Kansas data, shows how these functions behave in the range $-2 < z/L < 1$. Their forms may be approximated

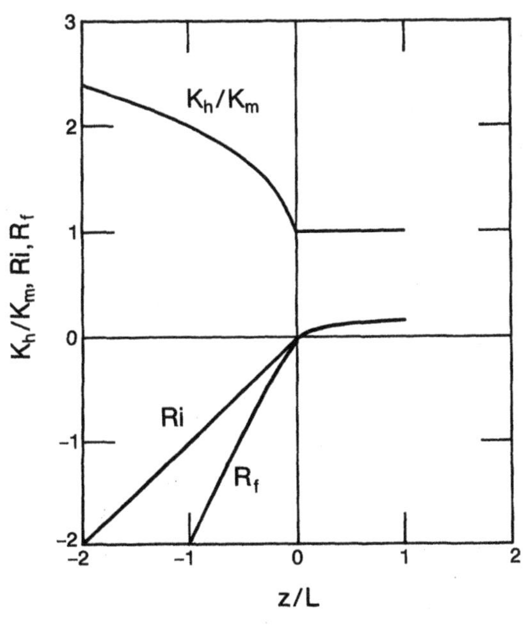

Fig. 1.8. Surface layer parameters K_h/K_m, R_i, and R_f shown as functions of z/L.

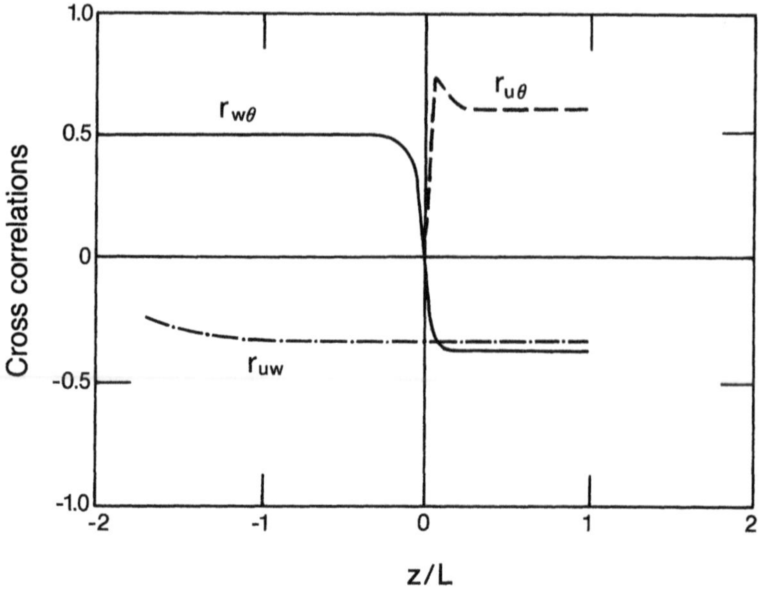

FIG. 1.9. Cross-correlation coefficients for uw, $w\theta$ and $u\theta$ in the surface layer.

as follows:

$$\begin{aligned} r_{uw} &= \overline{u'w'}/\sigma_u\sigma_w \\ &\simeq -0.35, (-1 < z/L < 1) \end{aligned} \tag{1.43}$$

$$r_{w\theta} \simeq \begin{cases} 0.5, & (-2 < z/L < 0) \\ 0, & (z/L = 0) \\ -0.4, & (0 < z/L < 1) \end{cases} \tag{1.44}$$

$$r_{u\theta} \simeq \begin{cases} 0, & (z/L = 0) \\ 0.6, & (0 < z/L < 1). \end{cases} \tag{1.45}$$

$r_{u\theta}$ is not M-O similar in unstable air. Its value is largest near neutral (where it might range anywhere from -0.2 to -0.6) but drops with increasing height, sometimes sharply and at other times more gradually.

Our similarity functions are shown as extending through $-2 \leq z/L \leq +1$. In reality, the curves begin to approach their free convection asymptotic forms at $z/L = -1$ (Wyngaard and Coté, 1971) as they become insensitive to surface stress τ_0/ρ. On the stable side, the similarity functions are well behaved only up to $z/L = +1$. Beyond that, the functions no longer respond to surface layer scaling as the layers decouple from the surface. On the basis of their conformity to M-O similarity, we define the surface layer depth at $z = |L|$ for both stable and unstable conditions. Some investigators (e.g., Holtslag and Nieuwstadt, 1986), however, prefer to define it as a fraction of the boundary layer depth, as $0.1z_i$ for the unstable surface layer and $0.1h$ for the stable surface layer.

1.4 Above the surface layer

1.4.1 *Convective mixed layer*

Above the unstable surface layer, roughly above $0.1z_i$, a different scaling law takes effect. This is the region where large-scale convective motions driven by surface heating force a near-uniform distribution of $\overline{u}$ and $\overline{\theta}$ and where the flow is clearly insensitive to the presence of the ground below. Neither τ_0/ρ nor z exerts any influence on the turbulence here. The controlling parameters are surface temperature flux $H_0/\rho c_p$ and CBL depth z_i.

In this *mixed layer* we have a new scaling velocity w_* and a new scaling temperature θ_*; mixed layer similarity implies that the statistical properties of turbulence, nondimensionalized with w_* and θ_*, should be functions only of z/z_i:

$$w_* = \left[\frac{g}{\overline{\theta}}\,(\overline{w'\theta'})_0\, z_i\right]^{1/3}, \tag{1.46a}$$

$$\theta_* = \frac{(\overline{w'\theta'})_0}{w_*}. \tag{1.46b}$$

The validity of this scaling law was established by Willis and Deardorff (1974) in their laboratory simulations of the CBL and was later confirmed for the atmosphere by Kaimal et al. (1976), Caughey and Palmer (1979), and Lenschow et al. (1980). Their findings are remarkably consistent. Plotted in Fig. 1.10 are the profiles of various turbulence parameters in their mixed layer forms. The curves are based on data from the Minnesota experiment (Kaimal et al., 1976), augmented by more measurements from Ashchurch (Caughey and Palmer, 1979); both were obtained with the same turbulence instrumentation, attached to the same type of balloon (Caughey and Palmer, 1979). The data points (not shown) do not exhibit the tight grouping observed in the surface layer plots of Businger et al. (1971) and Wyngaard and Coté (1971), but their distribution indicates a clear functional dependence that seems to hold in the mixed layer. Most studies of mixed layer turbulence describe results in terms of variances (not standard deviations, as in the surface layer), a practice we continue here.

The horizontal velocity variances, which did not follow M-O similarity in the surface layer, exhibit a rare consistency when plotted in the mixed layer framework. They show a slight maximum near the surface but otherwise stay relatively invariant with height over much of the mixed layer. In this layer we have, from Caughey and Palmer (1979),

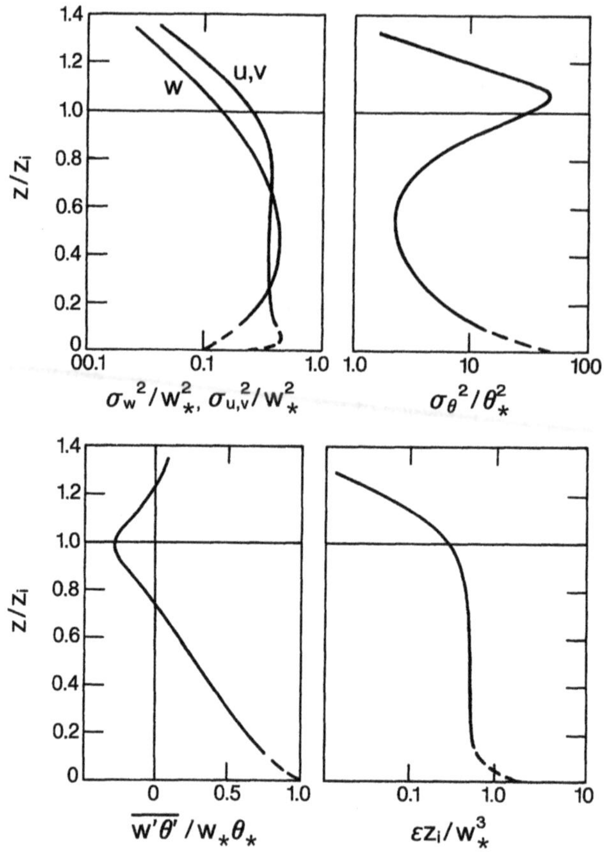

Fig. 1.10. Vertical profiles of nondimensional mixed layer parameters (after Caughey and Palmer, 1979). Dashed portions of curves imply extrapolations through the surface layer.

$$\frac{\sigma_u^2}{w_*^2} = \frac{\sigma_v^2}{w_*^2} \simeq 0.35. \tag{1.47}$$

Laboratory measurements of Adrian et al. (1986) are also in good agreement with (1.47).

The fluctuations in the horizontal, being controlled primarily by the large convective motions in the CBL, follow mixed layer scaling almost to the surface. A formulation in terms of u_*^2 is possible, but not as a function of z/L. Recognizing that

$$\frac{w_*}{u_*} = \left(\frac{z_i}{k|L|}\right)^{1/3}, \tag{1.48}$$

we can write

$$\frac{\sigma^2_{u,v}}{u^2_*} = \left(\frac{\sigma^2_{u,v}}{w^2_*}\right) \times \left(\frac{w^2_*}{u^2_*}\right)$$
$$= 0.35k^{-2/3}\left(\frac{z_i}{|L|}\right)^{2/3}$$
$$\simeq 0.6\left(\frac{z_i}{|L|}\right)^{2/3}. \tag{1.49}$$

The surface layer form for $\sigma^2_{u,v}/u^2_*$, proposed by Panofsky et al. (1977), in fact approaches (1.49) in the unstable limit from its value of 4 at neutral.

$$\frac{\sigma^2_{u,v}}{u^2_*} = 4 + 0.6\left(\frac{z_i}{|L|}\right)^{2/3}. \tag{1.50}$$

(z_i/L is often used as an indicator of the relative importance of convection over shear in the boundary layer, whereas z_i/w_* is viewed in diffusion studies as the characteristic time scale for mixing in the CBL.) The functional forms for σ^2_w/w^2_* as well as for $\overline{w'^3}/w^3_*$ established by aircraft (Lenschow et al., 1980) and tower (Hunt et al., 1988) measurements are

$$\frac{\sigma^2_w}{w^2_*} = 1.8\left(\frac{z}{z_i}\right)^{2/3}\left(1 - 0.8\frac{z}{z_i}\right)^2, \tag{1.51}$$
$$\frac{\overline{w'^3}}{w^3_*} = 1.0\left(\frac{z}{z_i}\right)\left(1 - 0.7\frac{z}{z_i}\right)^3. \tag{1.52}$$

In the mixed layer, Lenschow et al. (1980) found $\sigma^2_\theta/\theta^2_*$ and σ^2_q/Q^2_* (scaling parameter Q_* for humidity defined the same way as θ_*) to have similar profiles and to follow the same vertical distribution observed by Kaimal et al. (1976) in the range $0.1z_i < z < 0.5z_i$:

$$\frac{\sigma^2_\theta}{\theta^2_*} = \frac{\sigma^2_q}{Q^2_*} \approx 1.8\left(\frac{z}{z_i}\right)^{-2/3}. \tag{1.53}$$

This is, in fact, the asymptotic form for σ_θ/T_* in (1.34). Above $0.5z_i$, both the temperature and humidity variances rise to roughly their surface layer values, reach a maximum at $z \approx z_i$, then drop sharply above z_i, as in Fig. 1.10. This increase near z_i is associated with entrainment of warmer air downward from above the ABL and therefore does not follow mixed layer scaling. [For a discussion of the entrainment process, see Deardorff (1978).]

Also shown in Fig. 1.10 are the normalized profiles of $\overline{w'\theta'}$ and of ϵ, the dissipation rate of turbulent kinetic energy. $\overline{w'\theta'}$ decreases steadily with height, crossing zero at about $0.8z_i$ to reach a minimum value at about z_i. The strong negative heat flux at z_i is another manifestation of the entrainment process mentioned above. ϵ, on the other hand, remains fairly constant up to $z \simeq z_i$ and falls off rapidly above that height.

1.4.2 *Convective matching layer*

The layer of the CBL between $-L$ and $0.1z_i$ serves as a *matching layer* between the surface and mixed layers (Panofsky, 1978). The gradients of wind and temperature are not negligible in this layer; and although u_* is no longer significant, height z, not z_i, is the controlling length scale. It is considered a matching layer because relationships expressed as functions of z/L in the surface layer translate to equivalent forms for the mixed layer, as functions of z/z_i, with only an adjustment for the presence of k in the former. The scaling velocity u_f and scaling temperature T_f (Wyngaard et al., 1971) for this layer are

$$u_f = \left[\frac{g}{\theta}\,(\overline{w'\theta'})_0\, z\right]^{1/3}, \tag{1.54a}$$

$$T_f = \frac{(\overline{w'\theta'})_0}{u_f}. \tag{1.54b}$$

Turbulence parameters nondimensionalized with u_f and T_f should be constants in this framework. (The height dependence does not disappear; it is simply included in the scaling parameters.) Both σ_w/u_f and σ_θ/T_f assume values about 1.34 as we might expect from (1.51) and (1.53). The equivalence of dimensionless terms in the three frameworks in this layer is illustrated for the case of σ_w:

$$\frac{\sigma_w}{u_f} = 1.34, \tag{1.55}$$

$$\frac{\sigma_w}{w_*} = 1.34\left(\frac{z}{z_i}\right)^{1/3}, \tag{1.56}$$

$$\frac{\sigma_w}{u_*} = 1.34k^{-1.3}\left(\frac{z}{|L|}\right)^{1/3} \simeq 1.8\left(\frac{z}{|L|}\right)^{1/3}. \tag{1.57}$$

Note that (1.57) is the asymptotic form for the surface layer expression in (1.33). The scaling laws in the CBL, summarized in Table 1.2, show how the controlling parameters evolve with height as one progresses upward through the matching layer.

Although the presence of the matching layer is satisfying from a theoretical standpoint, its importance to modeling and other applications is minimal and therefore ignored in many discussions of CBL structure. Note that in CBLs where

Table 1.2. Evolution of Scaling Laws with Height in the CBL

Height Range	Layer Characteristic	Controlling Parameters	Scaling Parameters	Functional Dependence
$0.1z_i \leqslant z \leqslant z_i$	Mixed layer	z_i, $H_0/\rho c_p$, $g/\overline{\theta}$	w_*, θ_*	$f\,(z/z_i)$
$-L \leqslant z \leqslant 0.1z_i$	Matching layer	z, $H_0/\rho c_p$, $g/\overline{\theta}$	u_f, T_f	Constant
$z \leqslant -L$	Surface layer	z, L, τ_0/ρ, $H_0/\rho c_p$, $g/\overline{\theta}$	u_*, T_*	$f\,(z/L)$

$z_i < -10L$ (i.e., where the influence of buoyancy is weak compared with shear), the matching layer may be replaced by a deep layer where all three length scales, z, L, and z_i, exert control. Here the scaling laws are less clearly defined. This situation is typical for the marine boundary layer where we find $L \approx -200$ m and $z_i \approx 600$ m. On a convective day over land, in contrast, we often have $L \approx -20$ m and $z_i \approx 1$ km.

1.4.3 *Stable outer layer*

At heights greater than L in the SBL, in the layer we loosely refer to as the stable outer layer, we again encounter departures from surface layer scaling. Here it reflects a growing tendency for the turbulence to be decoupled from the surface. Height becomes irrelevant, hence the term z-less stratification (Wyngaard and Coté, 1972) is used to describe the flow in this layer. Caughey et al. (1979) showed how the stable outer layer statistics from the Minnesota experiment responded to scaling with h. Their h corresponded to the height at which heat flux dropped to 5% of its surface value. For those runs, that height was fairly close to the inversion top and to the jet maximum. All fluctuation statistics, normalized by u_* and T_*, dropped monotonically to 0 as z approached h (see Fig. 1.11).

Nieuwstadt (1984) and Sorbjan (1986) present arguments for a *local M-O similarity*, where u_*, T_*, and L are replaced by local values of those scaling parameters computed from $-\overline{u'w'}$ and $\overline{w'\theta'}$ measured at level z. When non-dimensionalized with these local scaling parameters, the variances and fluxes in the SBL display a certain invariance with height. This implies a strong internal consistency in the relationships among the gradients, variances, and fluxes. The local stability parameter z/L_z is essentially the flux Richardson number R_f discussed earlier. When z/L becomes much larger than unity, this local scaling breaks down giving way to buoyant oscillations in the form of gravity waves at frequencies equal to or less than the Brunt-Väisälä frequency N, where

$$N = \left[\frac{g}{\overline{\theta}}\left(\frac{\partial\overline{\theta}}{\partial z}\right)\right]^{1/2}, \tag{1.58}$$

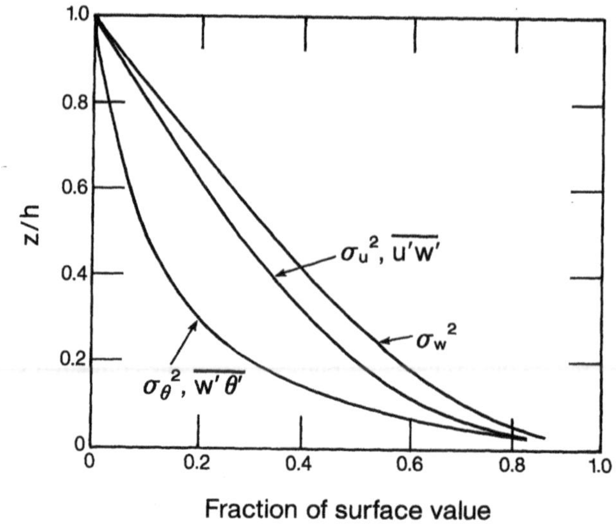

FIG. 1.11. Vertical profiles of variances and fluxes in the stable boundary layer normalized by their near-surface (4 m) values (after Caughey et al., 1979).

in units of radians/second. (Note the discussion of the buoyancy parameter $g/\overline{\theta}$ in Section 1.3.) These waves, with periods $(2\pi/N)$ ranging from roughly a minute to an hour, appear prominently in time traces of vertical velocity, temperature, and pressure. They also appear as separate peaks at the low-frequency end of the power spectra of those variables (Caughey, 1977; Finnigan et al., 1984).

1.5 Budgets of turbulent kinetic energy, temperature variance, and fluxes

An examination of the terms in the kinetic energy equation offers some important insights into the nature of turbulent production and destruction in the boundary layer. The behavior of the terms within the surface layer is especially interesting; there the terms are larger and thus easier to measure than in the layers above. In its simplified form, assuming horizontal homogeneity, the equation for the time rate of change of kinetic energy $D\overline{e}/dt$ can be written as in Panofsky and Dutton (1984):

$$\frac{D\overline{e}}{Dt} = \frac{\partial \overline{e}}{\partial t} = -\underset{\text{I}}{(\overline{u'w'})\left(\frac{\partial \overline{u}}{\partial z}\right)} + \underset{\text{II}}{\frac{g}{\overline{\theta}}(\overline{w'\theta'})} - \underset{\text{III}}{\frac{1}{\rho}\frac{\partial}{\partial z}(\overline{w'p'})} - \underset{\text{IV}}{\frac{\partial}{\partial z}(\overline{ew'})} - \underset{\text{V}}{\epsilon}, \tag{1.59}$$

where $e = (1/2)(u'^2 + v'^2 + w'^2)$, p is the atmospheric pressure, and ϵ is the dissipation rate of turbulent kinetic energy. The Eulerian derivative, $D/Dt =$

$\partial/\partial t + \overline{u}(\partial/\partial x) + \overline{v}(\partial/\partial y) + \overline{w}(\partial/\partial z)$, represents the time rate of change following a fluid element.

In (1.59), the first two terms on the right are the production rates of turbulence by shear and buoyancy. The *shear production* term (I) is always positive, a source for turbulence, whereas the *buoyant production* term (II) can be either a source or a sink, depending on the sign of $\overline{w'\theta'}$. Both terms can be measured directly. The third and fourth terms are transport terms and represent the import or export of turbulent kinetic energy by the turbulence itself. The third term, the *pressure transport* term (III), has never been accurately measured but is inferred indirectly as a residual after measurement of all other terms. The fourth term, the *turbulent transport* term (IV), is the rate at which turbulent kinetic energy is exported or imported by velocity fluctuations. This term can also be measured directly. The last term, the *viscous dissipation* term (V), is the rate at which turbulent kinetic energy is converted into internal energy (i.e., heat) by doing work against viscous stresses. As we will see in the next chapter, this term can be calculated from the spectral energy at the high-frequency end of the turbulence spectrum.

Assuming steady state (i.e., $\partial\overline{e}/\partial t = 0$) and multiplying all terms on the right by kz/u_*^3, we find that (1.59) in the surface layer reduces to a relationship between dimensionless terms, some of which are the similarity functions plotted in Fig. 1.7:

$$\phi_m - \frac{z}{L} - \phi_t - \phi_\epsilon + I = 0, \tag{1.60}$$

where I is the imbalance presumed to be the pressure term. Both ϕ_m and ϕ_ϵ are known functions of z/L. In the Kansas data, Wyngaard and Coté (1971) found the transport term

$$\phi_t = \begin{cases} -z/L, & z/L \leq 0 \\ 0, & z/L \geq 0. \end{cases} \tag{1.61}$$

In every case, in the unstable surface layer ϕ_t was positive, indicating turbulent transport is exporting energy upwards at the same rate as its production by buoyancy. The various terms in (1.60) are plotted as functions of z/L in Fig. 1.12. It appears that the imbalance term I in (1.60) is the difference between shear production and dissipation in the unstable surface layer. In the stable surface layer, with $\phi_m \simeq \phi_\epsilon$ and $\phi_t \simeq 0$, the imbalance term appears to be of the same order as the buoyant production term z/L, which is a loss in the budget. For the imbalance to be zero, ϕ_ϵ will have to be much smaller ($\approx 1 + 4\,z/L$) than observations indicate. The question of energy balance in the stable layer remains unresolved.

Energy budgets above the convective surface layer are much harder to measure accurately since all terms diminish in magnitude. Shear production is very small because $\partial\overline{u}/\partial z \approx 0$; buoyant production changes from being a source

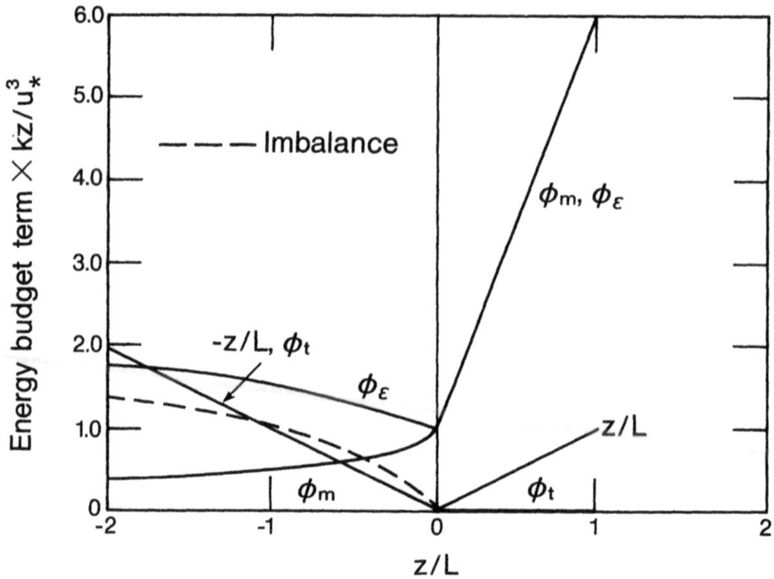

Fig. 1.12. Similarity functions representing terms in the turbulent kinetic energy budget in (1.59).

below $0.8z_i$ to a sink above, reflecting the change in the sign in the $\overline{w'\theta'}$ profile at that height. Turbulent transport changes from a loss to a gain above $0.4\,z_i$. The findings of Caughey and Wyngaard (1979) were inconclusive, but the aircraft measurements of Lenschow et al. (1980) confirm our expectations of a balance among buoyant destruction, turbulent transport, and dissipation through much of the mixed layer. In the SBL, turbulence levels are too small to measure with any degree of precision (see Fig. 1.11) and are often complicated by the presence of gravity waves. Energy budgets under the resulting unsteady conditions have been presented by Finnigan and Einaudi (1981), and Finnigan et al. (1984).

The budget of temperature variance (actually $\overline{\theta'^2}/2$ for consistency with the form for kinetic energy) was examined by Wyngaard and Coté (1971). The equation reduces to

$$\frac{D}{Dt}\left(\frac{\overline{\theta'^2}}{2}\right) = \frac{1}{2}\,\frac{\partial(\overline{\theta'^2})}{\partial t} = -\underbrace{(\overline{w'\theta'})\left(\frac{\partial\overline{\theta}}{\partial z}\right)}_{\text{I}} - \underbrace{\frac{1}{2}\,\frac{\partial}{\partial z}\,(\overline{w'\theta'^2})}_{\text{II}} - \underbrace{N_\theta}_{\text{III}}. \tag{1.62}$$

The terms on the right are the *production rate* (I), the *turbulent transpost rate* (II), and the *dissipation rate* (III) of half the temperature variance. N_θ can be calculated from the temperature spectrum if ϵ is known, as we will see in Chapter 2. All other terms can be measured. In the Kansas data, Wyngaard and Coté (1972) found the transport term to be negligible in the surface layer compared with the other two. Assuming steady state in the surface layer and multiplying the two remaining terms by $kz/u_*T_*^2$, and recognizing $\overline{w'\theta'} = -u_*T_*$, we have the nondimensional form of the budget

$$\phi_h - \phi_N \simeq 0, \tag{1.63}$$

which implies that most of the temperature variance produced is destroyed in the surface layer. (ϕ_N is the dimensionless temperature dissipation rate analogous to ϕ_ϵ.) Above the surface layer the terms in the temperature variance budget become extremely small, erratic, and difficult to evaluate.

The budgets of momentum flux and heat flux over a horizontally homogeneous surface reduce, respectively, to (Wyngaard et al., 1971)

$$\frac{D}{Dt}(\overline{u'w'}) = \frac{\partial}{\partial t}(\overline{u'w'}) = \underbrace{-\overline{(w')^2}\left(\frac{\partial \overline{u}}{\partial z}\right)}_{\text{I}} + \underbrace{\frac{g}{\overline{\theta}}(\overline{u'\theta'})}_{\text{II}} \underbrace{-\frac{\partial}{\partial z}(\overline{u'w'^2})}_{\text{III}}$$

$$\underbrace{-\frac{1}{\rho}\left[\overline{w'\left(\frac{\partial p'}{\partial x}\right)} + \overline{u'\left(\frac{\partial p'}{\partial z}\right)}\right]}_{\text{IV}} \tag{1.64}$$

and

$$\frac{D}{Dt}(\overline{w'\theta'}) = \frac{\partial}{\partial t}(\overline{w'\theta'}) = \underbrace{-\overline{(w')^2}\left(\frac{\partial \overline{\theta}}{\partial z}\right)}_{\text{I}} + \underbrace{\frac{g}{\overline{\theta}}\overline{(\theta')^2}}_{\text{II}} \underbrace{-\frac{\partial}{\partial z}(\overline{w'^2\theta'})}_{\text{III}}$$

$$\underbrace{-\frac{1}{\rho}\left[\overline{\theta'\left(\frac{\partial p'}{\partial z}\right)}\right]}_{\text{IV}}. \tag{1.65}$$

In both equations, the terms on the right represent *shear* or *gradient production* (I), *buoyant production* (II), *turbulent transport* (III), and *pressure destruction* (IV). Destruction of both these fluxes is brought about primarily by pressure forces and only minimally by viscous dissipation. This accounts for the absence of the dissipation terms in (1.64) and (1.65). Wyngaard et al. (1971) found the transport terms in both equations to be small in the surface layer compared to the shear and buoyant production terms. If we assume steady state, we are left with a balance between the pressure destruction term, which was not measured, and the two production terms.

References

Adrian, R. J., R. T. D. S. Ferreira and T. Boberg, 1986: Turbulent thermal convection in wide horizontal fluid layers. In *Experiments in Fluids,* Springer-Verlag, Berlin, 121–141.

Businger, J. A., 1988: A note on the Businger-Dyer profiles. *Bound.-Layer Meteor., 42,* 145–151.

Businger, J. A., J. C. Wyngaard, Y. Izumi and E. F. Bradley, 1971: Flux-profile relationships in the atmospheric surface layer. *J. Atmos. Sci., 28,* 181–189.

Caughey, S. J., 1977: Boundary-layer turbulence spectra in stable conditions. *Bound.-Layer Meteor., 11,* 3–14.

Caughey, S. J., 1982: Observed characteristics of the atmospheric boundary layer. In *Atmospheric Turbulence and Air Pollution Modeling* (F. T. M. Nieuwstadt and H. van Dop, Eds.). Reidel, Hingham, MA, 107–158.

Caughey, S. J., and S. G. Palmer, 1979: Some aspects of turbulence structure through the depth of the convective boundary layer. *Quart. J. Roy. Meteor. Soc., 105,* 811–827.

Caughey, S. J., and J. C. Wyngaard, 1979: The turbulent kinetic energy budget in convective conditions. *Quart. J. Roy. Meteor. Soc., 105,* 231–239.

Caughey, S. J., J. C. Wyngaard and J. C. Kaimal, 1979: Turbulence in the evolving stable layer. *J. Atmos. Sci., 36,* 1041–1052.

Deardorff, J. W., 1978: Progress in understanding entrainment at the top of a mixed layer. In *Workshop on Planetary Boundary Layer* (J. C. Wyngaard, Ed.). American Meteorological Society, Boston, MA, 36–66.

Dyer, A. J., 1974: A review of flux-profile relationships. *Bound.-Layer Meteor., 7,* 363–372.

Dyer, A. J., and E. F. Bradley, 1982: An alternative analysis of flux-gradient relationships at the 1976 ITCE. *Bound.-Layer Meteor., 22,* 3–19.

Finnigan, J. J., and F. Einaudi, 1981: The interaction between an internal gravity wave and the planetary boundary layer. Part II: Effect of the wave on the turbulence structure. *Quart. J. Roy. Meteor. Soc., 107,* 807–832.

Finnigan, J. J., F. Einaudi, and D. Fua, 1984: The interaction between an internal gravity wave and turbulence in the stably stratified nocturnal boundary layer. *J. Atmos. Sci., 41,* 2409–2436.

Haugen, D. A., J. C. Kaimal and E. F. Bradley, 1971: An experimental study of Reynolds stress and heat flux in the atmospheric surface layer. *Quart. J. Roy. Meteor. Soc., 97,* 168–180.

Högström, U., 1988: Nondimensional wind and temperature profiles. *Bound.-Layer Meteor., 42,* 55–78.

Holtslag, A. A. M., and F. T. M. Nieuwstadt, 1986: Scaling the atmospheric boundary layer. *Bound.-Layer Meteor., 36,* 201–209.

Hunt, J. C. R., J. C. Kaimal and J. E. Gaynor, 1988: Eddy structure in the convective boundary layer—new measurements and new concepts. *Quart. J. Roy. Meteor. Soc., 114,* 827–858.

Kaimal, J. C., J. C. Wyngaard, D. A. Haugen, O. R. Coté, Y. Izumi, S. J. Caughey and C. J. Readings, 1976: Turbulence structure in the convective boundary layer. *J. Atmos. Sci., 33,* 2152–2169.

LeMone, M. A., 1978: The marine boundary layer. In *Workshop on Planetary Boundary Layer* (J. C. Wyngaard, Ed.). American Meteorological Society, Boston, MA, 182–231.

Lenschow, D. H., J. C. Wyngaard and W. T. Pennell, 1980: Mean fields and second moment budgets in a baroclinic convective boundary layer. *J. Atmos. Sci., 37,* 1313–1326.

Monin, A. S., and A. M. Obukhov, 1954: Basic laws of turbulent mixing in the ground layer of the atmosphere. *Trans. Geophys. Inst. Akad. Nauk USSR, 151*, 163–187.

Nieuwstadt, F. T. M., 1984: Turbulence structure of the stable nocturnal boundary layer. *J. Atmos. Sci., 41*, 2202–2216.

Panofsky, H. A., 1978: Matching in the convective planetary boundary layer. *J. Atmos. Sci., 35*, 272–276.

Panofsky, H. A., and J. A. Dutton, 1984: *Atmospheric Turbulence*. Wiley-Interscience, New York, 397 pp.

Panofsky, H. A., H. Tennekes, D. H. Lenschow and J. C. Wyngaard, 1977: The characteristics of turbulent velocity components in the surface layer under convective conditions. *Bound.-Layer Meteor., 11*, 355–361.

Sorbjan, Z., 1986: On similarity in the atmospheric boundary layer. *Bound.-Layer Meteor., 34*, 377–397.

Sutton, O. G., 1953: *Micrometeorology*. McGraw-Hill, New York, 333 pp.

Tennekes, H., 1982: Similarity relations, scaling laws and spectral dynamics. In *Atmospheric Turbulence and Air Pollution Modeling* (F. T. M. Nieuwstadt and H. van Dop, Eds.). Reidel, Hingham, MA, 37–68.

Willis, G. E., and J. W. Deardorff, 1974: A laboratory model of the unstable planetary boundary layer. *J. Atmos. Sci., 31*, 1297–1307.

Wyngaard, J. C., 1990: Scalar fluxes in the planetary boundary layer—theory, modeling and measurement. *Bound.-Layer Meteor., 50*, 49–75.

Wyngaard, J. C., and O. R. Coté, 1971: The budgets of turbulent kinetic energy and temperature variance in the atmospheric surface layer. *J. Atmos. Sci., 28*, 190–201.

Wyngaard, J. C., and O. R. Coté, 1972: Cospectral similarity in the atmospheric surface layer. *Quart. J. Roy. Meteor. Soc., 98*, 590–603.

Wyngaard, J. C., O. R. Coté and Y. Izumi, 1971: Local free-convection, similarity and the budgets of shear stress and heat flux. *J. Atmos. Sci., 28*, 1171–1182.

Yaglom, A. M., 1979: Similarity laws for constant pressure and pressure gradient wall flows. *Ann. Rev. Fluid Mech., 11*, 505–540.

Additional Readings

Arya, S. P. S., 1988: *An Introduction to Micrometeorology*. Academic Press, San Diego, CA, 303 pp.

Haugen, D. A. (Ed.), 1973: *Workshop in Micrometeorology*. American Meteorological Society, Boston, MA, 392 pp.

Kaimal, J. C., and J. C. Wyngaard, 1990: The Kansas and Minnesota experiments. *Bound.-Layer Meteor., 50*, 31–47.

McBean, G. A. (Ed.), 1979: The Planetary Boundary Layer. WMO Technical Note No. 165, World Meteorological Organization, Geneva, Switzerland, 201 pp.

Priestley, C. H. B., 1959: *Turbulent Transfer in the Lower Atmosphere*. University of Chicago Press, Chicago, IL, 130 pp.

Sorbjan, Z., 1989. *Structure of the Atmospheric Boundary Layer*. Prentice-Hall, Englewood Cliffs, NJ, 317 pp.

Stull, R. B., 1988: *An Introduction to Boundary Layer Meteorology*. Kluwer Academic Press, Dordrecht, The Netherlands, 666 pp.

2

SPECTRA AND COSPECTRA OVER FLAT UNIFORM TERRAIN

Turbulent flows like those in the atmospheric boundary layer can be thought of as a superposition of eddies—coherent patterns of velocity, vorticity, and pressure—spread over a wide range of sizes. These eddies interact continuously with the mean flow, from which they derive their energy, and also with each other. The large "energy-containing" eddies, which contain most of the kinetic energy and are responsible for most of the transport in the turbulence, arise through instabilities in the background flow. The random forcing that provokes these instabilities is provided by the existing turbulence. This is the process represented in the production terms of the turbulent kinetic energy equation (1.59) in Chapter 1.

The energy-containing eddies themselves are also subject to instabilities, which in their case are provoked by other eddies. This imposes upon them a finite lifetime before they too break up into yet smaller eddies. This process is repeated at all scales until the eddies become sufficiently small that viscosity can affect them directly and convert their kinetic energy to internal energy (heat). The action of viscosity is captured in the dissipation term of the turbulent kinetic energy equation.

The second-moment budget equations presented in Chapter 1, of which (1.59) is one example, describe the summed behavior of all the eddies in the turbulent flow. To understand the conversion of mean kinetic energy into turbulent kinetic energy in the large eddies, the handing down of this energy to eddies of smaller and smaller scale in an "eddy cascade" process, and its ultimate conversion to heat by viscosity, we must isolate the different scales of turbulent motion and separately observe their behavior. Taking Fourier spectra and cospectra of the turbulence offers a convenient way of doing this. The spectral representation associates with each scale of motion the amount of kinetic energy, variance, or eddy flux it contributes to the whole and provides a new and invaluable perspective on boundary layer structure.

The spectrum of boundary layer fluctuations covers a range of more than five decades: millimeters to kilometers in spatial scales and fractions of a second to hours in temporal scales. The field experiments of the last two decades confirm that spectral representations in the wavenumber or frequency domain follow similarity laws much like the time-averaged statistics discussed in Chapter 1. When nondimensionalized with the appropriate scaling parameters ($u_*, T_*, w_*,$ and θ_*), the spectral and cospectral forms reduce to a set of universal curves that are functions only of z/L in the surface layer and z/z_i in the convective mixed layer. The existence of such order in the spectral domain is indeed gratifying to the boundary layer meteorologist. On a practical level, it provides engineers and modelers with the equations they need for a range of applications, from design of structures to air pollution predictions. In the ABL, the experimenter can use that information to determine the frequency requirements for sensors operating at a given height or, conversely, to adjust the observing height to suit the response characteristics of available sensors.

This chapter assumes familiarity with the general principles of the application of Fourier transforms to random processes. The essential formulas are presented in Appendix 2.1.

2.1 Spectral characteristics of boundary layer turbulence

The turbulent eddies we observe in the boundary layer are spatially extensive structures, and, ideally, their analysis requires information from many points in space. Such measurements are becoming increasingly available from aircraft and remote sensors, but the greater part of the data available to the micrometeorologist is still derived from point measurements in space as a function of time. To convert these temporal measurements into spatially distributed data, we commonly adopt Taylor's frozen turbulence hypothesis, which assumes that eddies change imperceptibly as they are convected by the mean wind $\overline{u}$ past an in situ sensor (Appendix 2.2). This assumption works best in the surface layer and worst in the high-intensity turbulence of a plant canopy or in the mixed layer. (In the recirculating flow behind a hill, it does not work at all.)

With two sensors a distance r apart, we can form the two-point covariance tensor R_{ij} (**x**,**r**) that provides the fundamental description of spatial structure in the turbulence

$$R_{ij}(\mathbf{x}, \mathbf{r}) = \overline{u_i'(\mathbf{x})u_j'(\mathbf{x}+\mathbf{r})}. \tag{2.1}$$

The Fourier transform of $R_{ij}(\mathbf{x},\mathbf{r})$ converts that covariance to a two-point spectrum tensor $E_{ij}(\mathbf{x},\boldsymbol{\kappa})$, where $\boldsymbol{\kappa}$ represents the wavenumber vector. $E_{ij}(\mathbf{x},\boldsymbol{\kappa})$ contains complete information on the distribution of turbulent variance and covariance over wavenumber space. Unfortunately, more information on the flow structure than we usually possess is required to form either $E_{ij}(\mathbf{x}, \boldsymbol{\kappa})$ or $R_{ij}(\mathbf{x}, \mathbf{r})$ and we usually work with simpler and more attainable descriptions.

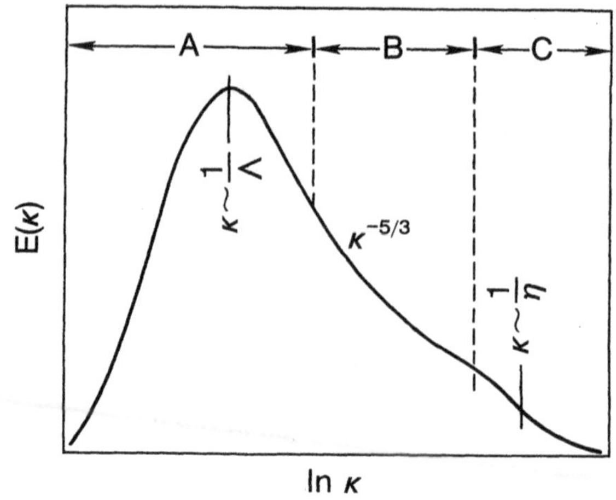

Fig. 2.1. Schematic of energy spectrum in the atmospheric boundary layer showing distinct regions of energy production (A) and dissipation (C) and the inertial subrange (B), where both energy production and dissipation are negligible. Λ is the integral scale of turbulence and η is the Kolmogorov microscale.

A particularly useful conceptual picture of the distribution of energy in wavenumber space, if we are dealing with turbulence that is homogeneous in all directions, is provided by the scalar energy spectrum $E(\kappa)$. $E(\kappa)$ represents the contribution to the total kinetic energy from Fourier modes with wavenumber magnitudes between κ and $\kappa + d\kappa$, where $\kappa = |\boldsymbol{\kappa}|$. For a precise definition of $E(\kappa)$, see Lumley and Panofsky (1964). $E(\kappa)$ peaks in the energy-containing region, by definition, and drops to zero at both ends of the spectrum. In its schematic representation in Fig. (2.1) we identify the three major spectral regions, A, B, and C, relevant to boundary layer flow.

- A The *energy-containing range*, which contains the bulk of the turbulent energy and where energy is produced by buoyancy and shear.
- B The *inertial subrange*, where energy is neither produced nor dissipated but handed down to smaller and smaller scales.
- C The *dissipation range*, where kinetic energy is converted to internal energy.

The energy-containing and dissipation ranges have their own characteristic length scales: In the former it is the Eulerian integral length scale Λ; in the latter, the so-called Kolmogorov microscale η. In Fig. 2.1 we show $E(\kappa)$ reaching its maximum at a wavenumber corresponding roughly to the Eulerian integral length scale[1]

[1] We cannot be more precise because $E(\kappa)$ is not easily measured in the atmosphere. Also, there is no simple relationship between $R_{ij}(\mathbf{x}, \mathbf{r})$ and $E(\kappa)$ except in fully isotropic turbulence (Batchelor, 1960), where the spectrum can be shown to peak at $\sqrt{\pi}/\Lambda$, close enough for our approximation of $1/\Lambda$.

($\kappa \sim 1/\Lambda$). With a single sensor and the use of Taylor's hypothesis, we are constrained to define that integral length scale in terms of its components Λ_u, Λ_v, and Λ_w, derived from the integral time scales $\mathcal{T}_u$, $\mathcal{T}_v$, and $\mathcal{T}_w$, available from our measurements of u, v, and w. These integral time scales actually represent the time scales over which the turbulence remains correlated (Fig. 2.2).

Taking Λ_u as an example we can write

$$\Lambda_u = \overline{u}\mathcal{T}_u = \overline{u}\int_0^\infty \rho_u(\xi) = \overline{u}\int_0^\infty \frac{\overline{u'(t)u'(t+\xi)}}{\sigma_u^2}\, d\xi, \tag{2.2}$$

where $\rho_u(\xi)$ is the autocorrelation function as defined above, ξ the time lag with respect to time t (Fig. 2.2), and $\overline{u}$ the mean wind velocity. If $\rho_u(\xi)$ is an exponential function, the integral time scale is the value of ξ at $\rho_u(\xi) = 1/e \simeq 0.37 (e \simeq 2.72$, the base for natural logarithms). For a discussion of autocorrelation functions and integral time scales, see Appendix 7.2 in Chapter 7.

In the dissipation range the scaling length η is given by

$$\eta = \left(\frac{\nu^3}{\epsilon}\right)^{1/4},$$

where ν is the kinematic viscosity of air and ϵ is the dissipation rate of turbulent kinetic energy. Whereas Λ_u varies typically from 10 to 500 m, η is of the order of 0.001 m.

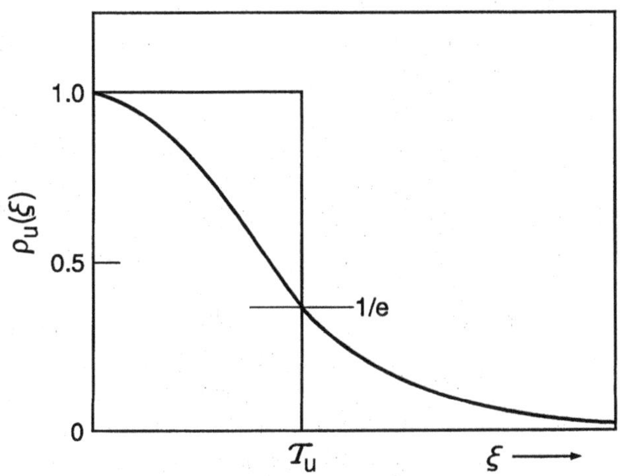

Fig. 2.2. Autocorrelation function and its relation to the integral time scale $\mathcal{T}_u$. The 1/e point on the curve is usually a good approximation of the integral time scale even when the correlation function is not strictly exponential. [The area under the rectangle should equal the area under $\rho_u(\xi)$.]

In the inertial subrange, energy is neither produced nor dissipated, and the transfer of energy from the energy-containing to the dissipation range is controlled entirely by ϵ, the rate at which energy is converted to heat in the dissipation range. Here, the form of the spectrum can be deduced from purely dimensional arguments (Kolmogorov, 1941).

Adopting Taylor's hypothesis implies a natural relationship between wavenumbers and frequencies. All our observations are henceforth defined in terms of the streamwise wavenumber κ_1 (subscript 1 for component in the streamwise direction), which corresponds to $2\pi/\lambda$, where λ is the wavelength approximated by $\overline{u}/f$, f being the cyclic frequency. The one-dimensional spectra $F_u(\kappa_1), F_v(\kappa_1)$, and $F_w(\kappa_1)$ of the three wind velocity components (available to us through spectrum analysis of measurements from fixed sensors) have forms somewhat different from $E(\kappa)$ but are predictable nevertheless in both the inertial subrange and the energy-containing region.

2.2 Inertial subrange

Kolmogorov, who first conceived the idea of an inertial subrange separating the energy-containing and dissipation ranges, argued from dimensional considerations that $E(\kappa)$ in this region should be proportional to $\epsilon^{2/3}\kappa^{-5/3}$ (Kolmogorov, 1941). Here the u spectrum, in its familiar one-dimensional form, becomes

$$F_u(\kappa_1) = \alpha_1 \epsilon^{2/3} \kappa_1^{-5/3}, \tag{2.3}$$

where α_1 is the Kolmogorov constant with a value estimated between 0.5 and 0.6. This is the well-known $-5/3$ power law for the inertial subrange. Theoretical arguments suggest that turbulence is isotropic in this range. Isotropy implies that the velocity field is independent of rotation and reflection about the spatial axes. Even though isotropy does not apply to the energy-containing eddies, we can assume that the small-scale ($\lambda \ll \Lambda_u$) structure is effectively isotropic. This *local isotropy* is important for the derivation of small-scale turbulence statistics. (*Local* in this context refers to wavenumber space not physical space.) If local isotropy exists in the inertial subrange, we have the following relationship among the u, v, and w spectra:

$$F_v(\kappa_1) = F_w(\kappa_1) = (4/3)F_u(\kappa_1), \tag{2.4}$$

which is illustrated in Fig. 2.3. In this log-log representation of the spectra, power laws appear as constant slopes and ratios as fixed separations. The spectra for v and w are placed higher than u in the inertial subrange, where they all fall off as $\kappa_1^{-5/3}$. Another consequence of local isotropy is the vanishing of all correlations between velocity components and between the velocity components and scalars;

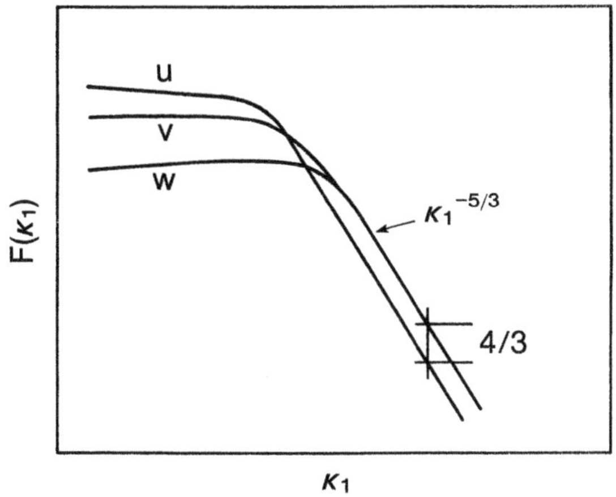

FIG.. 2.3. Idealized velocity spectra presented in log-log coordinates showing -5/3 slope in the inertial subrange and the 4/3 ratio between the transverse and streamwise velocity components.

this implies there can be no turbulent fluxes in the inertial subrange. The three conditions: $-5/3$ power law, $4/3$ ratio between transverse and longitudinal velocity components, and vanishing (or very low) cospectral levels are used as a test for the existence of an inertial subrange. [The ability to reach well into this region of the spectrum through the use of sonic anemometers (Chapter 6) to confirm these inertial subrange properties was critical to later development of universal forms for ABL spectra.] The $-5/3$ power law extends approximately to wavenumber $\kappa_1 = 0.1\eta^{-1}$ (Dubovikov and Tatarskii, 1987), above which it begins to fall off sharply with increasing κ_1.

Corrsin (1951) proposed an inertial subrange form for the temperature spectrum that appears to be valid for other scalars such as humidity:

$$F_\theta(\kappa_1) = \beta_1 \epsilon^{-1/3} N_\theta \kappa_1^{-5/3}, \tag{2.5}$$

where N_θ is the dissipation rate for half the temperature variance and β_1 is a universal constant with a value about 0.8 (Kaimal et al., 1972). [Note that at high wavenumbers approaching the dissipation range, the temperature spectrum exhibits a "bump" due to straining effects on the temperature eddies (Hill, 1978) not observed in the velocity spectra.]

2.3 Energy-containing range

The spectral forms in the energy-containing range tend to be different for each variable, since the integral scales they respond to are different. Yet they all seem to behave in a consistent manner when plotted in dimensionless similarity coordinates.

Implicit in the development of spectral forms in this range is the assumption that a spectral gap exists, separating boundary layer turbulence from external fluctuations. Fortunately for the meteorologist, a spectral gap of sorts can often be found in the CBL at a frequency f between 0.001 and 0.0001 Hz, between the convection-driven boundary layer scales and the synoptic scales (Van der Hoven, 1957). This end of the spectrum, however, is susceptible to contamination from long-term trends present in the data. These may be caused by gravity waves, diurnal variation, synoptically induced changes, or simply drift in the sensor. In the absence of such trends in the record, $F(\kappa_1)$ is seen to level off to a constant value as $\kappa_1 \to 0$, a consequence of the one-dimensional representation of the three-dimensional turbulence spectrum (see Appendix 2.3). Identifying energy peaks in this type of spectral plot is difficult. For a more realistic representation of peaks and valleys in the distribution of turbulent energy, meteorologists use the wavenumber-weighted form $\kappa_1 F(\kappa_1)$ of the spectrum. In Fig. 2.4 we see $\kappa_1 F(\kappa_1)$ for the u component peaking at $\kappa_1 \approx 1/\Lambda_u$ while falling off as κ_1^{+1} on the low side and $\kappa_1^{-2/3}$ on the high side.

The low-frequency spectrum can usually be approximated by one of two analytic forms:

$$\frac{\kappa_1 F_\alpha(\kappa_1)}{\sigma_\alpha^2} = \frac{A(\kappa_1/\kappa_{1m})}{1 + B(\kappa_1/\kappa_{1m})^{5/3}} \tag{2.6}$$

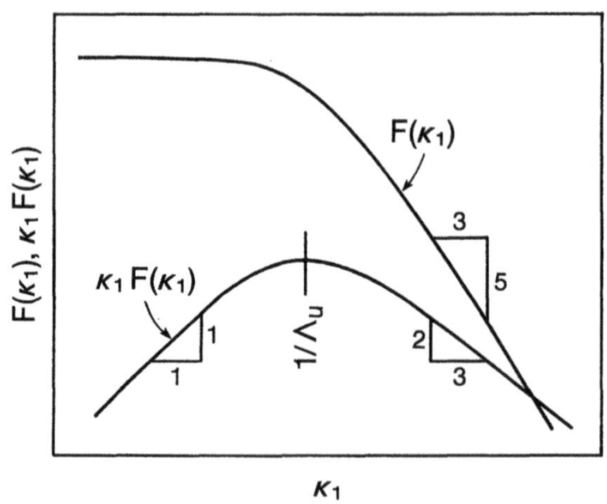

Fig. 2.4. Log-log representations of the frequency-weighted and unweighted power spectral densities. We show both plots in the same graph even though the two spectra have different dimensions (variance and variance/unit frequency interval) to compare their behavior as a function of κ_1.

or

$$\frac{\kappa_1 F_\alpha(\kappa_1)}{\sigma_\alpha^2} = \frac{C(\kappa_1/\kappa_{1m})}{[1 + D(\kappa_1/\kappa_{1m})]^{5/3}} \tag{2.7}$$

where $\alpha = u, v, w,$ or θ; $A, B, C,$ and D are adjustable constants, and the subscript m denotes κ_1 at the spectral peak. Equation (2.6) fits unstable w spectra and all the stable spectra. Equation (2.7) fits the unstable u, v, and θ spectra slightly better than (2.6).

This wavenumber-weighted spectrum is also referred to as a logarithmic spectrum since it represents the variance per unit logarithmic wavenumber interval. Its units are those of variance (e.g., $\mathrm{m}^2\ \mathrm{s}^{-2}$) instead of variance per wavenumber interval $\Delta\kappa_1$ (as in the unweighted spectrum). Yet, when plotted on a log-log scale, power relationships appear as straight lines as with the unweighted spectrum, a distinct advantage when dealing with atmospheric spectra. The log-log plot is, however, not area-preserving since

$$\sigma_\alpha^2 = \int_0^\infty F_\alpha(\kappa_1)\, d\kappa_1 = \int_0^\infty \kappa_1 F_\alpha(\kappa_1)\, d(\ln \kappa_1).$$

For an equal area spectrum, $\kappa_1 F_\alpha(\kappa_1)$ should be plotted on a linear scale and κ_1 on a log scale, but power laws will not be straight lines on that plot.

2.4 Conversion from wavenumber to frequency scales

Although spectral theories are formulated in wavenumber space, most measurements are performed in frequency space. To convert spatial scales to frequency scales we invoke Taylor's hypothesis, $\kappa_1 = 2\pi f/\overline{u}$, where f is the cyclic frequency as defined earlier.

Taking the u spectrum as an example, we have

$$\int_0^\infty F_u(\kappa_1)\, d\kappa_1 = \sigma_u^2 = \int_0^\infty S_u(f)\, df. \tag{2.8}$$

We use a different symbol to represent the frequency spectrum. If $\kappa_1 = 2\pi f/\overline{u}$, we have

$$\frac{2\pi}{\overline{u}} F_u\left(\frac{2\pi f}{\overline{u}}\right) = S_u(f), \tag{2.9}$$

or

$$\kappa_1 F_u(\kappa_1) = f S_u(f). \tag{2.10}$$

It can be shown that the relationship in (2.10) holds for any form of f, including the nondimensional frequency $n(= fz/\overline{u})$, used in surface layer work. Thus, $\kappa_1 F_u(\kappa_1)$, $fS_u(f)$, and $nS_u(n)$ all represent the same numerical value, allowing us considerable freedom in choosing frequency scales for the abscissa. For example, $fS(f)$ can be plotted as a function of κ_1, f, or n, without any conversion of units.

2.5 Surface layer spectra

Following conventions established for M-O scaling, we use u_* and T_* for nondimensionalizing velocity and temperature spectra in the surface layer. The appropriate dimensionless frequency scale for representing spectra would be $n = fz/\overline{u}$, which is the ratio of height z to wavelength λ. We also recall the dimensionless form for ϵ introduced in Chapter 1:

$$\phi_\epsilon = \frac{kz\epsilon}{u_*^3}, \tag{2.11}$$

which, according to similarity theory, should be a function only of z/L.

We start with the inertial subrange u spectrum expressed in the framework of surface layer similarity,

$$\begin{aligned} \frac{fS_u(f)}{u_*^2} &= \frac{\alpha_1}{(2\pi)^{2/3}} \left(\frac{\epsilon^{2/3} z^{2/3}}{u_*^2} \right) \left(\frac{fz}{\overline{u}} \right)^{-2/3} \\ &= \frac{\alpha_1}{(2\pi k)^{2/3}} \phi_\epsilon^{2/3} \left(\frac{fz}{\overline{u}} \right)^{-2/3} . \end{aligned} \tag{2.12}$$

Denoting the dimensionless frequency $fz/\overline{u}$ by n and using values $\alpha_1 = 0.55$ and $k = 0.4$ (see Appendix 2.4), we have (Kaimal et al., 1972)

$$\frac{fS_u(f)}{u_*^2 \phi_\epsilon^{2/3}} = 0.3\, n^{-2/3}. \tag{2.13}$$

When plotted on a log-log scale, the inertial subrange spectra collapse to a single straight line with a $-2/3$ slope. All the variability in the ordinate is absorbed in $\phi_\epsilon^{2/3}$. We have from Chapter 1 the functional form for $\phi_\epsilon^{2/3}$:

$$\phi_\epsilon^{2/3} = \begin{cases} 1 + 0.5|z/L|^{2/3}, & z/L \le 0 \\ (1 + 5z/L)^{2/3}, & z/L \ge 0 \end{cases} \tag{2.14}$$

The dimensionless spectra for the remaining velocity components and temperature can similarly be expressed as

$$\frac{fS_v(f)}{u_*^2 \phi_\epsilon^{2/3}} = 0.4\, n^{-2/3}, \tag{2.15}$$

$$\frac{fS_w(f)}{u_*^2\phi_\epsilon^{2/3}} = 0.4\,n^{-2/3}, \tag{2.16}$$

$$\frac{fS_\theta(f)}{T_*^2\phi_N\phi_\epsilon^{-1/3}} = 0.43\,n^{-2/3}. \tag{2.17}$$

For the temperature spectrum in (2.17) we assumed $\beta_1 = 0.8$. In Chapter 1 we defined the similarity function for N_θ as

$$\phi_N = \frac{kzN_\theta}{u_*T_*^2}. \tag{2.18}$$

Also, in the temperature variance budget, we found a local balance between the production and destruction of temperature variance ($N_\theta \simeq -\overline{w'\theta'}\partial\overline{\theta}/\partial z$), which suggests that $\phi_N \simeq \phi_h$. Hence (2.17) can be expressed in terms of ϕ_h as

$$\frac{fS_\theta(f)}{T_*^2\phi_h\phi_\epsilon^{-1/3}} \simeq 0.43n^{-2/3}. \tag{2.19}$$

With the velocity and temperature spectra anchored to the inertial subrange formulations of (2.13), (2.15), (2.16), and (2.17), we can now examine the rest of the spectrum to see if it will collapse into a narrow band (as in the inertial subrange) or progress in some orderly fashion as a function of z/L. The plots for the four variables from the Kansas experiment (Kaimal et al., 1972) are given in Figs. 2.5–2.8. Of the four, the w spectrum exhibits the most systematic variation with z/L;

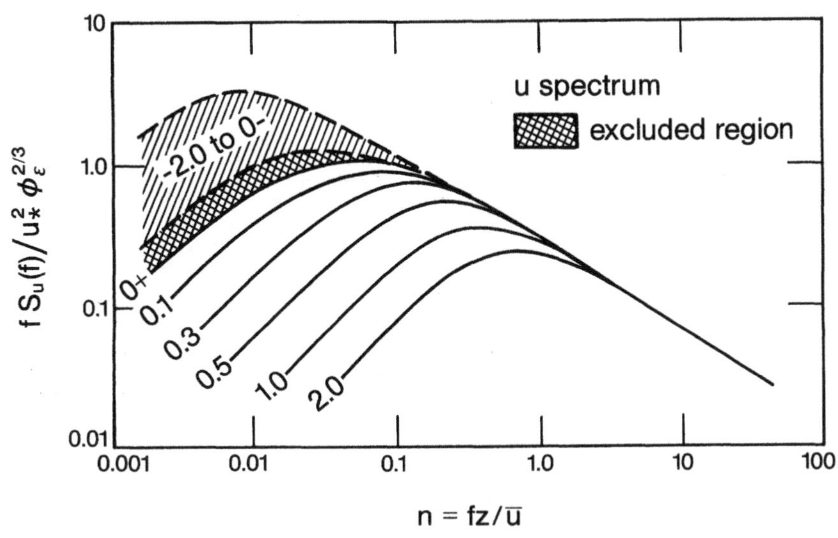

Fig. 2.5. Normalized surface layer u spectrum shown varying with z/L.

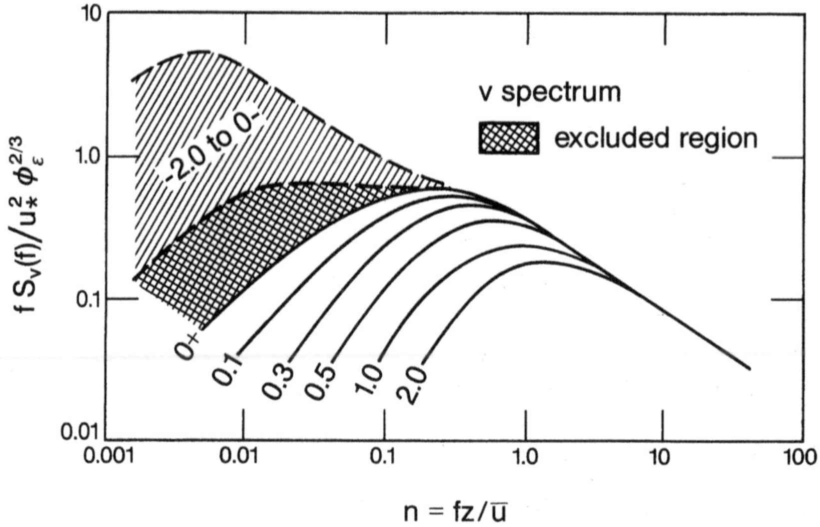

FIG. 2.6. Normalized surface layer v spectrum shown varying with z/L.

only in the range $-0.3 > z/L > -2$ (shown as a hatched area in Fig. 2.7) does the spectrum exhibit an insensitivity to z/L. This happens because the normalized spectral peak n_m stops shifting to lower frequencies with increasing z/L (see Fig. 2.9), an indication that the peak wavelength λ_m scales only with z. The u, v, and θ stable spectra also progress systematically, but their unstable spectra spread over a

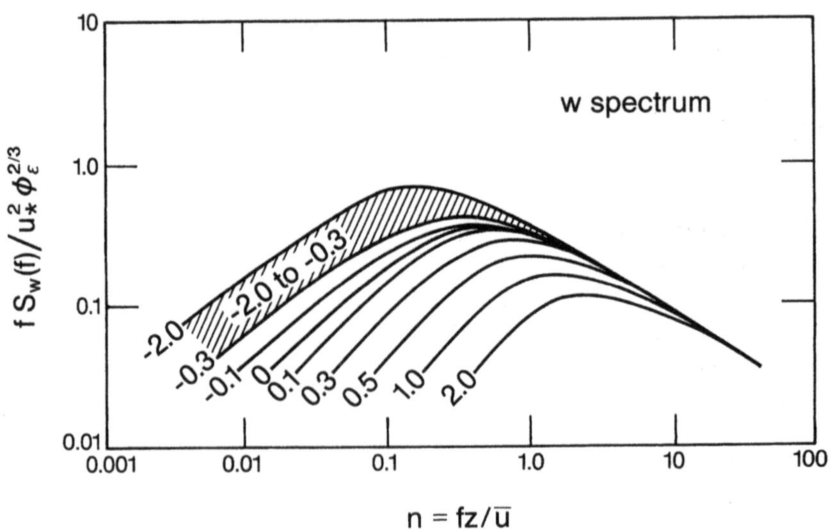

FIG. 2.7. Normalized surface layer w spectrum shown varying with z/L.

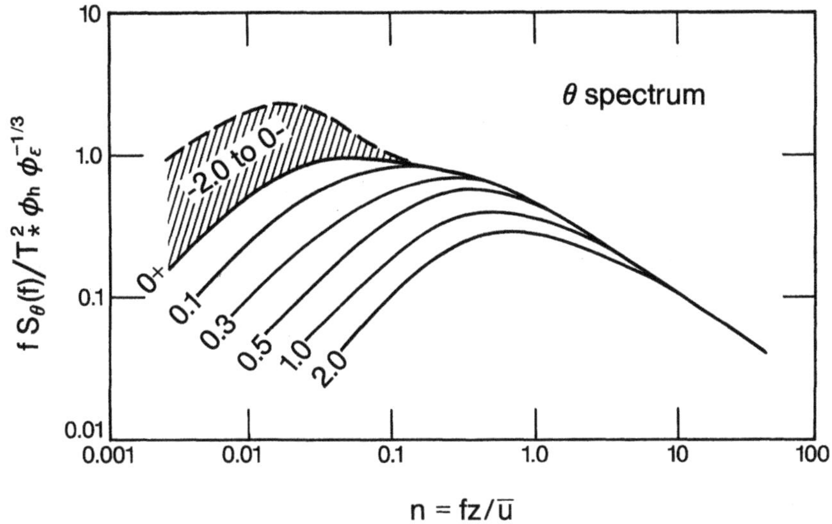

FIG. 2.8. Normalized surface layer θ spectrum shown varying with z/L.

larger area (hatched); in u and v, an "excluded" region (crosshatched) can be seen separating the stable and unstable spectra. The limiting curves for stability regimes approaching neutral from both sides are indicated by notations $z/L = 0+$ and $0-$ in the figures. Clearly, the unstable u and v spectra do not follow M-O similarity, and since no measurements of the boundary layer depth z_i were made in Kansas, it was not known at the time that their λ_m's scaled with z_i. Later, the Minnesota

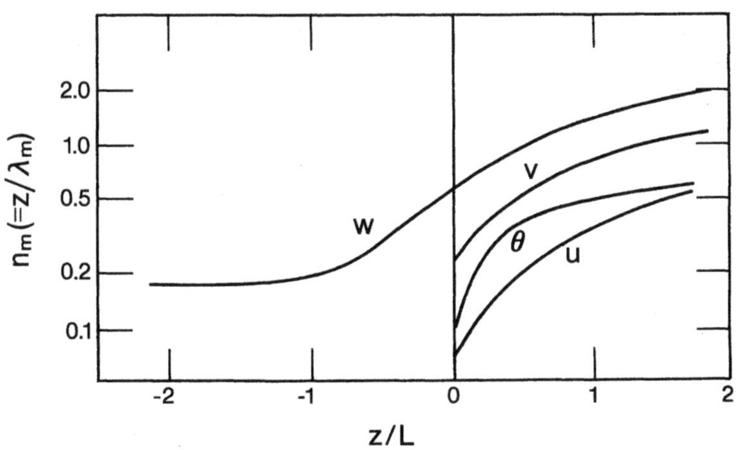

FIG. 2.9. Nondimensionalized frequency at spectral maxima shown as functions of z/L. For u, v, and θ, the curves are discontinuous at $z/L = 0$ and therefore valid only to $z/L =$ 0+, the neutral limit on the stable side.

experiment (Kaimal et al., 1976) and the laboratory work of Willis and Deardorff (1974) revealed λ_m to be approximately $1.5z_i$. The wavelength at the spectral peak λ_m is a length scale of greater importance to boundary layer meteorologists than the integral length scale Λ since it is representative of the size of the eddies with the most energy. In Fig. 2.9, λ_m is presented in its dimensionless form z/λ_m, or n_m. On the stable side, we find n_m for all variables increasing rapidly with z/L. On the unstable side, only n_m for w decreases systematically with $-z/L$, approaching its free-convection limit of $n_m = 0.17$ at $z > -L$. (The unstable θ spectral peaks tend to be less predictable than the w peaks.)

The behavior of $(\lambda_m)_w$ in the surface layer and in the layer immediately above may be expressed as follows:

$$(\lambda_m)_w = \begin{cases} z(0.55 - 0.38|z/L|)^{-1}, & 0 \le z \le -L \\ 5.9z, & -L \le z \le 0.1z_i \end{cases} \tag{2.20}$$

$$(\lambda_m)_w = \begin{cases} z(0.55 + z/L)^{-1}, & 0 \le z \le L \\ zL(0.45z + 1.1L)^{-1}, & L \le z \le 2L \\ L, & z \ge 2L. \end{cases} \tag{2.21}$$

The free-convection form for $(\lambda_m)_w$ leads to the familiar approximation $\Lambda_w \sim z$ used in the lower boundary layer, where $(\lambda_m)_w$ is typically six to seven times larger than Λ_w. The tendency for $(\lambda_m)_w$ to become independent of z as z exceeds $2L$ on the stable side is consistent with Wyngaard and Coté's (1972) "z-less stratification" concept mentioned in Chapter 1.

The stable spectra exhibit a common form when normalized by their variances and plotted against the dimensionless frequency f/f_0; f_0 is the value of f where the extrapolated inertial subrange meets the $fS_\alpha(f)/\sigma_\alpha^2 = 1$ line (see Fig. 2.10). The spectra for u, v, w, and θ conform to the shape (Kaimal, 1973)

$$\frac{fS_\alpha(f)}{\sigma_\alpha^2} = \frac{0.164 f/f_0}{1 + 0.164(f/f_0)^{5/3}}, \tag{2.22}$$

where $\alpha = u, v, w$, or θ.

By substituting for $fS_\alpha(f)$ from (2.13), (2.15), and (2.16) and assuming $\sigma_u/u_* = 2.17$, $\sigma_v/u_* = 1.78$, and $\sigma_w/u_* = 1.36$ from the Kansas stable data, Moraes (1988) showed

$$(f_0)_u = 0.012\phi_\epsilon, \tag{2.23}$$

$$(f_0)_v = 0.045\phi_\epsilon, \tag{2.24}$$

$$(f_0)_w = 0.094\phi_\epsilon. \tag{2.25}$$

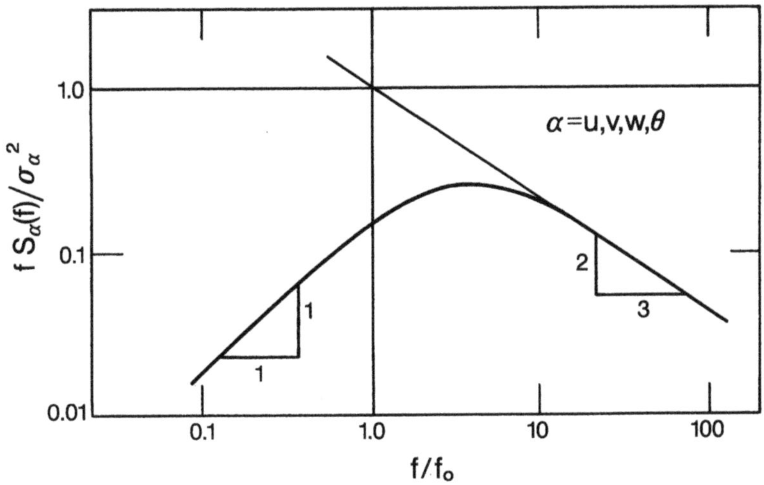

FIG. 2.10. Normalized stable surface layer spectrum representing u, v, w, and θ. The abscissa is normalized by the frequency where the inertial subrange slope intercepts the $fS(f) = 1$ line, as shown in the figure.

The frequency f_0 is also related to the integral time scale T_α of the variable α since the low-frequency behavior of (2.22) is controlled by T_α (see Appendix 2.3).

The forms for u, v, and w most commonly used in engineering applications are the neutral Kansas spectra (Kaimal et al., 1972) shown in Fig. 2.11 with minor adjustments to provide the expected 4/3 ratio in the inertial subrange. They can be represented by

$$\frac{fS_u(f)}{u_*^2} = \frac{102n}{(1+33n)^{5/3}}, \tag{2.26}$$

$$\frac{fS_v(f)}{u_*^2} = \frac{17n}{(1+9.5n)^{5/3}}, \tag{2.27}$$

$$\frac{fS_w(f)}{u_*^2} = \frac{2.1n}{(1+5.3n^{5/3})}. \tag{2.28}$$

2.6 Mixed layer spectra

In the mixed layer, which comprises the upper 9/10 of the CBL, we find M-O similarity being replaced by a different scaling law: with z_i replacing $-L, w_*$ replacing u_*, and θ_* replacing T_*, as we saw in Chapter 1. Since none of the new scaling parameters varies with height, we expect the spectra also to be invariant with height. In this new framework, λ_m, ϵ, and the magnitude of the $fS(f)$ peak will remain constant with height changing only from run to run in response to changes in $(\overline{w'\theta'})_0$ and z_i. This is indeed the case for u, v, and w spectra, as seen

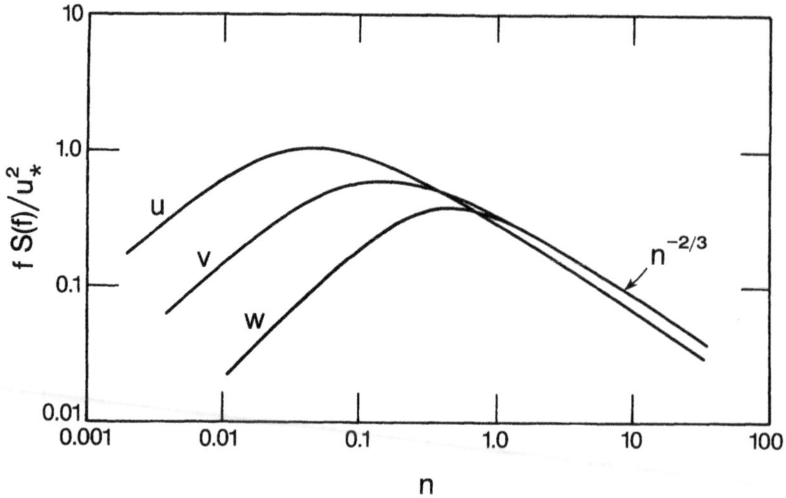

Fig. 2.11. Normalized neutral ($z/L = 0+$) spectra for u, v, and w in the surface layer.

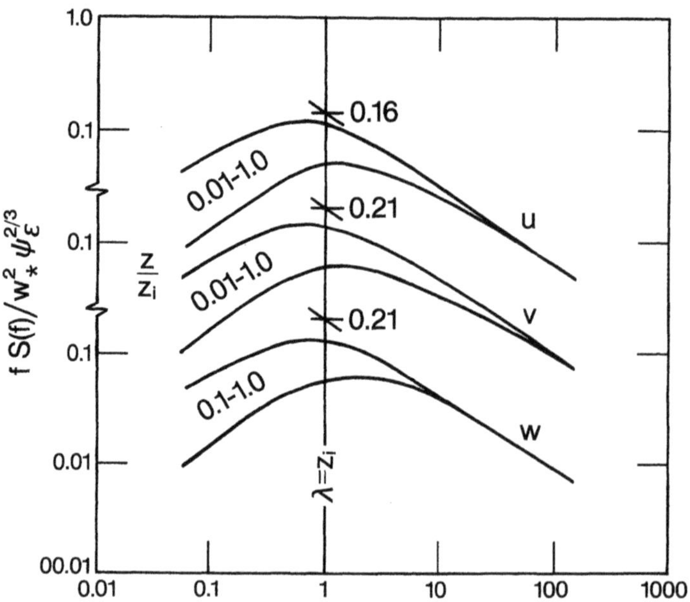

Fig. 2.12. Normalized mixed layer u, v, and w spectra. The two curves define the envelopes of spectra that fall within the z/z_i range indicated.

in their idealized representations in Fig. 2.12. Note that in the energy-containing range, w follows mixed layer similarity only down to $0.1z_i$, whereas u and v follow it almost all the way down.

The inertial subrange spectral form for u reduces to

$$\frac{fS_u(f)}{w_*^2} = \left[\frac{\alpha_1}{(2\pi)^{2/3}}\right] \psi_\epsilon^{2/3} \left(\frac{fz_i}{\overline{u}}\right)^{-2/3},$$

where $w_* = [(g/T)(\overline{w'\theta'})_0 z_i]^{1/3}$ and $\psi_\epsilon = \epsilon/(g/\overline{\theta})(\overline{w'\theta'})_0$, the ratio of dissipation rate to buoyant production rate near the surface. Representing $fz_i/\overline{u}$ by a new dimensionless frequency n_i and assuming $\alpha_1 = 0.55$, we can write (Kaimal et al., 1976)

$$\frac{fS_u(f)}{w_*^2 \psi_\epsilon^{2/3}} = 0.16\, n_i^{-5/3}, \tag{2.29}$$

$$\frac{fS_v(f)}{w_*^2 \psi_\epsilon^{2/3}} = 0.21\, n_i^{-5/3}, \tag{2.30}$$

$$\frac{fS_w(f)}{w_*^2 \psi_\epsilon^{2/3}} = 0.21\, n_i^{-5/3}. \tag{2.31}$$

ψ_ϵ should have a value between 0.4 and 0.5 in the mixed layer if we assume negligible wind shear and a linear heat flux profile in the layer. The Minnesota data show ψ_ϵ ranging from 0.5 to 0.7, perhaps because of larger than expected shear production rates in some runs (Kaimal et al., 1976). Note that ψ_ϵ is identically the dimensionless dissipation rate $\epsilon z_i/w_*^3$ shown in Fig. 1.10.

The spectral peaks for u, v, and w are approximated as follows:

$$(\lambda_m)_u = (\lambda_m)_v = 1.5z_i, \qquad 0.01z_i \leq z \leq z_i \tag{2.32}$$

$$(\lambda_m)_w = \begin{cases} 5.9z, & -L \leq z \leq 0.1z_i \\ 1.8z_i(1 - e^{-4z/z_i} - 0.0003e^{8z/z_i}), & 0.1z_i \leq z \leq z_i. \end{cases} \tag{2.33}$$

The mixed layer form for $(\lambda_m)_w$, shown in Fig. 2.13, was derived by Caughey and Palmer (1979) from a combined plot of data from the Minnesota and Ashchurch experiments. The mixed layer $(\lambda_m)_\theta$ does not show a consistent pattern because the temperature fluctuations are generally small above the surface layer and easily overwhelmed by the effects of entrainment and the diurnal trend in the temperature. The profile of σ_θ^2 presented in Chapter 1 shows the variance decreasing to a minimum at about $0.6z_i$ and rising to its near-surface value at z_i, a clear demonstration of the influence of entrainment in the temperature statistics.

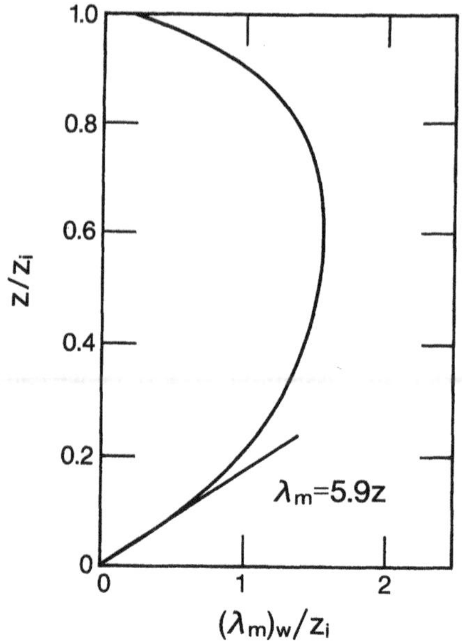

FIG. 2.13. Variation of the vertical velocity peak wavelength with height.

The evolution of u, v, w, and θ spectra with height in the ABL is shown schematically in Fig. 2.14. The u and v spectra appear flattened and stretched out in the surface layer as they adjust to conform to the mixed layer spectral forms of Fig. 2.12 in the energy-containing range and to the constraints of (2.13) and (2.15) in the inertial subrange. These spectra do not show much variation with height, except in the inertial subrange where the energy drops sharply with height in conformance with (2.14). The w spectrum, on the other hand, gains steadily in intensity as its peak moves down the frequency scale, consistent with the formulation in (2.20), and approaches the mixed layer form of Fig. 2.12 at $z = 0.2z_i$.

Two different approaches have been proposed to model this behavior of the u and v spectra. Kaimal (1978) used a simple interpolation formula linking the analytic form for the mixed layer spectrum

$$\frac{fS_u(f)}{\sigma_u^2} = n_i(1 + 3.1n_i^{5/3}) \tag{2.34}$$

to the inertial subrange forms of (2.29), (2.30), and (2.31). Højstrup (1982) treated the u and v spectra as the sum of two spectra—a low-frequency spectrum scaling with z_i and a high-frequency spectrum scaling with z:

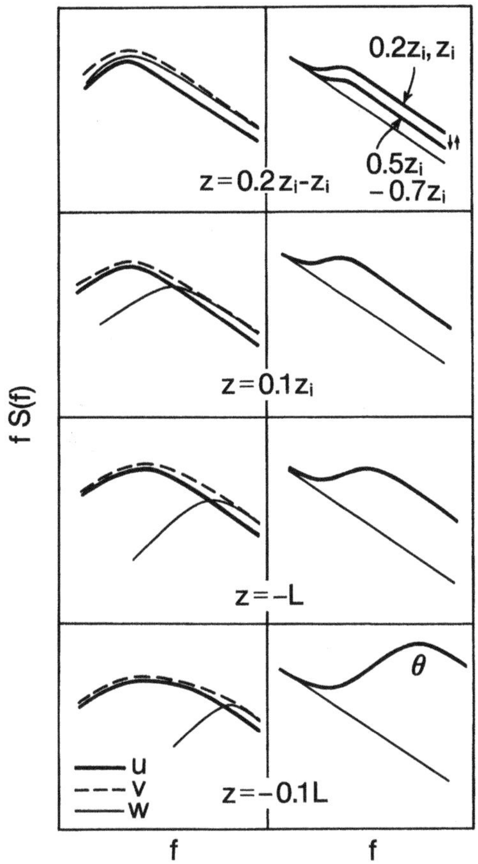

Fig. 2.14. Schematic representation of the evolution of spectra for velocity (left frames) and temperature (right frames) with height in the convective boundary layer. The thin reference line in the temperature plots is the spectrum of the diurnal trend.

$$\frac{fS_u(f)}{u_*^2} = \left[\frac{0.5n_i}{(1+2.2n_i^{5/3})}\right]\left(\frac{z_i}{|L|}\right)^{2/3} + \frac{102n}{(1+33n)^{5/3}}, \tag{2.35}$$

$$\frac{fS_v(f)}{u_*^2} = \left[\frac{0.95n_i}{(1+2n_i^{5/3})}\right]\left(\frac{z_i}{|L|}\right)^{2/3} + \frac{17n}{(1+9.5n)^{5/3}}. \tag{2.36}$$

The high-frequency contributions in (2.35) and (2.36) are identical to the Kansas forms in (2.26) and (2.27).

The modification of the θ spectrum with height is represented schematically in Fig. 2.14. The spectrum drops to its lowest point between $0.5z_i$ and $0.7z_i$, approaching spectral levels attributable to just the diurnal trend (used as reference in all the frames in Fig. 2.14); it rises again to its $z \approx 0.2z_i$ level at $z = z_i$. We can express these shifts in terms of the inertial subrange behavior of the θ spectrum which, in the mixed layer framework, becomes

$$\frac{fS_\theta(f)}{\gamma\theta_*^2} = \frac{\beta_1}{(2\pi)^{2/3}} n_i^{-2/3} = 0.24 n_i^{-2/3}. \tag{2.37}$$

$\theta_* = (\overline{w'\theta'})_0/w_*$, and γ is the nondimensional equivalent of $\psi_\epsilon^{2/3}$ in (2.29)–(2.31) given by

$$\gamma = \frac{N_\theta \epsilon^{-1/3}}{\theta_*^2 z_i^{-2/3}}. \tag{2.38}$$

In the CBL, γ exhibits a profile with a predictable minimum at about $0.5z_i$. The Minnesota results (Kaimal et al., 1976) have been approximated by

$$\gamma = \begin{cases} 0.83(z/z_i)^{-4/3}, & z \leq 0.5z_i \\ 2.1, & 0.5z_i \leq z \leq 0.7z_i \\ 6.1(z/z_i)^3, & 0.7z_i \leq z \leq z_i. \end{cases} \tag{2.39}$$

We attribute the rise in the spectral energy above $0.7z_i$ to entrainment of warmer air from above the capping inversion into the mixed layer.

2.7 Stable outer layer spectra

In the stable outer layer (i.e., $h > z > L$, where L often ranges from 1/10 to 1/3 of the inversion depth h), the energy due to turbulence in the spectrum decreases rapidly with height, whereas that of the wave components present does not. At $z = h$, only the wave energy remains unless nonlinear processes induce wave breaking, thereby producing turbulent kinetic energy; at frequencies where turbulent energy shows a peak at the lower heights, the spectrum is flat and significantly depressed. (Note that Taylor's hypothesis does not apply to gravity waves as they are not transported by the mean wind.) A flat $fS(f)$ spectrum appears to be characteristic of stable atmospheric flows at $Ri \approx 0.2$ (Kaimal and Izumi, 1965; Okamoto and Webb, 1970); it represents a background noise containing equal energy per octave, analogous to the "pink noise" used in audio testing.

The evolution of u, v, w, and θ spectra with height in a hypothetical stable layer is shown schematically in Fig. 2.15. It is apparent that the vertical variations of λ_m cannot be generalized in any way because the stable outer layer is continually evolving (Caughey et al., 1979) and never reaching equilibrium. The positions and relative magnitudes of the turbulent peaks and the wave peaks can be expected to change with time in this layer. Examples of combined wave and turbulence spectra can be found in Finnigan et al. (1984).

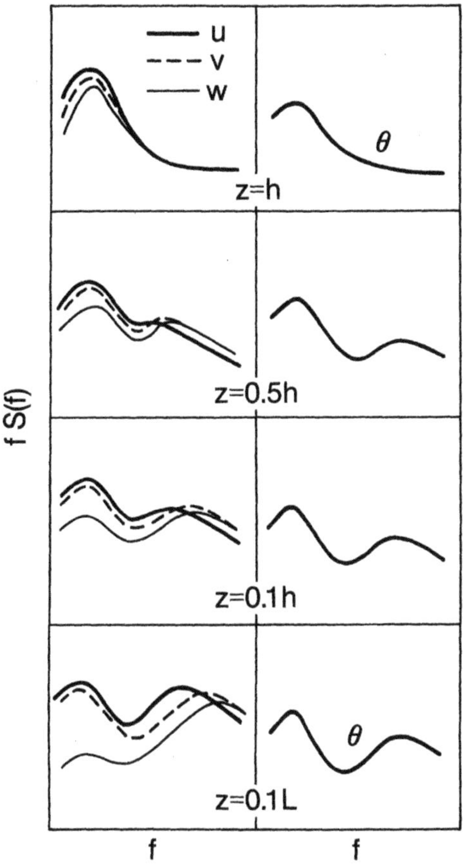

FIG. 2.15. Schematic representation of the evolution of spectra for velocity (left frames) and temperature (right frames) with height in the stable boundary layer.

2.8 Structure parameters and spectra

The structure parameter, a widely used indicator of the small-scale structure of turbulence in wave propagation studies, was originally defined in terms of the structure function $D(r)$ (Tatarski, 1961), where

$$D_\alpha(r) = \overline{[\alpha(x) - \alpha(x+r)]^2} = C_\alpha^2 r^{2/3}. \tag{2.40}$$

Here, C_α^2 is the structure parameter, α any variable (usually u, θ, or q), and r the distance separating two measurements of α along the x axis. It can be shown that C_α^2 is related to the one-dimensional spectrum of α through the relationship

$$F_\alpha(\kappa_1) = \frac{2C_\alpha^2}{3\Gamma(-1/3)}\kappa_1^{-5/3} \simeq 0.25 C_\alpha^2 \kappa_1^{-5/3}, \tag{2.41}$$

where Γ is the gamma function, provided the distance r is small enough to be within the range of length scales in the inertial subrange. The parameters, C_V^2, C_T^2, and C_Q^2 for u, θ, and q, respectively, are closely identified with C_n^2, the structure parameter of the refractive index for acoustic and electromagnetic waves propagating through the atmosphere. C_V^2 and C_T^2 are related to terms in the one-dimensional spectral forms for velocity and temperature as follows:

$$C_V^2 = 4\alpha_1 \epsilon^{2/3} \simeq 2.2\epsilon^{2/3}, \tag{2.42}$$

$$C_T^2 = 4\beta_1 N_\theta \epsilon^{-1/3} \simeq 3.2 N_\theta \epsilon^{-1/3}. \tag{2.43}$$

(The form for C_Q^2 is the same as for C_T^2, with the same numerical coefficient and the appropriate dissipation rate for humidity.) In the surface layer, the dimensionless forms for C_V^2 and C_T^2 may also be expressed in terms of ϕ functions defined earlier.

$$\frac{C_V^2 z^{2/3}}{u_*^2} = 2.2k^{-2/3}\phi_\epsilon^{2/3} = 4\phi_\epsilon^{2/3}, \tag{2.44}$$

$$\frac{C_T^2 z^{2/3}}{T_*^2} = 3.2k^{-2/3}\phi_N \phi_\epsilon^{-1/3} = 5.9\phi_N \phi_\epsilon^{-1/3}, \tag{2.45}$$

where $\phi_N = kzN_\theta/u_* T_*^2$ as in Chapter 1. For normalizing our temperature spectra we earlier assumed $\phi_N \simeq \phi_h$, but such a substitution in (2.45) would produce a form asymptotically incompatible with the $-4/3$ power law observed for γ at $z < 0.5z_i$. Instead we adopt the empirical form for C_T^2 derived directly from data (Wyngaard et al., 1971), adjusted to provide a good match with γ for $k = 0.4$. The form for $\phi_\epsilon^{2/3}$ in (2.14), in fact, evolves asymptotically to a constant in the lower mixed layer. Our formulations for the surface layer are

$$\frac{C_V^2 z^{2/3}}{u_*^2} = \begin{cases} 4(1+0.5|z/L|)^{2/3}, & z/L \le 0 \\ 4(1+5z/L)^{2/3}, & z/L \ge 0 \end{cases} \tag{2.46}$$

$$\frac{C_T^2 z^{2/3}}{T_*^2} = \begin{cases} 5(1+6.4|z/L|)^{-2/3}, & z/L \le 0 \\ 5(1+3z/L), & z/L \ge 0, \end{cases} \tag{2.47}$$

shown as functions of z/L in Fig. 2.16. Their mixed layer asymptotic forms are now

$$\frac{C_V^2 z_i^{2/3}}{w_*^2} = 2.52k^{2/3} \simeq 1.4, \tag{2.48}$$

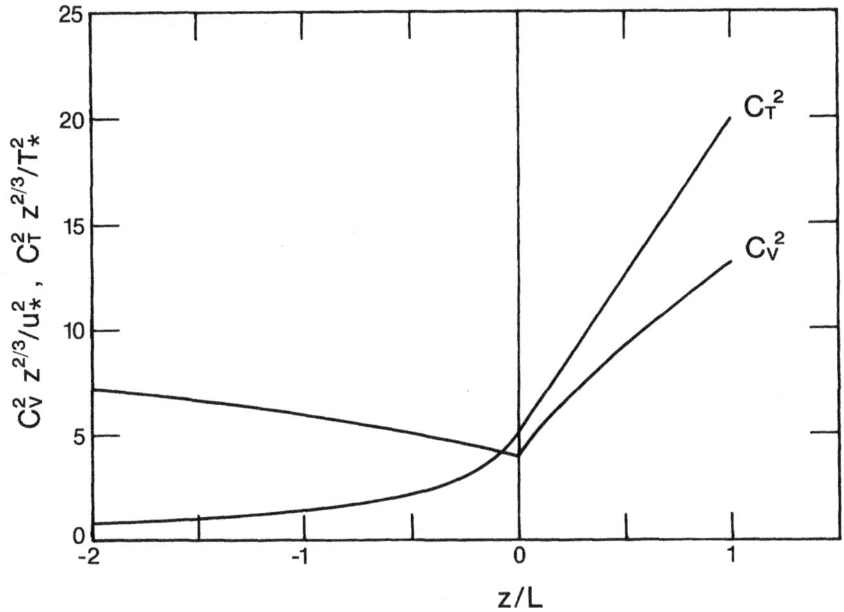

FIG. 2.16. Normalized structure parameters for velocity and temperature shown as functions of z/L.

$$\frac{C_T^2 z_i^{2/3}}{\theta_*^2} = 1.45k^{-2/3}(z/z_i)^{-4/3} \simeq 2.7\,(z/z_i)^{-4/3}. \tag{2.49}$$

The mixed layer profiles for C_V^2 and C_T^2 are presented in Fig. 2.17. Their evolution with height follows observations reported by Caughey and Palmer (1979). Of all the asymptotic power laws discussed so far, the form in (2.49) for C_T^2 shows the best fit with observations (Kaimal et al., 1976) over a very deep layer of the convective boundary layer. We do not have comparable asymptotic forms for C_V^2 and C_T^2 in the stable outer layer.

The functional forms for C_Q^2 are the same as for C_T^2 in the surface layer and in the lower mixed layer. The location of the minimum (Fig. 2.17) would, however, fluctuate greatly depending on the relative magnitude of moisture flux from above, through entrainment, and the surface flux (Fairall, 1987).

2.9 Cospectra of turbulence

The cospectra of uw and $w\theta$ give us valuable information on the averaging times and frequency responses needed for estimating momentum flux and heat flux (the fluxes are the integrals of the cospectra from $f = 0$ to ∞ as shown in Appendix 2.1). In the surface layer we have universal curves developed from field

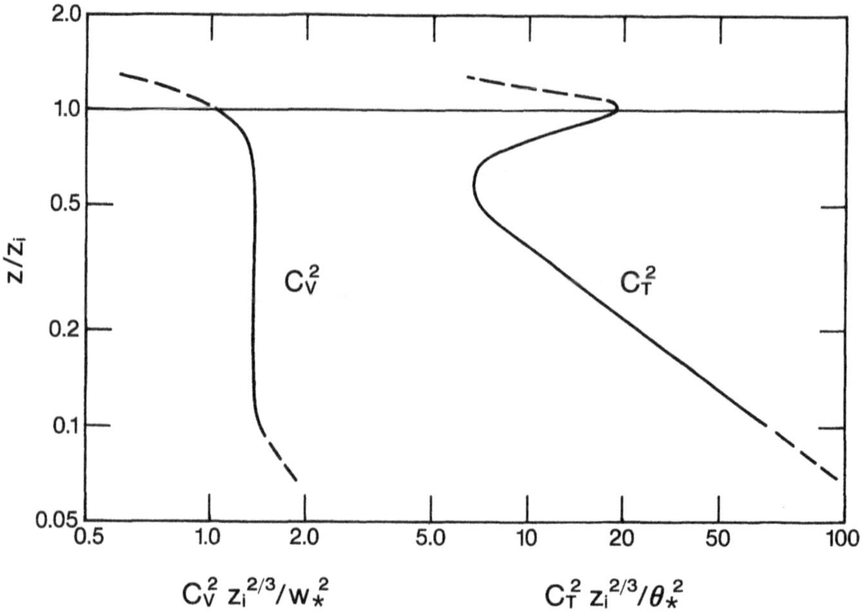

FIG. 2.17. Vertical profiles of the normalized structure parameters in the convective boundary layer. Note that the shape of the C_T^2 profile shown follows that of γ in (2.39).

measurements (Kaimal et al., 1972) that we can use as guides for establishing the sampling and sensor response requirements for any given application.

Inertial subrange cospectral forms for uw and $w\theta$ proposed by Wyngaard and Coté (1972) show them falling off as $n^{-7/3}$ (vanishing more rapidly than the power spectra of u, w, and θ, consistent with the requirement for isotropy). In this $-7/3$ range we would expect the normalized cospectral values to be functions only of z/L and n. Their logarithmic forms, when non-dimensionalized, become

$$-\frac{fC_{uw}(f)}{u_*^2} \propto G(z/L)n^{-4/3}, \tag{2.50}$$

$$-\frac{fC_{w\theta}(f)}{u_*T_*} \propto H(z/L)n^{-4/3}, \tag{2.51}$$

where $G(z/L)$ and $H(z/L)$ are functions of z/L, determined empirically from experiments. Note that $u_*T_* = -(\overline{w'\theta'})_0$. If we hold $G(0) = H(0) = 1$, from the Kansas experiment (Kaimal et al., 1972), we have

$$-\frac{fC_{uw}(f)}{u_*^2 G(z/L)} = 0.05n^{-4.3}, \tag{2.52}$$

$$-\frac{fC_{w\theta}(f)}{u_*T_*H(z/L)} = 0.14n^{-4/3}, \tag{2.53}$$

where

$$G(z/L) = \begin{cases} 1, & -2 \le z/L \le 0 \\ 1 + 7.9z/L, & 0 \le z/L \le 2 \end{cases} \tag{2.54}$$

$$H(z/L) = \begin{cases} 1, & -2 \le z/L \le 0 \\ 1 + 6.4z/L, & 0 \le z/L \le 2. \end{cases} \tag{2.55}$$

The universal curves for the two cospectra are combined in Fig. 2.18. In stable air the cospectral curves separate systematically according to z/L as the spectra did in earlier figures, whereas the unstable cospectra crowd into a narrow band that straddles the neutral cospectrum. For convenience, we take the $z/L = 0+$ curves for uw and $w\theta$ as the cospectral forms for the unstable surface layer ($0 \le z/L \le -2$). These curves can be approximated by

$$-\frac{fC_{uw}(f)}{u_*^2} = \frac{12n}{(1+9.6n)^{7/3}}, \tag{2.56}$$

$$-\frac{fC_{w\theta}(f)}{u_*T_*} = \begin{cases} \dfrac{11n}{(1\ +\ 13.3n)^{7/4}} & \text{for} \quad n \le 1.0 \\ \dfrac{4n}{(1\ +\ 3.8n)^{7/3}} & \text{for} \quad n \ge 1.0. \end{cases} \tag{2.57}$$

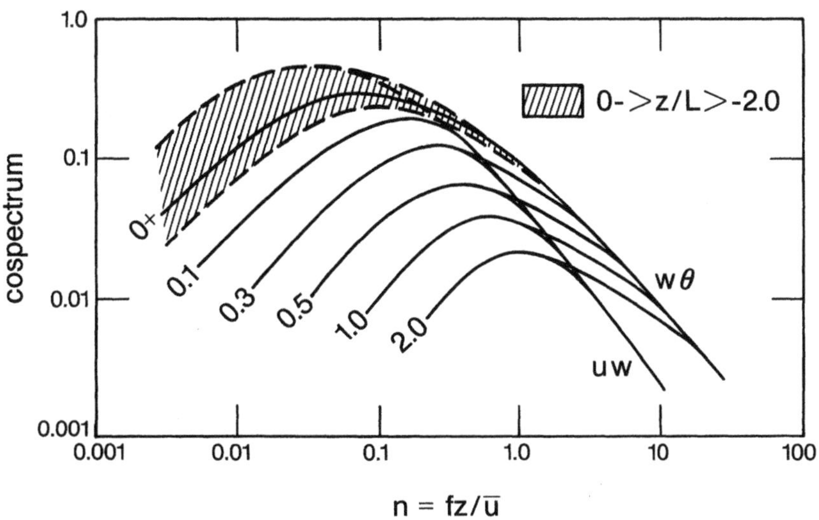

FIG. 2.18. Normalized surface layer cospectra of uw and $w\theta$, as represented in (2.52) and (2.53), shown varying with z/L. Note that the $w\theta$ cospectrum attains $-4/3$ behavior at a higher frequency than uw.

We can see from Fig. 2.18 that the two cospectra exhibit similar behavior at $n < 0.5$ (i.e., $\lambda > 2z$). They attain $-7/3$ behavior at different frequencies, however, about an octave higher in $w\theta$ than in uw. The extended range in $w\theta$ implies that the smaller eddies ($\lambda \sim z$) transport heat more effectively than momentum, pointing to the need for higher frequency response in surface layer measurements of heat flux compared to momentum flux. [On the basis of available evidence (e.g., Schmitt et al., 1979), we can assume that cospectra of w with scalars like moisture will exhibit the same shape and characteristics as the $w\theta$ cospectrum, and require sensors that are equally fast for flux measurement.]

The variation of the uw and $w\theta$ cospectral peaks with z/L is represented by a single curve in Fig. 2.19. In stable air it follows the same trend observed earlier in the velocity and temperature spectra (Fig. 2.9). In unstable air there is little, if any, variation with z/L, as one might surmise from Fig. 2.18.

As we move up into the mixed layer ($0.1z_i < z < z_i$) where the fluxes of momentum and heat are typically small and even change sign, as in the case of heat flux (Fig. 1.10), the cospectra become increasingly unpredictable. Often they show large excursions in both directions with no well-defined envelopes; the fluxes we measure are merely small differences between upward and downward transport located in different frequency bands (Kaimal et al., 1976). This is particularly true for the heat flux cospectra in which the balance shifts gradually with height until at about $0.8z_i$ the downward transport from entrainment begins to dominate. [Caughey and Kaimal (1977) found evidence of $(\lambda_m)_{w\theta}$ approaching $1.5z_i$ in the lower mixed layer.] Baroclinicity introduces similar uncertainties in stress cospectra. As a result, we have no universal forms to present for the mixed layer. Cospectral forms for the outer stable layer also tend to be unpredictable, but this

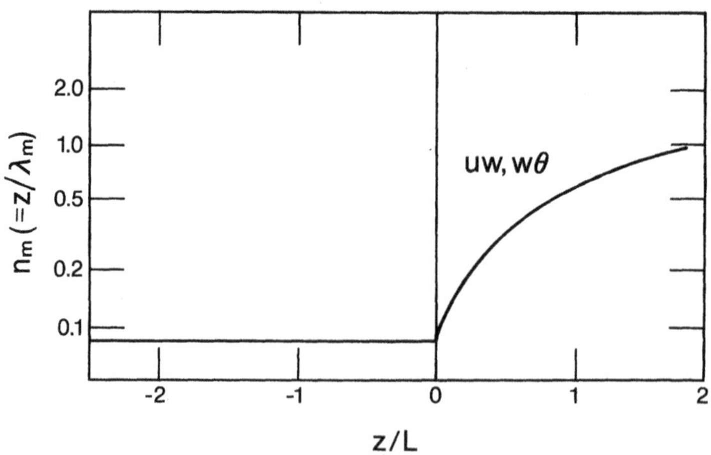

Fig. 2.19. Nondimensionalized frequency at the cospectral maxima shown as a function of z/L.

comes as no surprise given the very low levels of turbulence and the evolving nature of flow in that layer (Caughey et al., 1979).

Appendix 2.1 Fourier spectra: Essential formulas and relationships

A.2.1.1 *Fourier transform pair*

The three-dimensional energy spectrum $E_{ij}(\boldsymbol{\kappa})$ and the covariance tensor $R_{ij}(\mathbf{r})$ for homogeneous flow form a Fourier transform pair that is usually expressed as

$$E_{ij}(\boldsymbol{\kappa}) = \frac{1}{(2\pi)^3} \iiint\limits_{-\infty}^{\infty} R_{ij}(\mathbf{r}) e^{-i\boldsymbol{\kappa}\cdot\mathbf{r}} \, d\mathbf{r} \tag{2.58a}$$

$$R_{ij}(\mathbf{r}) = \iiint\limits_{-\infty}^{\infty} E_{ij}(\boldsymbol{\kappa}) e^{i\boldsymbol{\kappa}\cdot\mathbf{r}} \, d\boldsymbol{\kappa}. \tag{2.58b}$$

(For inhomogeneous flow, both the spectrum and the covariance tensor will be functions of the position vector $\mathbf{x}$ as well, as expressed in Section 2.1.)

A.2.1.2 *Energy spectrum*

The sum of the diagonal components of $R_{ij}(0)$ is twice the turbulent kinetic energy e:

$$R_{ii}(0) = \overline{u'^2} + \overline{v'^2} + \overline{w'^2} = 2e. \tag{2.59}$$

From (2.58b) we have

$$R_{ii}(0) = \iiint\limits_{-\infty}^{\infty} E_{ii}(\boldsymbol{\kappa}) \, d\boldsymbol{\kappa}, \tag{2.60}$$

which leads to the definition of spectral density as twice the contribution to turbulent kinetic energy from an element of wavenumber space $d\boldsymbol{\kappa}$.

The directional information in $E_{ii}(\boldsymbol{\kappa})$ can be removed by integrating over a spherical shell in wavenumber space of radius $\kappa = |\boldsymbol{\kappa}|$. This operation defines the energy spectrum $E(\kappa)$:

$$E(\kappa) = \frac{1}{2} \oiint E_{ii}(\boldsymbol{\kappa}) \, d\sigma, \tag{2.61}$$

where $d\sigma$ is an element of the shell and 1/2 is included to ensure that

$$\int\limits_{0}^{\infty} E(\kappa) \, d\kappa = e.$$

A.2.1.3 *One-dimensional covariances and spectra*

The most commonly measured covariances are those with separations in the x direction (r_1) since those can be replaced with a time delay (ξ) using Taylor's frozen turbulence hypothesis

$$r_1 = \overline{u}\xi.$$

The *autocovariance* function for variable α ($= u, v,$ or w)

$$R_\alpha(r_1) = \overline{\alpha'(x) \cdot \alpha'(x + r_1)} \tag{2.62a}$$

becomes

$$R_\alpha(\xi) = \overline{\alpha'(t) \cdot \alpha'(t + \xi)}. \tag{2.62b}$$

The Fourier transforms of $R_\alpha(r_1)$ and $R_\alpha(\xi)$ are the one-dimensional spectra of α' in the wavenumber and frequency domains, respectively. These are two-sided spectra we identify with a $\wedge$ to distinguish them from the one-sided power spectra we deal with in this chapter and elsewhere:

$$\hat{F}_\alpha(\kappa_1) = \frac{1}{2\pi} \int_{-\infty}^{\infty} R_\alpha(r_1) e^{-i\kappa_1 r_1}\, dr_1, \tag{2.63a}$$

$$\hat{S}_\alpha(\omega) = \frac{1}{2\pi} \int_{-\infty}^{\infty} R_\alpha(\xi) e^{-i\omega\xi}\, d\xi, \tag{2.63b}$$

where κ_1 ($= 2\pi f/\overline{u}$) is the wavenumber component in the streamwise (x) direction, and ω ($= 2\pi f$) is the angular frequency. Their inverse transforms have the form

$$R_\alpha(r_1) = \int_{-\infty}^{\infty} \hat{F}_\alpha(\kappa_1) e^{i\kappa_1 r_1}\, d\kappa_1, \tag{2.64a}$$

$$R_\alpha(\xi) = \int_{-\infty}^{\infty} \hat{S}_\alpha(\omega) e^{i\omega\xi}\, d\omega. \tag{2.64b}$$

The two spectra are related through Taylor's hypothesis

$$\hat{F}_\alpha(\kappa_1) = \overline{u}\hat{S}_\alpha(\omega). \tag{2.65}$$

The type of spectrum we deal with in atmospheric work and in many engineering applications is the one-sided power spectrum $S_\alpha(f)$ such that

$$\int_0^\infty S_\alpha(f)\,df = \sigma_\alpha^2 = \int_{-\infty}^\infty \hat{S}_\alpha(\omega)\,d\omega.$$

Since $S_\alpha(f)$ is the positive half of an even function, we are constrained, in the interest of preserving the variance, to have

$$S_\alpha(f) = 2\hat{S}_\alpha(f) = 4\pi\hat{S}_\alpha(\omega). \tag{2.66}$$

Note that the conversion of $\hat{S}_\alpha(f)$ to $\hat{S}_\alpha(\omega)$ requires multiplication by 2π.

A.2.1.4 *Eulerian integral length and time scales*

The autocovariance functions lead us directly to definitions of the Eulerian integral length scale Λ_α and time scale $\mathcal{T}_\alpha$:

$$\Lambda_\alpha = \frac{1}{\sigma_\alpha^2}\int_0^\infty R_\alpha(r_1)\,dr_1 = \int_0^\infty \rho_\alpha(r_1)\,dr_1, \tag{2.67a}$$

$$\mathcal{T}_\alpha = \frac{1}{\sigma_\alpha^2}\int_0^\infty R_\alpha(\xi)\,d\xi = \int_0^\infty \rho_\alpha(\xi)\,d\xi, \tag{2.67b}$$

where $\rho(r_1)$ and $\rho(\xi)$ are the familiar *autocorrelation* functions in space and time, the normalized versions of the autocovariance functions in (2.62a) and (2.62b). As in (2.65) we have, through Taylor's hypothesis,

$$\Lambda_\alpha = \overline{u}\mathcal{T}_\alpha. \tag{2.68}$$

We define the autocorrelation only for positive r_1's and ξ's because the autocovariance is an even function, and we have

$$\int_0^\infty R_\alpha(r_1)\,dr_1 = \frac{1}{2}\int_{-\infty}^\infty R_\alpha(r_1)\,dr_1. \tag{2.69}$$

The integral scales are related to the one-dimensional spectra through (2.63a,b), (2.67a,b), (2.66), and (2.69):

$$\Lambda_\alpha = \frac{1}{\sigma_\alpha^2}\int_0^\infty R_\alpha(r_1)\,dr_1$$

$$
\begin{aligned}
&= \frac{\pi}{\sigma_\alpha^2}\left[\frac{1}{2\pi}\int_{-\infty}^{\infty} R_\alpha(r_1)e^{-i\kappa_1 r_1}\,dr_1\right] \qquad \text{at } \kappa_1 = 0 \\
&= \frac{\pi}{\sigma_\alpha^2}\hat{F}_\alpha(0) \qquad (2.70)
\end{aligned}
$$

$$
\begin{aligned}
\mathcal{T}_\alpha &= \frac{\pi}{\sigma_\alpha^2}[\hat{S}(f)]_{f=0} \\
&= \frac{1}{4\sigma_\alpha^2}[S(\omega)]_{\omega=0}. \qquad (2.71)
\end{aligned}
$$

(In Appendix 7.2, Chapter 7, we discuss the consequences of high-pass filtering of data which essentially removes all energy at zero frequency, forcing Λ_α and $\mathcal{T}_\alpha$, as defined here, to zero.)

A.2.1.5 *Cross covariances and cross spectra*

The cross covariance between variables α and β can be expressed as

$$R_{\alpha\beta}(r_1) = \overline{\alpha'(x)\cdot\beta'(x+r_1)}. \qquad (2.72)$$

This function, unlike the autocovariance function, is not an even function. We find, in general,

$$\overline{\alpha'(x)\cdot\beta'(x+r_1)} \neq \overline{\alpha'(x+r_1)\cdot\beta'(x)},$$

and its Fourier transform has both real and imaginary parts. The cross spectrum $\hat{C}r_{\alpha\beta}(\kappa_1)$ [transform of $R_{\alpha\beta}(r_1)$] separates into real and imaginary parts:

$$
\begin{aligned}
\hat{C}r_{\alpha\beta}(\kappa_1) &= \hat{C}_{\alpha\beta}(\kappa_1) - i\hat{Q}_{\alpha\beta}(\kappa_1) \\
&= \frac{1}{2\pi}\int_{-\infty}^{\infty} R_{\alpha\beta}(r_1)e^{-i\kappa_1 r_1}\,dr_1. \qquad (2.73)
\end{aligned}
$$

The real part of the cross spectrum is the cospectrum and the imaginary part the quadrature spectrum. If $R_{\alpha\beta}(r_1)$ is split into its even and odd parts, $\hat{E}_{\alpha\beta}(r_1)$ and $\hat{O}_{\alpha\beta}(r_1)$, respectively,

$$R_{\alpha\beta}(r_1) = \hat{E}_{\alpha\beta}(r_1) + \hat{O}_{\alpha\beta}(r_1), \qquad (2.74)$$

it can be shown that $\hat{C}_{\alpha\beta}(\kappa_1)$ is the Fourier transform of $\hat{E}_{\alpha\beta}(r_1)$ and $\hat{Q}_{\alpha\beta}(\kappa_1)$ the transform of $\hat{O}_{\alpha\beta}(r_1)$:

$$\hat{C}_{\alpha\beta}(\kappa_1) = \frac{1}{2\pi}\int_{-\infty}^{\infty} \hat{E}_{\alpha\beta}(r_1)e^{-i\kappa_1 r_1}\,dr_1, \qquad (2.75a)$$

$$\hat{Q}_{\alpha\beta}(\kappa_1) = \frac{i}{2\pi}\int_{-\infty}^{\infty} \hat{O}_{\alpha\beta}(r_1)e^{-i\kappa_1 r_1}\,dr_1. \qquad (2.75b)$$

Their transforms are, respectively,

$$\hat{E}_{\alpha\beta}(r_1) = \int_{-\infty}^{\infty} \hat{C}_{\alpha\beta}(\kappa_1) e^{i\kappa_1 r_1}\, d\kappa_1, \tag{2.76a}$$

$$\hat{O}_{\alpha\beta}(r_1) = -i \int_{-\infty}^{\infty} \hat{Q}_{\alpha\beta}(\kappa_1) e^{i\kappa_1 r_1}\, d\kappa_1. \tag{2.76b}$$

$\hat{O}_{\alpha\beta}(r_1)$ makes no contribution to $R_{\alpha\beta}(0)$. Hence, from (2.76b) we have

$$\overline{\alpha'\beta'} = R_{\alpha\beta}(0) = \int_{-\infty}^{\infty} \hat{C}_{\alpha\beta}(\kappa_1)\, d\kappa_1. \tag{2.77}$$

The one-sided cospectrum $C_{\alpha\beta}(f)$ used in our discussions is defined such that

$$\int_0^{\infty} C_{\alpha\beta}(f)\, df = \overline{\alpha'\beta'}.$$

Appendix 2.2 Taylor's hypothesis in the atmospheric boundary layer

Taylor's frozen turbulence hypothesis enables us to convert temporal measurements at a point to spatial patterns in space through the transformation $x = \overline{u}t$. Implicit is the assumption that the turbulent field is frozen in time and transported horizontally past the observer. Frequency scales become wavenumber scales ($\kappa_1 = 2\pi f/\overline{u}$), but the spectra remain unchanged in their shapes as well as their magnitudes.

We know, however, that atmospheric turbulence is neither frozen (it evolves with time) nor transported precisely at local mean wind speeds. The former fact we ignore because, typically, the eddy life times are long compared to their travel time across the sensor. The latter we usually ignore on the basis of correlation studies conducted by early investigators (Lumley and Panofsky, 1964). The high degree of consistency in the time-averaged statistics and spectra observed over the years has served to reinforce that practice.

There is, however, ample evidence of eddy convection velocities departing from $\overline{u}$, the local mean wind speed, in the CBL. Wilczak and Businger (1984), in their comprehensive study of eddy transport in the convective surface layer ($z < 150$ m) using tower measurements, tracked ramplike temperature structures (plumes and thermals) moving at speeds that varied from ramp to ramp; the larger structures, as a rule, moved faster than the smaller ones. The large thermals, along with their roots in the surface layer, are convected at mean mixed layer wind velocities (Davison, 1974), whereas the smaller (and shorter) plumes, sustained by local sources of warm air near the surface, travel at speeds smaller than $\overline{u}$ (Kaimal, 1974). Taken as a whole, the effective convection velocity, according to Wilczak and Businger (1984), is between 0.7 and 0.8 times the mixed layer mean wind. This is consistent with the findings of Kaimal et al. (1982), which showed 150-m tower u, v, and w spectral peaks shifted to slightly lower (less than one octave) wavenumbers compared to their aircraft counterparts. Whereas the aircraft and tower u spectra are virtually indistinguishable in their shapes and placement, the aircraft v and w spectra showed more rapid rolloff on the low-frequency side of their peaks compared to tower spectra.

Wind tunnel turbulence (neutral air) shows somewhat different behavior, as pointed out by Perry and Li (1990). In the near-wall region, they find small eddies traveling with the local mean wind, whereas larger "attached" eddies moved faster, causing u spectra from a flying hot-wire anemometer probe to shift slightly to lower frequencies. Within plant canopies, this behavior is exaggerated. Finnigan (1979), for example, measured the mean convection velocity of energy-containing eddies in a wheat canopy as 1.8 times the mean wind speed at canopy top. There is strong evidence of similar behavior over a wide range of natural and model plant canopies (Raupach et al., 1989). The behavior of large eddies in plant canopies is discussed in detail in Chapter 3.

Despite these findings, we continue to use $\overline{u}$ as the convection velocity in most boundary layer applications. This practice will continue as long as investigators find reasonably good agreement between atmospheric measurements made with moving and stationary probes.

Appendix 2.3 Low-frequency behavior of one-dimensional spectra

One-dimensional spectra of turbulence, available to us through measurements in the atmosphere, give misleading information on the behavior of three-dimensional turbulence at very low frequencies (Tennekes and Lumley, 1972) because modes of wavenumber κ traveling nearly normal to the κ_1 direction (the direction that defines the one-dimensional spectrum) appear as very low wavenumber contributions in $F_u(\kappa_1)$, $F_v(\kappa_1)$, and $F_w(\kappa_1)$. The phenomenon is analogous to measuring wave separation along the direction parallel to the shoreline when ocean waves are approaching the shoreline almost at right angles to it. As a result, the one-dimensional spectrum may show finite energy at $\kappa_1 = 0$ when, in fact, there is no energy at zero wavenumber.

Typically, the one-dimensional spectra of u, v, and w level off to a constant value at the low wavenumber end, as shown in Fig. 2.4. Interestingly enough, it is their intercept at $\kappa_1 = 0$ that transforms into the integral time scale $\mathcal{T}_\alpha$ (or Λ_α).

For the one-dimensional, one-sided power spectrum of α, the relationship between this apparent energy at $f = 0$ and $\mathcal{T}_\alpha$ can be derived directly from the relationship

$$\frac{S_\alpha(f)}{\sigma_\alpha^2} = 4 \int_0^\infty \rho_\alpha(\xi) \cos(2\pi f \xi)\, d\xi, \tag{2.78}$$

where σ_α^2 is the variance and $\rho_\alpha(\xi)$ the autocorrelation function of α (Fig. 2.2). At $f = 0$, we have

$$S_\alpha(0) = 4\sigma_\alpha^2 \int_0^\infty \rho_\alpha(\xi)\, d\xi = 4\sigma_\alpha^2 \mathcal{T}_\alpha, \tag{2.79}$$

which is the same as the relationship (2.71) derived from the two-sided spectrum. From (2.78) and (2.79) we can derive the asymptotic form for the low-frequency behavior of $f S_\alpha(f)/\sigma_\alpha^2$ in Fig. 2.10:

$$fS_\alpha(f)/\sigma_\alpha^2 = 4\mathcal{T}_\alpha f. \tag{2.80}$$

Combining (2.80) and the asymptotic form for (2.22) as $f \to 0$, we get

$$4\mathcal{T}_\alpha f = 0.164 f / f_0$$

or

$$f_0 \simeq \frac{1}{24\mathcal{T}_\alpha}. \tag{2.81}$$

In the spectrum defined by (2.22), the peak occurs at frequency $f_m \simeq 3.8 f_0$, which leads us to the relationship

$$(f_m)_\alpha \simeq \frac{1}{6.3\mathcal{T}_\alpha}$$

or

$$(\lambda_m)_\alpha \simeq 6.3\Lambda_\alpha. \tag{2.82}$$

For the hypothetical spectrum with $\rho_\alpha(\xi) = e^{-\xi/\mathcal{T}_\alpha}$ (Lumley and Panofsky, 1964), the peak frequency is exactly $1/2\pi\mathcal{T}_\alpha$.

Appendix 2.4 Relationship between Kolmogorov and von Karman constants

In the formulation of (2.12) we have an implicit relationship between the Kolmogorov constant α_1 and the von Karman constant k that dictates the value of one when the other is known. In neutrally stable air, where we can assume $\phi_\epsilon = 1$, (2.12) reduces to

$$\begin{aligned} \frac{fS_u(f)}{u_*^2} &= \frac{\alpha_1 k^{-2/3}}{(2\pi)^{2/3}} \left(\frac{fz}{\overline{u}}\right)^{-2/3} \\ &= \frac{\alpha_1 k^{-2/3}}{3.4} (n)^{-2/3} \end{aligned} \tag{2.83}$$

or

$$\alpha_1 k^{-2/3} = 3.4 \left[\frac{fS_u(f)}{u_*^2}\right] n^{2/3}. \tag{2.84}$$

In the inertial subrange, at say $n = 4$, the Kansas data (Kaimal et al., 1972) yield

$$\left[\frac{fS_u(f)}{u_*^2}\right] = 0.12 \qquad \text{for} \quad z/L = 0.$$

Then

$$\alpha_1 k^{-2/3} = 3.4 \times 0.12 \times (4)^{2/3} \simeq 1.0. \tag{2.85}$$

We have in (2.85) the rationale for choosing $\alpha_1 = 0.55$ to match the now accepted value of $k = 0.4$ in place of the $\alpha_1 = 0.5, k = 0.35$ combination used with the Kansas data. The formulations of the inertial subrange relationships in (2.13), (2.15), and (2.16) remain unchanged. The choice of $\alpha_1 = 0.55$ does, however, alter the coefficient for the structure function of velocity in (2.42). The new value is 2.2 instead of the generally used value of 2.0.

Frenzen (1973), who pointed out the connection between the two constants, used a different formulation. In place of u_*^2 he used the wind profile equivalent $(kz\,\partial\overline{u}/\partial z)^2$ to get $\alpha_1 k^{4/3} = 0.136$ in the logarithmic layer under neutral conditions. Frenzen's present estimate (personal communication) for $\alpha_1 k^{4/3}$, based on a reexamination of his wind profile measurements, falls between 0.16 and 0.17. A value of 0.162 is consistent with choosing $\alpha_1 = 0.55$ and $k = 0.4$.

References

Batchelor, G. K., 1960: *The Theory of Homogeneous Turbulence.* Cambridge University Press, Cambridge, 197 pp.

Caughey, S. J., and J. C. Kaimal, 1977: Vertical heat flux in the convective boundary layer. *Quart. J. Roy. Meteor. Soc., 103,* 811–815.

Caughey, S. J., and S. G. Palmer, 1979: Some aspects of turbulence structure through the depth of the convective boundary layer. *Quart. J. Roy. Meteor. Soc., 105,* 811–827.

Caughey, S. J., J. C. Wyngaard, and J. C. Kaimal, 1979: Turbulence in the evolving stable boundary layer. *J. Atmos. Sci., 36,* 1041–1052.

Corrsin, S., 1951: On the spectrum of isotropic temperature fluctuations in an isotropic turbulence. *J. Appl. Phys., 22,* 469–473.

Davison, D. S., 1974: The translation velocity of convective plumes. *Quart. J. Roy. Meteor. Soc., 100,* 572–592.

Dubovikov, M. M., and V. I. Tatarskii, 1987: Calculation of the asymptotic form of spectrum of locally isotropic turbulence in the viscous range. *Soviet Phys. J. Exp. Theor. Phys., 66* (English translation by the American Institute of Physics, 1988, 1136–1141).

Fairall, C. W., 1987: A top-down and bottom-up diffusion model for C_T^2 and C_Q^2 in the entraining convective boundary layer. *J. Atmos. Sci., 44,* 1010–1017.

Finnigan, J. J., 1979: Turbulence in waving wheat. I. Mean statistics and honami. *Bound.-Layer Meteor., 16,* 181–211.

Finnigan, J. J., F. Einaudi, and D. Fua, 1984: The interaction between an internal gravity wave and turbulence in the stably stratified nocturnal boundary layer. *J. Atmos. Sci., 41,* 2409–2436.

Frenzen, P., 1973: The observed relation between the Kolmogorov and von Kármán constants in the surface boundary layer. *Bound.-Layer Meteor., 3,* 348–358.

Hill, R. J., 1978: Models of the scalar spectrum for turbulent advection. *J. Fluid Mech., 88,* part 3, 541–562.

Højstrup, J., 1982: Velocity spectra in the unstable boundary layer. *J. Atmos. Sci., 39,* 2239–2248.

Kaimal, J. C., 1973: Turbulence spectra, length scales and structure parameters in the stable surface layer. *Bound.-Layer Meteor., 4,* 289–309.

Kaimal, J. C., 1974: Translation speed of convective plumes in the atmospheric surface layer. *Quart. J. Roy. Meteor. Soc., 100,* 46–52.

Kaimal, J. C., 1978: Horizontal velocity spectra in an unstable surface layer. *J. Atmos. Sci.*, *35*, 18–24.

Kaimal, J. C., and Y. Izumi, 1965: Vertical velocity fluctuations in a nocturnal low-level jet. *J. Appl. Meteor.*, *4*, 576–584.

Kaimal, J. C., R. A. Eversole, D. H. Lenschow, B. B. Stankov, P. H. Kahn, and J. A. Businger, 1982: Spectral characteristics of the convective boundary layer over uneven terrain. *J. Atmos. Sci.*, *39*, 1098–1114.

Kaimal, J. C., J. C. Wyngaard, Y. Izumi, and O. R. Coté, 1972: Spectral characteristics of surface layer turbulence. *Quart. J. Roy. Meteor. Soc.*, *98*, 563–589.

Kaimal, J. C., J. C. Wyngaard, D. A. Haugen, O. R. Coté, Y. Izumi, S. J. Caughey, and C. J. Readings, 1976: Turbulence structure in the convective boundary layer. *J. Atmos. Sci.*, *33*, 2152–2169.

Kolmogorov, A. N., 1941: The local structure of turbulence in incompressible viscous fluid for very large Reynolds numbers. *Doklady ANSSSR*, *30*, 301–304.

Lumley, J. L., and H. A. Panofsky, 1964: *The Structure of Atmospheric Turbulence.* Wiley-Interscience, New York, 239 pp.

Moraes, O. L. L., 1988: The velocity spectra in the stable surface layer. *Bound.-Layer Meteor.*, *43*, 223–230.

Okamoto, M., and E. K. Webb, 1970: Temperature fluctuations in stable stratification. *Quart. J. Roy. Meteor. Soc.*, *96*, 591–600.

Perry, A. E., and J. D. Li, 1990: Experimental evidence for the attached-eddy hypothesis in zero pressure gradient turbulent boundary layers. *J. Fluid Mech.*, *218*, 405–438.

Raupach, M. R., J. J. Finnigan, and Y. Brunet, 1989: Coherent eddies in vegetation canopies. *Proc. Fourth Australian Conference on Heat and Mass Transfer*, Christchurch, New Zealand, 9–12 May 1989, 75–90.

Schmitt, K. F., C. A. Friehe, and C. H. Gibson, 1979: Structure of marine surface layer turbulence. *J. Atmos. Sci.*, *36*, 602–618.

Tatarski, V. I., 1961: *Wave Propagation in a Turbulent Medium* (translation from Russian by R. A. Silverman). McGraw-Hill, New York, 285 pp.

Tennekes, H., and J. L. Lumley, 1972: *A First Course in Turbulence.* MIT Press, Cambridge, 300 pp.

Van der Hoven, I., 1957: Power spectrum of horizontal wind speed in the frequency range from 0.0007 to 900 cycles per hour. *J. Meteor.*, *14*, 160–164.

Wilczak, J. M., and J. A. Businger, 1984: Large-scale eddies in the unstably stratified atmospheric surface layer. Part II: Turbulent-pressure fluctuations and budgets of heat flux, stress, and turbulent kinetic energy. *J. Atmos. Sci.*, *41*, 3551–3567.

Willis, G. E., and J. W. Deardorff, 1974: A laboratory model of the unstable planetary boundary layer. *J. Atmos. Sci.*, *31*, 1297–1307.

Wyngaard, J. C., and O. R. Coté, 1972: Cospectral similarity in the atmospheric surface layer. *Quart. J. Roy. Meteor. Soc.*, *98*, 590–603.

Wyngaard, J. C., Y. Izumi, and S. A. Collins, 1971: Behavior of the refractive-index-structure parameter near the ground. *J. Opt. Soc.*, *61*, 1646–1650.

Additional Reading

Panofsky, H. A., and J. A. Dutton, 1984: *Atmospheric Turbulence.* Wiley-Interscience, New York, 397 pp.

3

FLOW OVER PLANT CANOPIES

Any land surface that receives regular rainfall is almost certain to be covered by vegetation. Most of the inhabitable regions of the globe fall into this category. Often the vegetation is tall enough to call into question the assumption, implicit in the discussion of the first two chapters, that the roughness elements on the ground surface are much lower than any observation height of interest to us. In fact, if we venture to make measurements too close to tall vegetation, we discover significant departures from many of the scaling laws and formulas that seem to work in the surface layer above the canopy.

To take one example, momentum is absorbed from the wind not just at the ground surface but through the whole depth of the canopy as aerodynamic drag on the plants. Consequently, although we still observe a logarithmic velocity profile well above the canopy, its apparent origin has moved to a level $z = d$ near the top of the plants. The precise position of this "displacement height," d, depends on the way the drag force is distributed through the foliage and this in turn depends on the structure of the mean wind and turbulence within the canopy.

Our interest in the nature of within-canopy turbulence, however, is not motivated solely by its influence on the surface layer above. The understanding of turbulent transfer within foliage canopies provides the intellectual underpinning for the physical aspects of agricultural meteorology. As such it has a history almost as venerable as investigations of the surface layer itself. The landmark study of *Weather in Wheat* by Penman and Long (1960) was the first of a series of seminal papers to establish the quantitative link between the turbulent fluxes in a canopy and the physiological sources and sinks of heat, water vapor, and carbon dioxide (CO_2). Prominent and influential among these early publications were those by Uchijima (1962), Denmead (1964), Brown and Covey (1966), and Lemon and Wright (1969).

Whereas these authors were motivated by curiosity about plant physiology and the transfer of water and other scalars through the soil-plant-air continuum, other workers forged the link between the classical surface layer studies detailed in

Chapter 1 and the structure of within-canopy turbulence. Some early contributors to this facet of canopy work were Inoue (1963), Uchijima and Wright (1964), and Wright and Lemon (1966). The many wind tunnel and field studies conducted over the last two decades have shed considerable light on the nature of turbulent transport in and above plant canopies, but most of what we know, in the micrometeorological sense, comes from studies of crops and forests on level ground. We shall therefore limit the scope of this chapter to uniform plant canopies on flat land, making the assumption that on the scale of a few or more plants, the canopy structure can be viewed as one-dimensional, varying only in the vertical (z) direction.

Characterizing and measuring the flow within a canopy is made particularly difficult by the very high turbulence intensities encountered there. These exceed by far the turbulent intensities encountered elsewhere in the surface layer and result, predominantly, from energy production in the intense shear layer at the top of the plants, although smaller scale turbulence produced in plant wakes also makes a significant contribution. Furthermore, the quasirandom distribution of air spaces between the plants introduces spatial as well as temporal complexity to the turbulence there. Add to that the presence of three-dimensional arrays of sources and sinks of momentum and scalars such as heat and water vapor, and it is not surprising that many of the theoretical concepts that form the mainstay of surface layer theory do not apply here. For example, the failure of eddy diffusivities to describe turbulent transfer in canopies was established a decade ago. As a direct consequence, there is to date no general similarity scheme with the unifying power of Monin-Obukhov theory.

The plant canopy is a biologically active entity with water vapor and CO_2 transfers across leaf surfaces. These exchanges are controlled by a complex biomechanical system that opens and closes tiny pores called stomata on the leaf surface in response to a variety of stimuli. Among the foremost of these is the amount of water in the plant and the soil, which depends in turn on the history of evaporation and precipitation. The canopy can be viewed, therefore, as an interface between essentially rapid physical processes such as solar irradiation and turbulent exchange in the boundary layer and slower-acting biological factors. Consequently, a proper treatment of the canopy recognizes that it does not simply form a rough lower boundary to the surface layer but plays a dynamic role in the partitioning of radiant energy into sensible and latent heat.

3.1 Flow above the canopy

There are many situations where, although we wish to take account of the presence of the foliage and its influence on the flow above, we prefer not to consider the detailed distribution of turbulence in the canopy air space. In such cases, we turn to single-layer models that treat the canopy as a notional plane at the bottom of the boundary layer at which momentum and scalars are absorbed (or emitted) and where radiant energy is partitioned into sensible and latent heat. Such models are of two kinds: those deriving directly from surface layer similarity (Chapter 1) and those involving application of the "combination equation" to the whole canopy. The first group includes the logarithmic law for velocity and scalars and bulk transfer formulas, which are essentially applications of the logarithmic law at two levels, the ground and a reference level above. The combination equation takes a different approach, which we shall discuss later.

3.1.1 *Logarithmic law*

In near-neutral conditions, the logarithmic portion of the wind profile above the canopy extrapolates downward to a height $(d + z_0)$ at $\overline{u} = 0$ (Fig. 3.1). The height d, referred to as the displacement height, typically falls within 70% to 80% of the canopy height h_c. We write, therefore,

$$\overline{u}(z) = \frac{u_*}{k} \ln\left(\frac{z-d}{z_0}\right). \tag{3.1}$$

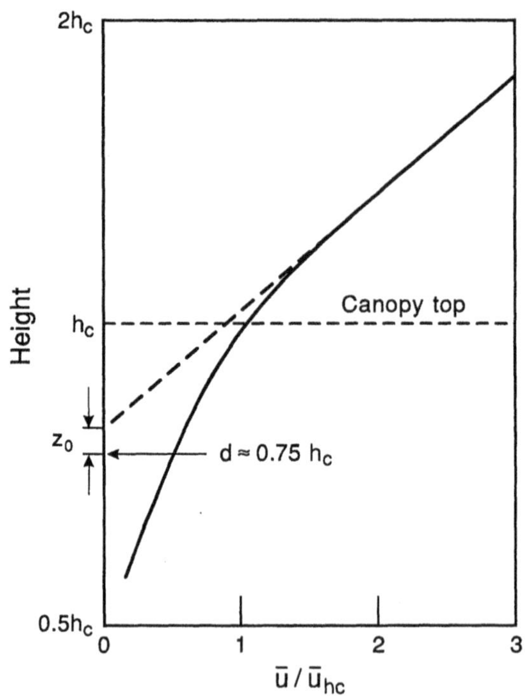

FIG. 3.1 Mean wind profile over a plant canopy showing relationships between canopy height (h_c), displacement height (d), and roughness length (z_0). Here, height is plotted on a logarithmic scale and mean wind speed, normalized by $\overline{u}_{hc}$ (the mean speed at canopy top), on a linear scale.

We now have a new origin for the logarithmic law at $z = d$, which also coincides with the mean level of momentum absorption (Jackson, 1981). Plotted as a function of $\ln(z-d)$, the $\overline{u}$ profile will show a zero intercept at the roughness length z_0 and a slope equal to u_*/k, as in (1.11). Corresponding expressions for heat and humidity, assuming the same d for both, are

$$\overline{\theta}(z) - \overline{\theta}_0 = \frac{T_*}{k} \ln\left(\frac{z-d}{z_h}\right), \tag{3.2}$$

$$\overline{q}(z) - \overline{q}_0 = \frac{q_*}{k} \ln\left(\frac{z-d}{z_q}\right), \tag{3.3}$$

where z_h and z_q are new roughness lengths for heat and humidity (analogous to z_0), $\overline{\theta}_0$ and $\overline{q}_0$ are the values of $\overline{\theta}$ and $\overline{q}$ at heights $(d+z_h)$ and $(d+z_q)$, respectively, and q_* is the humidity scale analogous to T_* defined as

$$q_* = \frac{-\overline{w'q'}}{u_*}. \tag{3.4}$$

Over canopies, the traditional nondimensional surface layer functions become

$$\phi_m = \frac{k(z-d)}{u_*}\frac{\partial \overline{u}}{\partial z}, \tag{3.5}$$

$$\phi_h = \frac{k(z-d)}{T_*}\frac{\partial \overline{\theta}}{\partial z}, \tag{3.6}$$

$$\phi_q = \frac{k(z-d)}{q_*}\frac{\partial \overline{q}}{\partial z}, \tag{3.7}$$

where ϕ_m, ϕ_h, and ϕ_q are functions now of $(z-d)$ instead of z. It is readily seen that (3.5), (3.6), and (3.7) are equivalent to flux-gradient expressions

$$\frac{\tau}{\rho} = u_*^2 = \left[k(z-d)u_*\phi_m^{-1}\right]\frac{\partial \overline{u}}{\partial z} = K_m \frac{\partial \overline{u}}{\partial z}, \tag{3.8}$$

$$\frac{H}{\rho c_p} = -u_* T_* = -\left[k(z-d)u_*\phi_h^{-1}\right]\frac{\partial \overline{\theta}}{\partial z} = -K_h \frac{\partial \overline{\theta}}{\partial z}, \tag{3.9}$$

$$\frac{E}{\rho} = -u_* q_* = -\left[k(z-d)u_*\phi_q^{-1}\right]\frac{\partial \overline{q}}{\partial z} = -K_q \frac{\partial \overline{q}}{\partial z}. \tag{3.10}$$

In neutral conditions, $K_m = K_h = K_q = k(z-d)u_*$. This is an assumption supported by a great deal of evidence collected over short vegetation and embodied in the ϕ functions (1.26) and (1.27) presented in Chapter 1 (Dyer, 1974; Högström, 1988).

3.1.2 *Bulk transfer coefficients and roughness lengths*

A convenient way to characterize the capacity of the canopy to absorb momentum, heat, and water vapor is in terms of transfer coefficients C_f, C_h, and C_q, where

$$C_f(z) = \frac{u_*}{\overline{u}(z)}, \tag{3.11}$$

$$C_h(z) = \frac{u_* T_*}{\overline{u}(z)[\overline{\theta}(z) - \overline{\theta}_0]}, \tag{3.12}$$

$$C_q(z) = \frac{u_* q_*}{\overline{u}(z)[\overline{q}(z) - \overline{q}_0]}. \tag{3.13}$$

We then have from (3.8) to (3.10)

$$\frac{\tau}{\rho} = C_f^2 \overline{u}^2, \tag{3.14}$$

$$\frac{H}{\rho c_p} = -C_h \overline{u}(\overline{\theta} - \overline{\theta}_0), \tag{3.15}$$

$$\frac{E}{\rho} = -C_q \overline{u}(\overline{q} - \overline{q}_0). \tag{3.16}$$

Combining (3.1), (3.2), and (3.3) with (3.11), (3.12), and (3.13), we can express z_0, z_h, and z_q in neutral conditions as

$$z_0 = (z - d)e^{-k/C_f}, \tag{3.17}$$

$$z_h = (z - d)e^{-kC_f/C_h}, \tag{3.18}$$

$$z_q = (z - d)e^{-kC_f/C_q}. \tag{3.19}$$

Hence, while z_0 expresses the capacity of the canopy to absorb momentum, z_h and z_q characterize its capacity to absorb (or emit) heat and water vapor *relative* to its ability to absorb momentum. Thom (1975) estimated z_h and z_q to be equal to $z_0/5$, but their precise values will be different for different canopies and for different values of $\overline{\theta}_0$ and $\overline{q}_0$. The numerical value of this ratio illustrates, nevertheless, that canopies are far more efficient at absorbing momentum than at absorbing scalars. This discrepancy is called the "bluff body effect" (Thom, 1975) and arises because momentum is transferred to the vegetation mainly by pressure forces that have no counterpart in the transfer of scalars.

In the absence of direct measurement of the turbulent fluxes in the constant flux layer above the vegetation, the usual way of estimating d, z_0, z_h, z_q, u_*, T_*, and q_* is to fit equations (3.1), (3.2), and (3.3) to measurements of mean concentrations and wind speed above the canopy. Unfortunately, over even moderately tall vegetation such as cereal crops, let alone over forests, the multiple regression analysis required is fraught with difficulty. At best, it can lead to errors of $\pm 25\%$ in d, z_0, and u_* (Bradley and Finnigan, 1973); at worst, it can produce profiles that exhibit completely spurious diabatic forms. The best way to avoid errors is to fix d independently by measuring the velocity profile within the canopy, finding the height of the centroid of the drag force and equating it to d (Thom, 1971; Jackson, 1981). This approach has been found to produce a fairly consistent value of $d \simeq 0.75h_c$ over a wide range of natural canopies.

The dependence of z_0 on canopy densities has been studied by Seginer (1974). As surface roughness increases with the addition of plants to bare soil, z_0 and C_f also increase until a point is reached when the plants crowd together so closely that the "shelter effect" reduces their individual capacity to absorb momentum. The physiological requirements of each plant to grow and absorb sunlight place an upper limit on this process, but the maximum aerodynamic roughness of the canopy occurs at an intermediate density where each plant can use its area to maximum advantage in intercepting the wind. The corresponding value of z_0 is close to $h_c/5$.

3.1.3 *Roughness sublayer*

The roughness sublayer is the region at the bottom of the boundary layer where the presence of the canopy impinges directly on the character of the turbulence. It extends from the ground to a height of about $3h_c$ and includes the canopy air space. If we make measurements above the canopy but within the roughness sublayer, we encounter further problems in applying the formulas of the last section. The first departure from standard behavior that we find is an increase in the eddy diffusion coefficients, K_m, K_h, and K_q above their logarithmic values of $[ku_*(z-d)\phi_m^{-1}]$, $[ku_*(z-d)\phi_h^{-1}]$, and $[ku_*(z-d)\phi_q^{-1}]$, respectively.

We can examine these departures by comparing measured values of the inverse diabatic influence functions, ϕ_m^{-1}, ϕ_h^{-1}, and ϕ_q^{-1} as defined in (3.5), (3.6), and (3.7) with the standard surface layer formulas given in Chapter 1. We compare the inverses because their ratios are directly proportional to the ratios of the eddy diffusivities. To clarify the distinction we identify the *measured* ϕ functions over canopies with a star ϕ_m^*, ϕ_h^*, and ϕ_q^*; the unstarred ϕ functions denote the formulas of Chapter 1.

We expect that as we move outside the roughness sublayer to heights $z \gtrsim 3h_c$, (ϕ_m^{*-1}, ϕ_h^{*-1}, and ϕ_q^{*-1}) will approach (ϕ_m^{-1}, ϕ_h^{-1}, and ϕ_q^{-1}). An increasing number of experiments where fluxes and gradients have been measured independently in the roughness sublayer allows us to examine how this approach occurs. Thus, we see in Fig. 3.2 that, over closed canopies, ϕ_m^{*-1} scatters around unity at neutral stability and moderate instabilities, whereas ϕ_h^{*-1} and ϕ_q^{*-1} reach values ranging from 2 to 5 under similar conditions (Shuttleworth, 1989). Garratt (1978) found that over a scattered, open, savannah-type canopy, ϕ_m^{*-1} at neutrality was close to 2. Figure 3.2 includes data points from a variety of experiments with measurement heights at varying distances above the canopy top. The scatter in the data points,

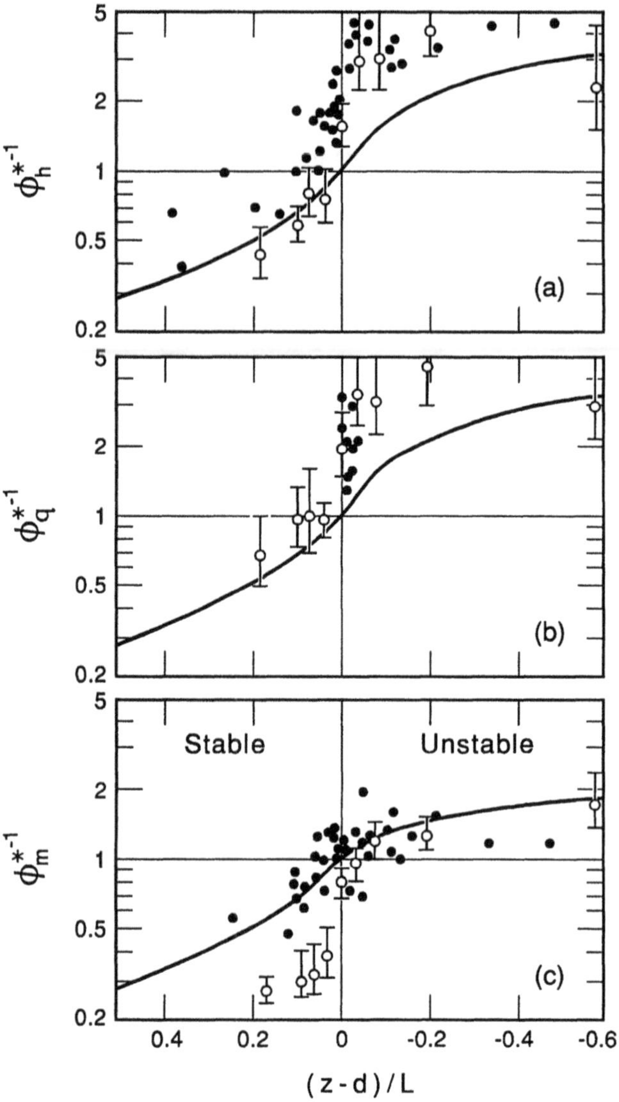

FIG. 3.2. Diabatic influence functions ϕ^{*-1} shown as a function of stability for (a) heat, (b) humidity, and (c) momentum. Data points are measured values; solid lines follow standard surface layer formulas for ϕ_h^{-1}, ϕ_q^{-1}, and ϕ_m^{-1} from Chapter 1 (after Shuttleworth, 1989).

therefore, reflects not only the variation between canopies but also the height dependences of ϕ_m/ϕ_m^*, ϕ_h/ϕ_h^*, and ϕ_q/ϕ_q^*. This feature has been explored in more detail by Cellier (1986) and Fazu and Schwerdtfeger (1989). We shall return to the reasons for this kind of behavior when we examine turbulent transfer in the canopy later in this chapter.

These roughness sublayer effects present another obstacle to inferring eddy fluxes over tall vegetation by fitting measurements of mean concentrations to equations (3.1), (3.2), and (3.3). Over very tall vegetation such as forests it may be a practical impossibility to make measurements above two or three canopy heights; furthermore, over canopies of any size, scalar gradients may become extremely small outside the roughness sublayer. This is particularly true for CO_2.

3.1.4 *Combination equation*

The surface layer similarity relationships we have just discussed treat the transfer of different scalars and of momentum independently. An important relationship that, in contrast, emphasizes the link between sensible and latent heat fluxes is the combination equation. The combination equation was first developed to describe evaporation from an open water surface or from short green vegetation amply supplied with water (Penman, 1948; Ferguson, 1952). It was extended by Monteith (1965) to situations in which the physiological state of the plants restricted the availability of water for evaporation.

To obtain the combination equation we first recast (3.15) and (3.16) in terms of transfer resistances r_{ah} and r_{aq}:

$$-\frac{H}{\rho c_p} = \frac{\overline{\theta}(z) - \overline{\theta}_0}{r_{ah}}, \tag{3.20}$$

$$-\frac{E}{\rho} = \frac{\overline{q}(z) - \overline{q}_0}{r_{aq}}, \tag{3.21}$$

where the aerodynamic resistances r_{ah} and r_{aq} describe the resistance to diffusion between plant surfaces and the reference level z. The resistances, which have dimensions of inverse velocity, can be written immediately in terms of the logarithmic law. In neutral conditions the relationships are

$$r_{ah} = \frac{C_f}{C_h u_*} = \frac{1}{ku_*} \ln\left(\frac{z-d}{z_h}\right), \tag{3.22}$$

$$r_{aq} = \frac{C_f}{C_q u_*} = \frac{1}{ku_*} \ln\left(\frac{z-d}{z_q}\right). \tag{3.23}$$

Unless the plants are wet, water is not available for evaporation on the surface of the canopy but instead diffuses as water vapor from the interior of the leaves via the tiny stomatal pores on the leaf surface. It is usually a good approximation to take thin leaves to be isothermal and to assume that the substomatal air cavities in the leaves are saturated at the leaf temperature. Consequently, we can introduce a canopy resistance r_c by the expression

$$-\frac{E}{\rho} = \frac{\overline{q}_0 - \overline{q}_{\text{sat}}(\overline{\theta}_0)}{r_c}, \tag{3.24}$$

where $\overline{q}_{\text{sat}}(\overline{\theta}_0)$ is the saturated specific humidity at $\overline{\theta}_0$.

The term r_c is intended to capture the physiological response of a canopy to vapor transfer from the leaf interior to the surface, which is controlled by the stomatal apertures. These respond through complex biomechanical feedbacks to sunlight, the amount of water in the plant, the ambient water vapor, CO_2 levels, and other stimuli. At the present time, these are far from fully understood (Zeiger et al., 1987). In fact, the canopy resistance r_c cannot be a completely accurate reflection of the combined stomatal resistances of all the leaves in a canopy for a variety of reasons. These have been explored and quantified by Raupach and Finnigan (1988). The most obvious reason is that for this to be so, $\overline{\theta}_0$ and $\overline{q}_0$ should be properly weighted averages of $\overline{\theta}$ and $\overline{q}$ on the surfaces of all the transpiring leaves. In fact, the values of $\overline{\theta}_0$ and $\overline{q}_0$ that fit the logarithmic expressions (3.2) and (3.3) do not correspond to such properly weighted quantities (except by accident) because they are affected by nonbiological factors such as turbulent transfer within the canopy and the distribution of wind speed and thermal radiation. Another aspect of the same problem is that r_{ah} and r_{aq} are defined in typical single-layer fashion, as if all the sources of heat and water vapor were concentrated close to $z = d$, whereas the spatial distribution of sources and sinks of momentum heat and moisture is often quite different (Denmead, 1984).

The thermal energy balance at the bottom of the surface layer is conventionally written as

$$A = R_n - G_s - Pi = H + \lambda_e E, \tag{3.25}$$

where R_n is the net radiation, made up of incoming short and long waves minus outgoing short and long waves, G_s is the heat flux going into storage in the soil or vegetation, Pi is the small amount (usually less than 2% of R_n) consumed in photosynthesis, and λ_e is the latent heat of vaporization. A is called the available energy because it is available for conversion to sensible and latent heat at the surface. Equations (3.22), (3.23), (3.24), and (3.25) can be combined, using a set of now standard procedures, to yield the combination equation

$$\lambda_e E = \frac{\epsilon_s A + \lambda_e \rho \Delta / r_{ah}}{\epsilon_s + r_{aq}/r_{ah} + r_c/r_{ah}}, \tag{3.26}$$

where $\Delta = \overline{q}(z) - \overline{q}_{\text{sat}}(\overline{\theta})$ is the specific saturation deficit at level z and

$$\epsilon_s = \frac{\lambda_e}{c_p} \frac{dq_{\text{sat}}}{d\theta}$$

is the dimensionless rate of change of saturated specific humidity with temperature ($\epsilon_s = 2.2$ at 20°C).

Equation (3.26) is often written in a more compact way by defining the climatological resistance r_i

$$r_i = \frac{\lambda_e \rho \Delta}{A} \tag{3.27}$$

and noting that, since r_{aq} and r_{ah} seldom vary by more than a few percent, we can write

$$r_{ah} \simeq r_{aq} = r_a. \tag{3.28}$$

Then,

$$\frac{\lambda_e E}{A} = \frac{1}{1+\beta} = \frac{\epsilon_s + r_i/r_a}{\epsilon_s + 1 + r_c/r_a}, \tag{3.29}$$

where $\beta = H/\lambda_e E$ is known as the Bowen ratio.

The combination equation neatly encapsulates the two essential physical controls on evaporation: the supply of energy and the diffusion of water vapor from the surface. It also separates the physical constraint on diffusion, r_a, from the physiological one, r_c.

Equation (3.26) or (3.29) can be used in several different ways. If r_a and r_c are known or assumed, the combination equation furnishes a diagnostic model of evaporation. Widely used approaches such as the Priestley-Taylor model of evaporation (Priestley and Taylor, 1972) can be shown to be equivalent to inserting particular expressions for r_a and r_c into (3.26). With knowledge of the behavior of Δ, the specific saturation deficit, which might come from the output of a boundary layer model, the combination equation forms the core of prognostic models of evaporation. For example, it is the basis of land surface parameterizations in global climate models; see, for example, Raupach (1991) and references therein. Alternatively, inserting measurements of $A, \lambda_e E$, and r_a into (3.26) or (3.29) allows inferences to be drawn about the physiological state of the canopy as embodied in r_c. Used in this way, the combination equation is an important tool of agronomy and agricultural meteorology, one that allows, for example, the water use efficiency of different plant species to be compared (Dunin et al., 1978). In Table 3.1, typical values of r_a and r_c for well-watered cereal crops or grasses and forests are compared with an open water body to show the range of variation commonly encountered.

We can infer useful information about the partition of radiant energy by considering the limiting cases of (3.29). This has been done in a complete and instructive way by Thom (1975). Here we give three examples.

Table 3.1. Typical Values of Aerodynamic and Canopy Resistance for Well-Watered Plants

Surface	r_a **(s m^{-1})**	r_c **(s m^{-1})**
Water	200	0
Grass/cereal	50	50
Forest	20	200

1. If r_c is negligible or actually zero, as occurs after rain when the vegetation is wet, we have

$$\frac{1}{1+\beta} = \frac{\lambda_e E}{A} = \frac{\epsilon_s + r_i/r_a}{\epsilon_s + 1}. \tag{3.30}$$

This is called "potential evaporation," the evaporation you get when the stomata exert no control.

2. If $r_a \gg r_c, r_i$, a situation corresponding to very light winds and humid conditions, we have

$$\frac{1}{1+\beta} = \frac{\lambda_e E}{A} = \frac{\epsilon_s}{\epsilon_s + 1} (\simeq 0.7 \text{at } 20°\text{C}) \tag{3.31a}$$

or

$$\lambda_e E = \epsilon_s H. \tag{3.31b}$$

This is known as "equilibrium evaporation." In this case also, the canopy resistance exerts no control on evaporation but for quite a different reason: not because r_c is negligible but because the inability of water vapor to diffuse away from the leaves is the limiting factor. Equilibrium evaporation has been proposed as the asymptotic state to be expected far downwind over extensive regions of well-watered, transpiring crops (Denmead and McIlroy, 1970). More complete and recent analysis (McNaughton and Spriggs, 1986) has shown that diffusion of dry air from above the capping inversion at z_i, and hence the dynamics of the entire convective boundary layer, must be considered to obtain the proper limiting behavior in such situations. Nevertheless, equation (3.31) usually provides a useful first estimate of energy partition over extensive vegetation free of water stress.

3. Finally, consider $r_i/r_a \to \infty, r_c/r_a \to \infty$, the situation that occurs in dry windy conditions over irrigated vegetation, what is sometimes called the 'oasis' situation. We see now, by combining (3.27), (3.28), and (3.29), that

$$E = \frac{\rho\Delta}{r_c}. \tag{3.32}$$

Evaporation is maintained by the downward flux of dry air, and stomatal control is paramount.

To relate the "bulk" parameters that have appeared in the logarithmic and combination equations to the morphology of the underlying canopy in any but an empirical way, we must look in detail at the structure of turbulent flow within the canopy itself. As we pointed out in the introduction, the motivation for a closer investigation of the canopy air space extends much further than providing underpinnings for single layer models; there are many hydrological, agronomic, and biological applications for which the proper characterization of the canopy microclimate is indispensable. It is this requirement that we now address.

3.2 Flow within the canopy

Measurements of canopy turbulence both in the field and in the wind tunnel pose difficulties because of the very high turbulence intensities present and the need for some kind of spatial averaging. From a practical point of view it means that much of the data collected more than 15 years ago cannot give reliable indications of all the components of the wind vector. The earliest data set where confidence could be placed in the instantaneous values of u, v, and w was obtained in a cornfield by Shaw et al. (1974). Their mean velocity data $\overline{u}(z)$, normalized with $\overline{u}_{hc}$ (the value of $\overline{u}$ at $z = h_c$), is plotted against z/h_c in Fig. 3.3a, together with data from seven other uniform canopies where we have confidence in the measurements. These range from two plantations of Pinus (Uriarra, Bordeaux), through a forest of eucalyptus and banksia (Moga), two fields of corn at different stages of growth (Shaw's corn and Wilson's corn), down to three wind tunnel (WT) model crops: rigid strips (WT strips), rods (WT rods), and an aeroelastic model of wheat (WT wheat). The vital statistics of the canopies and references to the published results are given in Table 3.2.

What is striking is the good collapse of these data when normalized by h_c and $\overline{u}_{hc}$, given that the range of heights varied by a factor of 500, whereas the type and distribution of "foliage" ranged from simple rods in a wind tunnel to complex combinations of crown, trunk space, and understory in forests. The only significant variation in the form of the profiles is the secondary maximum that occurs in the two forest canopies with clearly marked trunk spaces, Uriarra and Bordeaux.

Other recent data sets of comparable accuracy that fall within the range of scatter of Fig. 3.3a (and the subsequent Figs. 3.3b, 3.3c, and 3.3d) have been presented by Amiro (1990) from three natural boreal forest canopies and by Shaw et al. (1988) and Maitani and Shaw (1990) from a deciduous forest. Statistics collected in an almond orchard by Baldocchi and Hutchison (1987) and in a deciduous forest of oak and hickory by Baldocchi and Meyers (1988) fall somewhat outside the range—the first probably because of its unnaturally uniform row structure, the

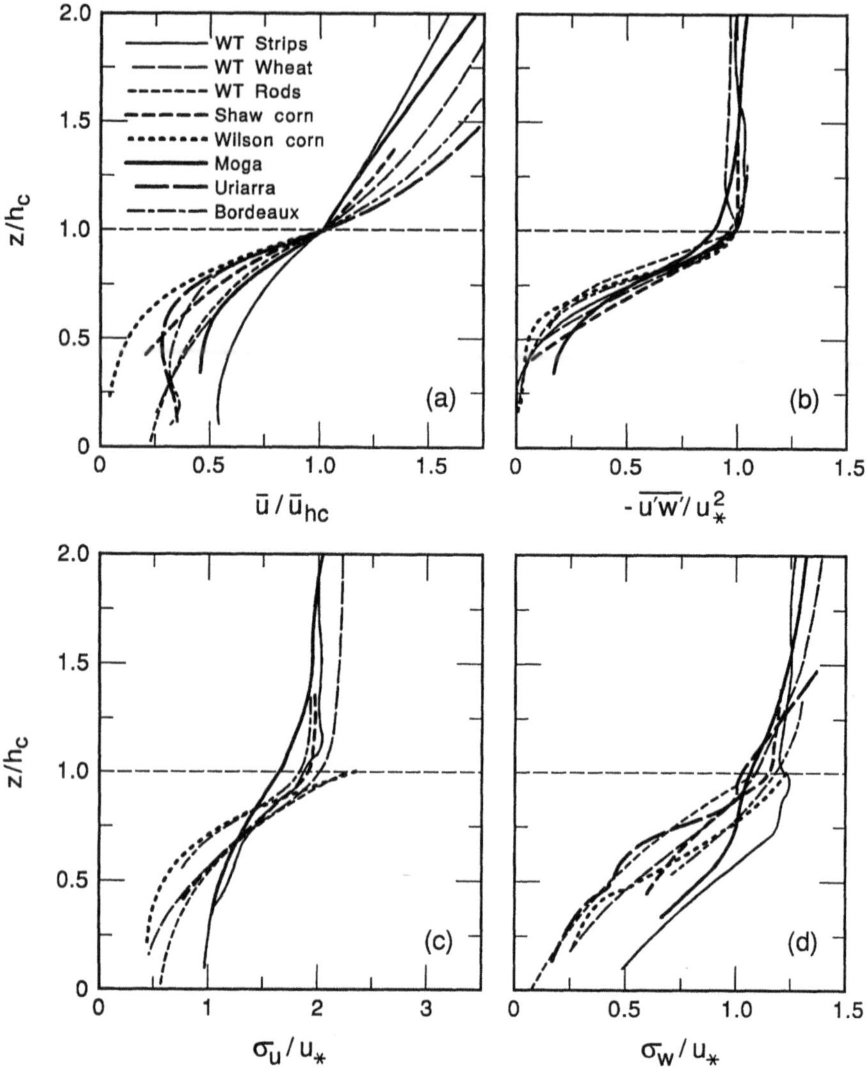

FIG. 3.3. Observed profiles of (a) mean wind speed, (b) shear stress, (c) standard deviations of u, and (d) standard deviations of w, within and immediately above the canopy, for a range of real and model experiments.

second because of the exaggerated concentration of foliage in the crown space. If the deciduous data of Baldocchi and Meyers are plotted against $(z-d)/(h_c-d)$, however, the correspondence with the curves of Fig. 3.3a is restored. This sounds a note of warning that in dense canopies, pronounced vertical variation in foliage density may override the simple picture of Fig. 3.3a and a more sophisticated vertical scaling may be called for. Nevertheless, the composite curves of Fig. 3.3

Table 3.2. Physical and Aerodynamic Properties of Eight Canopies for Which High-Quality Turbulence Data Exist

Canopy	Site	Reference	h_c	LAI	u_{hc}/u_*	ν_e	Sensors: Mean	Sensors: Turbulence
Strips	WT	Raupach et al. (1986)	60 mm	0.23	3.3	0.84	T	T
Wheat	WT	Brunet, Finnigan, and Raupach[a]	47 mm	0.47	3.4	1.6	T	T
Rods	WT	Seginer et al. (1976)	19 cm	1.0	5.0	1.7	X	X
Corn	Elora	Shaw et al. (1974)	260 cm	3.0	3.6	2.4	C	F
Corn	Elora	Wilson et al. (1982)	225 cm	2.9	3.2	4.1	C, F	F
Forest	Moga	Raupach, Finnigan and Brunet[a]	12 m	1.0	2.9	1.7	C, S3	S3
Forest	Uriarra	Denmead and Bradley (1987)	20 m	4.0	2.5	1.7	C	S1
Forest	Bordeaux	Brunet[a] (personal communication)	13.5 m	3.0	2.2	3.2	C	S3

Note: WT denotes wind tunnel. Sensors: C, cup anemometer; F, split-film servo-driven anemometer; X, X-configuration hot-wire anemometer; T, coplanar triple hot-wire anemometer; S1, single-axis (vertical) sonic anemometer; S3, three-axis sonic anemometer.
[a]Papers in preparation

form a remarkably robust guide to a wide range of canopies. Let us examine them in more detail.

3.2.1 *Velocity statistics*

The mean wind shear attains a maximum at the canopy top with both $\overline{u}$ and $\partial\overline{u}/\partial z$ attenuating within the canopy at a rate determined by the density of the foliage. Foliage area density $a(z)$ is the area of plant surface per unit volume of space. The integral of $a(z)$ through the whole canopy depth is called the leaf area index (LAI), where

$$\text{LAI} = \int_o^{h_c} a(z)\,dz. \tag{3.33}$$

LAI is the simplest useful measure of the canopy area density. The upper part of the $\overline{u}$ profile is fairly well approximated by

$$\frac{\overline{u}}{\overline{u}_{hc}} = e^{-\nu_e(1-z/h_c)}, \tag{3.34}$$

where the extinction coefficient ν_e tends to increase with LAI as seen in Table 3.2.

The data collapse in Fig. 3.3b, where we plot $-\overline{u'w'}/u_*^2$ as a function of z/h_c, is even more impressive. We identify u_* with $(-\overline{u'w'})^{1/2}$ at h_c, as all the profiles display an excellent constant stress layer down to the top of the canopy. Here we encounter one of the principal ways in which canopy turbulence is different from that in the layer above; it is strongly inhomogeneous in the vertical. In Fig.

3.3b we see, at least in the denser canopies (LAI >1), that all the momentum is absorbed in the upper part of the canopy so the shearing stress transmitted to the ground surface is essentially zero.

Vertical heterogeneity is prominent again in the measurements of σ_u/u_* and σ_w/u_*. These are shown in Figs. 3.3c and 3.3d, respectively. Although both σ_u and σ_w scale with u_*, they approach typical surface layer values of $2.5u_*$ and $1.25u_*$ only above the roughness sublayer. Both quantities decrease through the depth of the roughness sublayer to attain values of $2.0u_*$ and $1.1u_*$, respectively, at $z = h_c$.

Although we observe a constant stress layer above the canopy, the correlation coefficient r_{uw}, which has a value of -0.3 in the surface layer, increases through the roughness sublayer to reach a value of -0.45 at $z = h_c$. This coefficient, defined as

$$r_{uw} = \frac{\overline{u'w'}}{\sigma_u \sigma_w}, \tag{3.35}$$

is a measure of the efficiency of the turbulence in transferring momentum relative to the absolute amount of turbulence present. Within the canopy, r_{uw} attenuates rapidly. In other words, turbulence at the top of the canopy is efficient at downward momentum transfer, but this efficiency decays even more rapidly than the variance; in fact, $-\overline{u'w'}$ decreases more rapidly than σ_u^2 or σ_w^2.

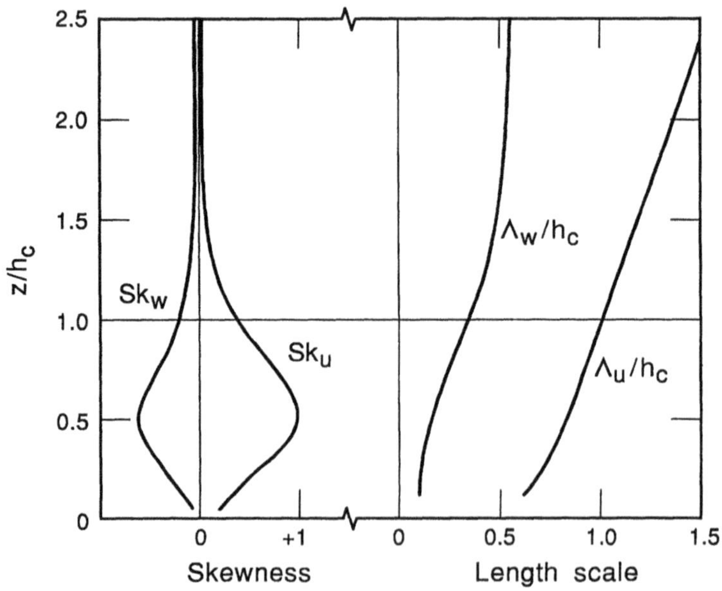

FIG. 3.4. Typical profiles of skewness and normalized integral length scales for u and w within and above the same range of canopies as in Fig 3.3.

Another characteristic property of canopy turbulence, its intermittency, emerges when we consider higher velocity moments. In the surface layer, the skewnesses of u and w, both measures of intermittency in the flow, are close to the Gaussian value of zero:

$$\mathrm{Sk}_u = \frac{\overline{u'^3}}{(\sigma_u)^3}, \tag{3.36a}$$

$$\mathrm{Sk}_w = \frac{\overline{w'^3}}{\sigma_w^3}. \tag{3.36b}$$

As the canopy is approached from above, Sk_u increases through the roughness sublayer to reach a value of 1.0 in the canopy, whereas Sk_w changes over the same distance to -0.6 (Fig. 3.4). There is considerable scatter in published results caused primarily by morphological differences between canopies, but this combination of strongly positive u skewness and strongly negative w skewness tells us that the turbulence is dominated by intermittent downward moving gusts.

In the description of turbulence, length scales and time scales are just as important as velocity moments. We can obtain the Eulerian integral length scales, Λ_u and Λ_w of Chapter 2, for streamwise and vertical components by applying Taylor's *frozen turbulence* hypothesis to the respective integral time scales. Despite the high turbulence intensities, Λ_u and Λ_w are good indicators of the size of the eddies that dominate the turbulence spectrum (Raupach et al., 1989), the so-called energy-containing eddies. The profiles of length scale normalized by h_c that are shown in Fig. 3.4 are reasonable approximations for the canopies of Fig. 3.3. Around $z = h_c$, we have $\Lambda_u \approx h_c$ and $\Lambda_w \approx h_c/3$. The realization that these eddies are typically of canopy size, coupled with the picture of a canopy turbulence dominated by large-scale intermittent incursions of fast-moving air, has profound consequences for the way we parameterize or describe canopy transport.

3.2.2 *Scalar statistics*

In Fig. 3.5a we illustrate typical midday and nighttime profiles of potential temperature $\overline{\theta}$ for the natural forest canopies of Fig. 3.3. Note the strong maximum in $\overline{\theta}$ at about $z/h_c \simeq 0.75$ in the daytime profile. The peak coincides with the level of strongest absorption of solar radiation in the foliage. Above this maximum we observe a conventional lapse profile, but below it the airspace is stably stratified. This inversion is a prominent feature of forest canopies with distinct crown and trunk spaces but is also observed, albeit less dramatically, in uniform canopies such as corn and rice (Uchijima, 1976). The nighttime profile is almost a mirror image of midday behavior with a weak lapse profile in the lower canopy and stable stratification above. The numerical values of $(\overline{\theta} - \overline{\theta}_{hc})$, the departure from the canopy top temperature, are typical of summertime conditions in temperate forests with light-to-moderate winds.

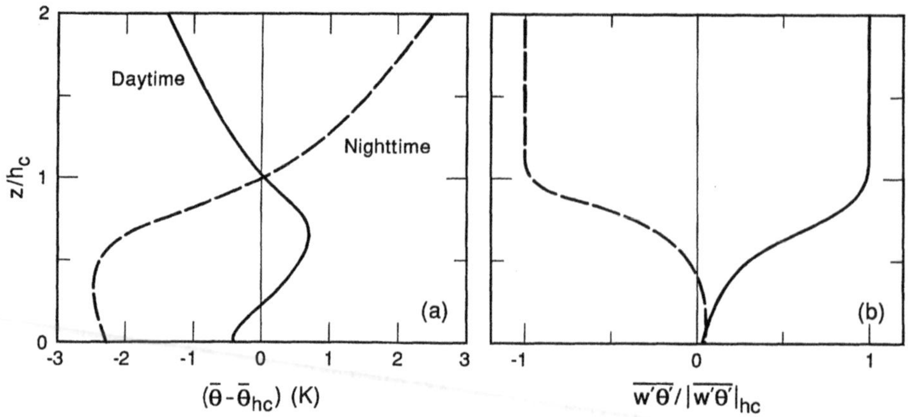

Fig. 3.5. Typical mean potential temperature and heat flux profiles in and above the canopy for daytime and nighttime conditions.

The corresponding flux profiles are shown in Fig. 3.5b. To preserve the sign of the flux, they are normalized with the modulus of $\overline{w'\theta'}$ at the top of the canopy. The most striking feature is the region of counter-gradient flux between the ground and the maximum in the daytime $\overline{\theta}$ profile. The reasons for this surprising behavior lie in the large size of the eddies responsible for most of the transport, as we noted in Fig. 3.4b. We shall return to this point in the next section. Measured nighttime flux profiles generally display no counter-gradient behavior, although few simultaneous eddy flux and profile data have been gathered at such times.

Figures 3.6a and 3.6b display daytime specific humidity concentration and flux profiles for the same average canopy as in Fig. 3.5. The profile of $\overline{q}$ has a

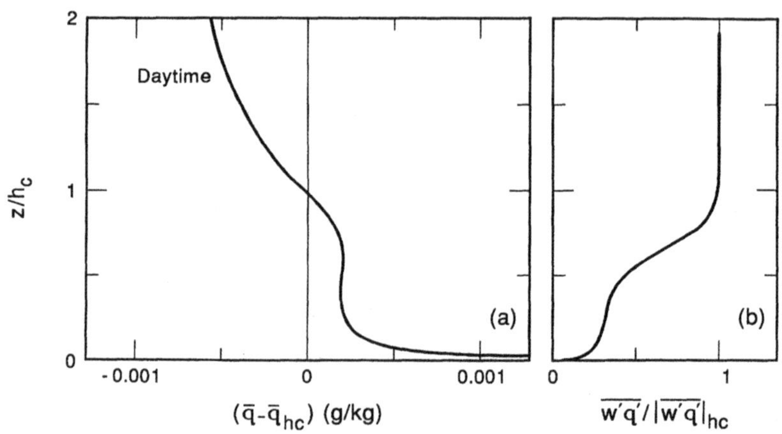

Fig. 3.6. Typical daytime mean humidity and humidity flux profiles in and above the canopy when the ground is moist.

less pronounced secondary maximum in the crown space and attains its greatest value just above the ground. A profile of this shape is observed when the ground and overlying leaf litter are moist, a condition that may persist beneath closed forest canopies for several days after rain. The flux profile once again displays counter-gradient behavior in the midcanopy space as well as a rapid flux increase at the lowest level.

Figures 3.3, 3.5, and 3.6 demonstrate the rich vertical structure of wind and scalar concentrations and fluxes that even simply structured plant canopies conceal. As we shall see in more detail in Sections 3.3 and 3.4, this structure is closely linked to the distribution of sources and sinks of momentum and scalar quantities, the latter a direct consequence of the canopies' biological function.

3.2.3 *Stability effects in the roughness sublayer*

The potential temperature profile of Fig. 3.5a makes it obvious that local stability within the canopy air space varies strongly with height. [Because of the importance of water vapor transport in canopy dynamics, it is conventional to express stability in terms of virtual potential temperature $\overline{\theta}_v$ (see Chapter 1) as we shall do henceforth.] During the day, in the lower canopy, the Richardson number Ri is positive, denoting stability, but the flux Richardson number R_f is negative, denoting instability. In fact, the stability of the lower canopy space often appears to be bimodal in time also with intermittent incursions of large eddies that effect the positive heat flux (negative R_f) interspersed with quiescent periods where Ri is strongly positive. In these intervening times it is not unusual to record periodic gravity wave oscillations on surface pressure sensors. These waves have frequencies close to the average Brunt-Väisälä frequency of the inversion. Such phenomena have been clearly observed in Uriarra and Moga forests (Table 3.2) by Finnigan and at Camp Borden by Leclerc (personal communication).

At night, the lower canopy space may be unstably stratified, promoting turbulent mixing there, whereas the upper crown space is stable with little turbulence. This activity can lead to dew formation in the crown, which is cooled by radiation, while the trunk space remains dry. Once again it must be emphasized that these features are not observed with such clarity in shorter, more uniform canopies such as cereals and grasses.

Above the midcrown both Ri and R_f have the same sign as in the surface layer, but neither parameter has been used consistently to collapse data from a range of stabilities. Attempts to use a simpler measure h_c/L (where L is the Obukhov length) have met with some success (Shaw et al., 1988; Leclerc et al., 1990), but no universal behavior has yet emerged. For h_c/L to be a true measure of canopy response to stability, it would imply that the canopy turbulence structure was determined by the same parameters that govern the behavior of the surface layer above the roughness sublayer. In fact, as we shall see in the next section, canopy turbulence is dominated by the large eddies that form in the intense shear layer confined to the crown or upper part of the canopy. The character of this shear layer and the turbulence within it is set by the foliage distribution both in the vertical and the horizontal and also by the distribution of temperature on the plant surfaces. The velocity shear itself has a maximum value roughly equal to $\overline{u}_{hc}/(h_c - d)$ (Raupach et al., 1989), where $\overline{u}_{hc}$ is the mean wind at canopy top.

Above $z = d$, where fluxes are generally cogradient, the flux Richardson number R_f is, by definition, a good measure of stability effects. In Chapter 1 we saw that it is possible to write $R_f = z/L\phi_m$. Use of h_c/L as a stability parameter, therefore, is tantamount to representing the shear at the canopy top by $u_*/k(h_c - d)$ just as in the surface layer above, rather than by the correct and somewhat larger quantity $\overline{u}_{hc}/(h_c - d)$. Hence, while h_c/L may be generally indicative of the canopy's response, it is unlikely to exhibit the success enjoyed by z/L above the roughness sublayer in ordering a wide range of phenomena. Clearly a good deal more work is required before we can speak with confidence about this particular facet of canopy flow.

3.3 Dynamics of canopy turbulence

3.3.1 *The momentum balance*

The essential differences between turbulence in the canopy air space and that in the boundary layer above result from the sources and sinks of momentum and scalars that are spread through the canopy. The distribution of these plant surfaces dictates a corresponding variation in the time-averaged moments of velocity and scalars. If we want average moments to reflect the same dependence on z as does the average canopy morphology, we have to average them in space as well as in time. The appropriate spatial average is taken over a thin slab parallel to the ground and extending over many plants in the horizontal direction. The practical lower limit on the thickness of this slab is the resolution of the measuring instrument. A schematic view of such an averaging volume is provided in Fig. 3.7. Note that because plant parts intercept the averaging volume (which is taken to consist only of the air, not the solid plants), the averaging volume is a "multiply connected" space (Courant, 1959). This is a mathematical concept that has profound consequences for the derivation of the canopy flow equations.

Applying this spatial average to the conservation equations for momentum and for turbulent kinetic energy is a convenient way of introducing the extra physical processes that occur in the canopy. The now standard form in which we present them evolved through the efforts of various workers. The essential steps may be found in Raupach and Shaw (1982), Finnigan (1985), and Raupach et al. (1986).

If we assume steady flow with no evolution in the streamwise direction, at least on scales larger than the spatial average, the conservation of streamwise momentum can be written as

$$\frac{D}{Dt}\langle\overline{u}\rangle = 0 = -\frac{\partial}{\partial z}\langle\overline{u'w'}\rangle - \frac{\partial}{\partial z}\langle\overline{u}''\overline{w}''\rangle - \frac{\langle\overline{D}\rangle(z)}{\rho}, \tag{3.37}$$

where $D(z)$ is the aerodynamic drag on the canopy per unit volume of space. The angle brackets denote volume averages, whereas time mean quantities have been split into the sum of their volume average and the local departure from it, the latter denoted by a double prime. For example,

$$\overline{u} = \langle\overline{u}\rangle + \overline{u}''. \tag{3.38}$$

In discussing canopy statistics such as $\overline{u}$ up to this point, we have been tacitly assuming the volume-averaging operations that have now been formally introduced. In particular, reinterpret $\overline{u}_{hc}$ as $\langle\overline{u}\rangle_{hc}$, and so on.

The first term on the right-hand side of (3.37) is the vertical gradient of volume-averaged turbulent shear stress. The next term is less familiar; it represents the contribution to momentum transfer from correlations between point-to-point variations in the time-averaged flow. It is called the *dispersive flux* (of momentum) and on the relatively few occasions that it has been measured, has proved to be only about 1% of the first term, the Reynolds stress (Coppin et al., 1986).

The last term is the drag exerted by the canopy on the mean air flow. It is minus $\overline{D}(z)$, the average of viscous plus pressure drag forces exerted on each

FIG. 3.7. Schematic view of an averaging volume V in a forest. The solid plant parts are excluded from the average, causing V to be a "multiply connected" space.

plant part in the averaging volume. Equation (3.37) shows immediately why we observe large gradients of $\langle\overline{u'w'}\rangle$ in the canopy. In horizontally homogeneous conditions, a vertical gradient in shearing stress is necessary to sustain a steady air flow against the aerodynamic drag of the foliage. Just above the canopy, in the surface layer, the aerodynamic drag disappears, the gradient of shearing stress is zero, and we recover the familiar constant stress layer.

Equation (3.37) also illustrates the two extra processes we encounter in the canopy: the transport by correlations between spatial variations in time-averaged quantities and the contribution of spatially distributed sources and sinks to the conservation of some measurable quantity. The equations for conservation of heat and humidity under the same horizontally homogeneous conditions, therefore, become

$$\frac{D}{Dt}\langle\overline{\theta}\rangle = 0 = -\frac{\partial}{\partial z}\langle\overline{w'\theta'}\rangle - \frac{\partial}{\partial z}\langle\overline{w}''\overline{\theta}''\rangle + \langle\overline{F}_h\rangle, \tag{3.39a}$$

$$\frac{D}{Dt}\langle\overline{q}\rangle = 0 = -\frac{\partial}{\partial z}\langle\overline{w'q'}\rangle - \frac{\partial}{\partial z}\langle\overline{w}''\overline{q}''\rangle + \langle\overline{F}_q\rangle, \tag{3.39b}$$

where $\langle\bar{F}_h\rangle$ and $\langle\bar{F}_q$ represent the average transport of heat and humidity across the surface of every plant part in the averaging volume.

3.3.2 *Turbulent kinetic energy*

We see both extra processes at work again when we consider the budget of turbulent kinetic energy in the canopy. The equation is

$$\begin{aligned}\frac{D}{Dt}\langle\overline{e}\rangle = 0 = &-\underset{\text{I}}{\langle\overline{u'w'}\rangle\frac{\partial}{\partial z}\langle\overline{u}\rangle} - \underset{\text{II}}{\frac{\partial}{\partial z}\langle\overline{w'e}\rangle} - \underset{\text{III}}{\frac{1}{\rho}\frac{\partial}{\partial z}\langle\overline{w'p'}\rangle} + \underset{\text{IV}}{\frac{g}{\overline{\theta}_{v0}}\langle\overline{w'\theta'_v}\rangle}\\ &-\underset{\text{V}}{\langle\epsilon\rangle} - \underset{\text{VI}}{\frac{\partial}{\partial z}\langle\overline{e}''\overline{w}''\rangle} - \underset{\text{VII}}{\langle(\overline{u'_iu'_j})''\frac{\partial}{\partial x_j}\overline{u}''_i\rangle} + \underset{\text{VIII}}{\frac{1}{\rho}\langle\overline{v'_id'_i}\rangle},\end{aligned} \tag{3.40}$$

where d'_i is the fluctuating part of the drag on the foliage and v'_i the component of foliage velocity. When the canopy is not moving $v'_i = 0$. In the last two terms we have resorted to tensor notation since *wake production* (VII) and *waving production* (VIII) receive significant contributions from all three axial directions. The other terms above are *shear production* (I), *turbulent transport* (II), *pressure transport* (III), *buoyant production* (IV), *viscous dissipation* (V), and *dispersive transport* (VI). Equation (3.40) has the same form as the turbulent kinetic energy budget given in Chapter 1, but angle brackets now remind us that all mean moments are averaged spatially as well as in time and three new terms have appeared that are peculiar to canopy flow.

The first new term is the dispersive transport (VI), which is precisely analogous to the dispersive flux terms of equations (3.37), (3.39a), and (3.39b). It is the counterpart under volume averaging of the standard turbulent transport term. The other additional terms, wake and waving production, represent the conversion of mean to turbulent kinetic energy that occurs as the wind interacts directly with the foliage.

The wake production term has the same form as shear production but depends on local variations in shear stress doing work against local variations in mean strain rates. We associate these variations with the wakes of individual plants, hence, wake production. If we assume the canopy is horizontally homogeneous on the large scale and the flow steady, the wake production term reduces to $-\langle\overline{u}\rangle\, \partial/\partial z\big(\langle\overline{u'w'}\rangle + \langle\overline{u}''\overline{w}''\rangle\big)$. Furthermore, if we can assume $\langle\overline{u}''\overline{w}''\rangle$, the dispersive stress term, to be negligible, as measurements so far have shown it to be in the upper part of uniform canopies (Coppin et al., 1986), we have a form that enables the wake production term to be evaluated in practice.

The waving production term is associated specifically with plant motion and is present in varying degrees in all real canopies. It arises because work must be done to move the plants. This process is a net sink for turbulent kinetic energy since the plant waves in response to fluctuating aerodynamic forcing, while the damping of plant motion through internal friction or as adjacent plants rub together converts this motion to heat. This is not to say that leaf fluttering or plant waving is not an effective way of converting aerodynamic forcing at one frequency to eddy motion at the plant's preferred (resonant) frequencies, but this transfer of energy across the frequency spectrum cancels out in the total budget equations when contributions from all frequencies are summed.

Figure 3.8 presents profiles of the scaled budget terms representative of three of the canopies in Table 3.2: Moga, WT strips, and WT wheat, of Camp Borden (Leclerc et al., 1990) and of the deciduous forest of Meyers and Baldocchi (1991). Each term is multiplied by h_c/u_*^3, where $u_*$is determined in the constant stress region above the canopy. The forest data are taken in neutral conditions whereas those from the wind tunnel are adiabatic. Individual budgets are surprisingly similar over a 500-fold disparity in size, from the forest canopies to the wind tunnel models.

Shear production peaks near the canopy top where $\partial\langle\overline{u}\rangle/\partial z$ is at a maximum but $-\langle\overline{u'w'}\rangle$ has not fallen below its constant value of u_*^2. Wake production is highest in the upper third of the canopy, but the most unexpected behavior is exhibited by the turbulent transport term, which appears as a significant sink for turbulent kinetic energy at the top of the canopy and up to about $z = 2h_c$, and an equally strong source in the lower canopy, where both shear and shear production have fallen to very low values. Hence, transport is instrumental in maintaining

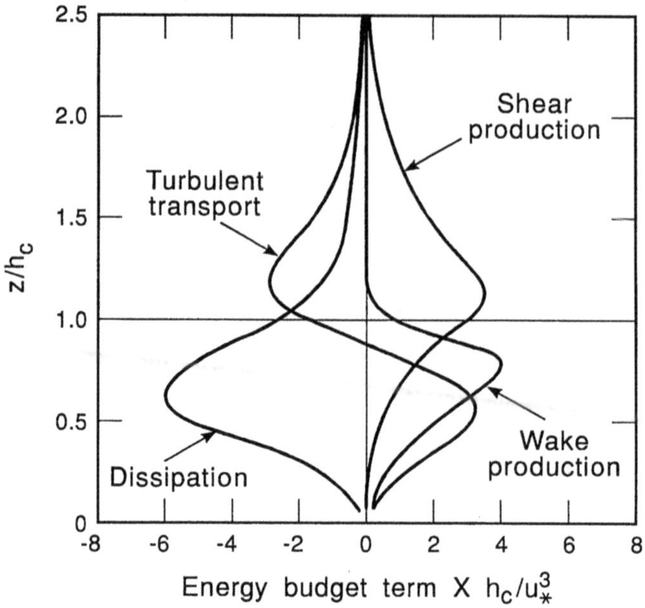

FIG. 3.8. Normalized profiles of terms in the kinetic energy budget within and above the canopy.

turbulent kinetic energy levels in the lower canopy, where most of the turbulence is not locally produced but imported from above. This is consonant with the large values of skewness observed in the lower canopy. Dispersive transport has been measured in one forest canopy (Moga) and one wind tunnel model (WT strips) and was found to be negligible in both cases.

In the WT strips, WT wheat, and Moga forest canopies, viscous dissipation has been estimated using the standard Kolmogorov relationship (Chapter 2); and values obtained from u and w spectra show surprising consistency. Both the wake and waving production terms, however, have the capacity to "short circuit" the spectral energy cascade by converting eddy motion with length and time scales in the energy-containing range to smaller eddies much closer to the dissipation range. This process should produce dissipation rates much higher than those observed in the free stream.

We see in Fig. 3.8, for example, that the dimensionless dissipation rate $\epsilon h_c/u_*^3$ is about 6.0, which is close to the imbalance in the budget terms, whereas, according to surface layer scaling, $\epsilon z/u_*^3 = 1/k = 2.5$. Consequently, dissipation values obtained from Kolmogorov's relationship should be taken as merely indicative of the actual values. In particular, they cannot be used to infer the other major unknown component of the budget, the pressure transport term. That term is even harder to measure in a canopy than in the free surface layer because of the high turbulence levels. Present estimates suggest that pressure transport is roughly proportional to turbulent transport but opposite in sign.

Not enough data exist concerning the effect of buoyancy on the turbulent kinetic energy budget to state firm conclusions. The most complete investigation to date is that by Leclerc et al. (1990), who noted not only the expected contribution of buoyant production as a source or sink term, depending on the sign of the local heat flux, but strong influences on the other terms in the budget. For example, when h_c/L was positive or stable, they observed large increases in shear production at the top of the canopy.

We shall not present budgets of shear stress or scalar fluxes here. Examples of these for the type of canopies we are examining can be found in Raupach et al. (1986), Raupach (1988), Coppin et al. (1986), Leclerc et al. (1990), and Meyers and Baldocchi (1991). They too show, as a central feature, the maintenance of stress or eddy flux within the canopy by turbulent transport, which appears as a sink in the range $2.5h_c > z > 0.8h_c$ and as a source below $0.8h_c$. In contrast to the turbulent kinetic energy budget, there are no important extra source terms in the $\langle\overline{u'w'}\rangle$ or $\langle\overline{w'c'}\rangle$ budgets (where c is the arbitrary scalar), although source terms similar to wake and waving production do appear in the budgets of turbulent normal stresses and horizontal scalar fluxes.

3.3.3 *Turbulent transport in the canopy*

We saw in (3.8), (3.9), and (3.10) that within the surface layer, Monin-Obukhov similarity is equivalent to describing turbulent transport by flux-profile or "K-theory" relationships. That is, eddy fluxes and mean gradients are linked by an eddy diffusivity. Adopting these parameterizations is tantamount to assuming that turbulent transport and source terms make negligible contributions to the eddy fluxes of momentum, heat, and moisture. [This was demonstrated by Wyngaard (1983) for the free boundary layer and Finnigan (1985) in the canopy context.] Absence of transport terms connotes a state of *local equilibrium* in the turbulence, where production of turbulent energy or eddy flux is in balance with destruction (by viscous dissipation or interaction with pressure fluctuations).

Much earlier than this, Corrsin (1974) showed that the eddy diffusivity concept can only be applied if the random motions that do the mixing, the large eddies in this case, are much smaller than the scale over which the mean gradients change. Neither this nor the condition of local equilibrium is satisfied in canopy flow. As we have seen, turbulent transport is vital in maintaining turbulent kinetic energy and eddy flux within the canopy, whereas the characteristic length scales of the turbulent motion, Λ_u and Λ_w, are of order h_c, which is also the distance over which mean gradients change most rapidly.

Despite arguments such as these and earlier rumblings of disquiet, eddy diffusivity models formed the basis of canopy transport prediction for many years and such efforts are still appearing in the literature. The excellent two-volume work, *Vegetation and the Atmosphere*, edited by Monteith (1975, 1976), and the review by Raupach and Thom (1981) are watershed volumes in the understanding of this particular aspect of canopy flow as they separate the time of universal acceptance of K-theory from the present, when it has been abandoned by serious students.

Early evidence for the failure of K-theory came from measurements of secondary maxima in forest wind speed profiles. Combined with necessarily monotonic $-\langle\overline{uw}\rangle$ profiles, these would have required counter-gradient momentum transfer in the lower crown space. Compare, for example, the curves for $\overline{u}/\overline{u}_{hc}$ and $-\overline{u'w'}/u_*^2$ for the Uriarra and Bordeaux forest in Fig. 3.3. Earlier examples of secondary velocity maxima were usually explained away as a consequence of cup anemometer overspeeding (Chapter 6) in the high-intensity, trunk-space turbulence, until the pivotal example of counter-gradient behavior, from simultaneous measurements of heat, water vapor, and CO_2 eddy fluxes in Uriarra forest by Denmead and Bradley (1985) made it impossible to ignore the phenomenon. Their observations of fluxes in the canopy (Fig. 3.9) demonstrate the coexistence of counter gradient (or zero gradient) fluxes within the trunk space and cogradient behavior above.

It is most important to realize that the mere existence of co-gradient fluxes does *not* mean that the mechanisms that cause pathological, counter-gradient behavior in extreme cases have switched off. Simple K-theory, in fact, fails throughout the roughness sublayer, leading to the breakdown of the classical surface layer

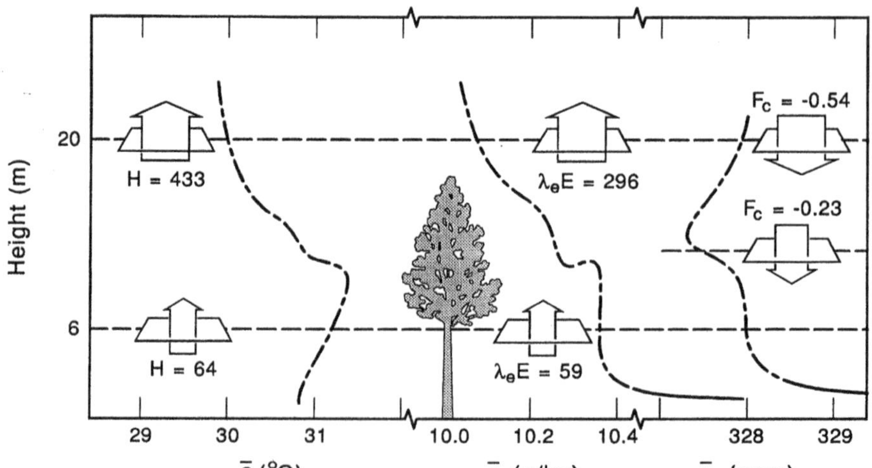

FIG. 3.9. Profiles of mean potential temperature ($\overline{\theta}$), mixing ratio ($\overline{r}$), and CO_2 concentration ($\overline{c}$) observed in Uriarra forest over a period of 1 hour, near noon, together with directly measured eddy fluxes (in W m^{-2}) at two levels (after Denmead and Bradley, 1985).

forms for ϕ_m, ϕ_h, and ϕ_q that we discussed in Section 3.1. In this context, we must mention the role of Professor George Thurtell of Guelph University, who contributed seminal ideas to the reappraisal of K-theory. The physical reason for the failure of K-theory, the domination of transport by large gusts, was also demonstrated by Denmead and Bradley (1985) as they graphed the penetration of the canopy by cool, dry gusts that displace the warm moist canopy air. Similar but more extensive pictures of heat, moisture, and momentum transfer by large eddies have now been obtained in the Camp Borden forest by Gao et al. (1989) and in a pine forest by Bergström and Högström (1989).

Another striking visual demonstration of the large eddies, this time in their role as transporters of momentum, is afforded by the coherent waves that appear on flexible cereal crops such as rice, wheat, or barley on windy days. Finnigan (1979a) showed that this phenomenon, which is often called by its Japanese name "honami," results from the coupling between the natural elastic frequency of the stalks and the characteristic frequency of the energy-containing eddies of the turbulence. This work provided one of the early demonstrations that the dominant canopy eddies were typically of canopy size.

An interesting discussion of the rise and fall of K-theory may be found in Raupach (1988), who also develops the third theoretical objection to such models, which relies on a Lagrangian formulation. This Lagrangian formulation also makes it very clear that the *apparent* diffusivity (i.e., the flux divided by the gradient) is strongly influenced by the source distribution of the particular entity under consideration. It is worth taking some time to explore the Lagrangian viewpoint since it provides the most direct refutation of the eddy diffusivity concept as well as forming the basis of a physically valid alternative (Raupach, 1989a, 1989b).

The Lagrangian or fluid-particle-following formulation assumes that the transfer of a scalar released from a point source is statistically equivalent to the dispersion of an ensemble of *marked* fluid particles that pass through the point and thereafter carry the scalar. This assumption is an extremely good one in the high-Reynolds-number turbulence of a natural plant canopy. The scalar sources and sinks that spread over the surfaces of plants throughout the canopy are regarded as a distribution of such point sources, each producing a plume of marked particles that is advected downwind by the mean flow and spread by the turbulence. The concentration of the scalar at any point is the ensemble average of all the particles arriving at that point from all the sources or, equivalently, the average of the concentrations in all the overlapping plumes that envelop the point. This process is shown schematically in Fig. 3.10a. Note that most of the sources influencing a given measurement point will be located upwind but that in the high-intensity turbulence of the canopy air space, nearby downwind sources will also contribute.

To understand the behavior of any given plume, we can focus on a single typical particle; the Lagrangian treatment assumes that each marked particle moves

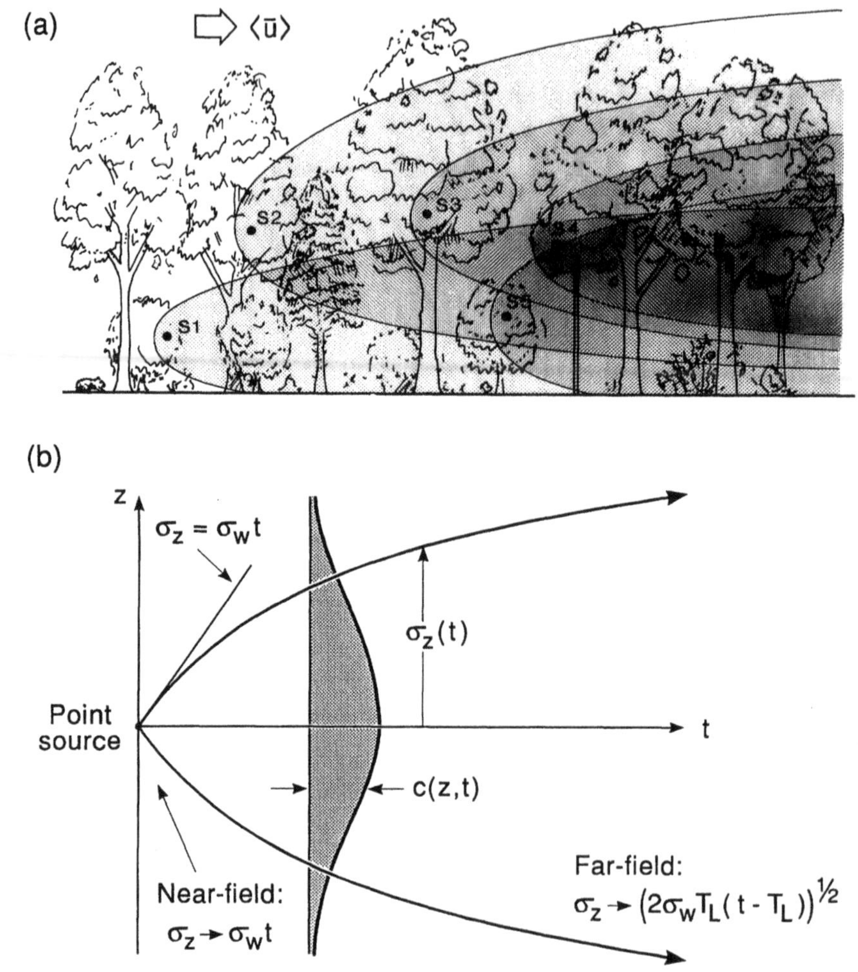

FIG. 3.10. Figure 3.10 (a) Scalar plumes from just a few of the infinite number of point sources scattered through the canopy. The plumes overlap at the measuring point M, and the measured concentration is the sum of the contributions from all the sources. (b) Schematic of the distribution of concentration across the plume.

independently of all others. If $Z(t)$ is the height of the typical particle and t is the travel time from the source, then $W(t) = dZ/dt$ is its Lagrangian (particle-following) vertical velocity. The depth of the scalar plume is identified with σ_z, the standard deviation of $Z(t)$, given by

$$\sigma_z(t) = [\overline{Z'^2(t)}]^{1/2}, \tag{3.41}$$

where the square brackets denote an ensemble average over all particles with the same travel time.

In homogeneous turbulence, Taylor (1921) derived the classic result:

$$\frac{d\sigma_z^2}{dt} = 2\sigma_w^2 \int_0^t R_L(\xi)\, d\xi, \tag{3.42}$$

where $R_L(\xi)$ is the Lagrangian correlation function of W at time lag ξ defined by

$$R_L(\xi) = \frac{\overline{[W'(t)W'(t+\xi)]}}{\sigma_w^2}. \tag{3.43}$$

The most important characteristic of $R_L(\xi)$ is its integral time scale $\mathcal{T}_L$, the integral of $R_L(\xi)$ from $\xi = 0$ to $\xi = \infty$. This *Lagrangian time scale* $\mathcal{T}_L$ is related to the vertical Eulerian integral length scale Λ_w of Fig. 3.4b; for our present purposes we can assume that $\mathcal{T}_L \sim \Lambda_w/\sigma_w$ (Raupach, 1989b). Raupach (1989a) described $\mathcal{T}_L$ as a measure of the "persistence" of the turbulent eddies or, equivalently, of the "memory" of $W(t)$.

According to (3.42), the depth of the scalar plume has two different types of behavior in the limits of small and large relative travel time $t/\mathcal{T}_L$. These are known as the near field and far field limits, respectively:

$$\sigma_z(t) \to \sigma_w t \qquad \text{for } t/\mathcal{T}_L \to 0, \quad \text{(near field)}, \tag{3.44a}$$

$$\sigma_z(t) \to (2\sigma_w^2 \mathcal{T}_L t)^{1/2} \qquad \text{for } t/\mathcal{T}_L \to \infty, \quad \text{(far field)}. \tag{3.44b}$$

This behavior is illustrated in Fig. 3.10b. To quote Raupach (1989b): "In the near field, persistence causes $W(t)$ to differ only a little from $W(0)$ and the particle trajectories $Z(t)$ to depart only a little from straight lines so that the cloud depth grows linearly with *t*. In the far field, on the other hand, the effects of persistence are negligible and $Z(t)$ behaves as a random walk just as in a diffusive process." In fact, the Lagrangian analysis makes it clear that, for travel times from the source that are large compared to $\mathcal{T}_L$, the spread of the plume and the consequent relationship between the vertical flux of the scalar and its concentration gradient can be described by an eddy diffusivity.

The forms that we have given for (3.42) and (3.44) are only precisely true in homogeneous turbulence. We have seen that canopy turbulence is far from this state, but the homogeneous formulas are sufficient to illustrate the crucial point we wish to make; indeed, surprisingly, Raupach (1989a) shows that much of the paradoxical behavior of canopy transport, including counter-gradient diffusion, can be reproduced by assuming that canopy turbulence is homogeneous.

Returning to our picture of a single plume in Fig. 3.10b, the time axis can be turned into a distance axis x by the transformation $x = \langle\overline{u}\rangle t$. (We will ignore the complications caused by streamwise diffusion.) Then, because of the large size of σ_w and of Λ_w (Figs. 3.3d and 3.4b), we see that the near field of any point source extends a distance of order h around the source. Hence, any measurement point will be in the near field of sources, less than about a canopy height away and in the far field of more distant sources. The concentration measured will be the superposition of near-field and, therefore, more concentrated plumes from nearby sources and more diffuse far-field plumes from distant sources. These more concentrated, near-field plumes dominate the concentration profile, impressing the character of the local source distribution upon it, whereas the more numerous far-field plumes dominate the total rate of spread and therefore the scalar flux. In essence this near-field, far-field dichotomy decouples the flux from the gradient and invalidates the functional dependence implied by an eddy diffusivity.

Raupach (1989a, 1989b) treats these concepts in exhaustive detail and shows how to incorporate the mechanics of near-field, far-field diffusion into a practical and physically valid model of canopy dispersion that is equivalent to K-theory and can be used in its stead. The only drawback of the Lagrangian approach is that there is no obvious way to apply it to momentum transport, primarily because the existence of pressure forces makes it invalid to assume that fluid particles "marked" with momentum move independently.

3.4 Sources and sinks in the canopy

3.4.1 *Momentum sinks*

We said in the last section that the aerodynamic drag term $\langle\overline{D}\rangle(z)$ represents the mean value of the pressure and viscous forces on all the canopy elements in an averaging volume. Focusing on a single leaf,[1] its total aerodynamic drag F_{di} can be expressed in terms of its area A_{Li} and the mean wind speed U_i measured just outside the leaf boundary layer through a dimensionless drag coefficient C_{di}:

$$F_{di} = C_{di}\rho A_{Li}U_i^2, \tag{3.45}$$

where the subscript i distinguishes the ith leaf in the averaging volume. (Note that the micrometeorological convention does not include a factor of 1/2 in the definition of C_d.)

Pressure forces on the ith leaf vary as U_i^2 and the frontal area of the leaf, whereas, viscous forces vary as U_i and the leaf's total surface or "wetted" area. At Reynolds numbers typical of most leaves in the upper part of a canopy, however, which is where the greater part of the momentum is absorbed, the pressure force is at least four times larger than the viscous force so C_{di} is only a weak function of U_i.

It is not a practical proposition to sum (3.45) for every leaf in an averaging volume. We write instead

[1]In this section for convenience we shall use *leaf* to refer to any plant part. Remember, however, that although all of the plant can be active in heat and momentum exchange, transpiration occurs almost entirely at the surface of a genuine leaf.

$$\langle\overline{D}\rangle(z) = C_d(z)\,\rho\, a(z)\,\langle\overline{u}\rangle^2(z), \tag{3.46}$$

where $\langle\overline{D}\rangle(z)$ is the momentum sink or the drag term in (3.37). The variations in leaf area, the random orientations and mutual interference of leaves, as well as the effects of turbulence are all subsumed in C_d, the *effective* drag coefficient, and $a(z)$, the leaf area per unit volume of space.

C_{di}, the drag coefficient of any individual leaf, depends not only on how A_i is chosen but also on physical effects such as leaf fluttering, orientation, streamlining (which emphasizes the viscous component of drag), and particularly on the turbulence intensity and scale. There have been many wind tunnel measurements of the drag coefficients of individual leaves and groups of leaves, with and without turbulence; their comparison with the drag of a similar leaf in situ in the real canopy is generally poor.

The net effect of these influences is to reduce C_d below what is measured for an isolated leaf in a wind tunnel. At Reynolds numbers typical of a leaf in a canopy, the wind tunnel value of C_{di} is roughly equal to 0.5, but in field conditions C_{di} falls to about 0.2. This is the so-called shelter effect mentioned earlier; it is discussed more thoroughly in Raupach and Thom (1981).

The connection between the distribution of drag in the canopy and bulk parameters such as u_*, z_0, and C_f that we dealt with in earlier sections is readily seen by integrating (3.37) from just below the surface 0− (so that the ground is counted as part of the canopy) to just above canopy height, h_c+:

$$\begin{aligned} u_*^2 &= -\int_{0-}^{h_c+} \frac{\partial}{\partial z}\left[\langle\overline{u'w'}\rangle + \langle\overline{u}''\overline{w}''\rangle\right] dz = \int_{0-}^{h_c+} \frac{\langle\overline{D}\rangle(z)}{\rho}\, dz \\ &= \int_{0-}^{h_c+} C_d(z)\, a(z)\,\langle\overline{u}\rangle^2(z)\, dz. \end{aligned} \tag{3.47}$$

3.4.2 *Scalar sources*

The transfer of a scalar to or from a leaf is precisely analogous to the transfer of momentum by viscosity, but there is no counterpart of pressure drag in scalar transport. This leads to significant differences between the distribution of momentum sinks and scalar sources, differences often referred to as the bluff body effect or the failure of "Reynolds analogy." We encountered this phenomenon earlier in the context of differences between z_0, z_h, and z_q in Section 3.1.

As we did for the single layer model [equations (3.20) and (3.21)], we follow convention and parameterize the scalar flux from an individual leaf through a "boundary layer resistance," r_b.

Hence, using subscript i to distinguish an individual leaf, we have

$$-\frac{H_i}{\rho c_p} = \frac{\overline{\theta}_i - \overline{\theta}_{0i}}{r_{bhi}}, \tag{3.48a}$$

$$-\frac{E_i}{\rho} = \frac{\overline{q}_i - \overline{q}_{0i}}{r_{bqi}}, \tag{3.48b}$$

where the flux densities are taken as positive along the outward normal to the leaf surface, $\overline{q}_i$ and $\overline{\theta}_i$ mark conditions just outside the leaf boundary layer, while $\overline{\theta}_{0i}$ and $\overline{q}_{0i}$ denote surface values, and all the variables are averages across the leaf surface.

As in the case of leaf drag, we do not attempt to obtain F_h and F_q, the scalar source terms of equation (3.39a,b) by summing (3.48a,b) for every leaf but rather by defining average resistances r_{bh} and r_{bq}.

$$-\langle \overline{F}_h \rangle = a(z)\, \frac{\langle \overline{\theta} \rangle (z) - \langle \overline{\theta}_{0i} \rangle}{r_{bh}}, \tag{3.49a}$$

$$-\langle \overline{F}_q \rangle = a(z)\, \frac{\langle \overline{q} \rangle (z) - \langle \overline{q}_{0i} \rangle}{r_{bq}}. \tag{3.49b}$$

r_{bh} and r_{bq} must now accommodate a similar range of environmental factors as C_d, the effective drag coefficient, and there are similar discrepancies between r_{bhi} and r_{bqi}, measured on single leaves in laboratory conditions, and the values that apply to the average leaf in a canopy. In contrast to what we saw for momentum transfer, these influences tend to *increase* the transfer rate so that within-canopy resistances are lower than those measured on leaves in isolation or deduced directly from engineering formulas. On most occasions, leaves in the canopy experience a regime of forced convection when their boundary layer resistances vary directly as the square root of a characteristic leaf dimension, such as the width of a deciduous leaf or the diameter of a pine needle, and inversely as $U_i^{1/2}$. Formulas for these resistances in both forced and free convection together with a discussion of the effects of turbulence intensity and scale may be found in Finnigan and Raupach (1987).

The connection between the distribution of scalar source strength and the bulk parameters T_*, q_*, C_h, and C_q, follows in the same way as it did for momentum [equation (3.47)] by integrating (3.49a) and (3.49b) from $0-$ to h_c+.

3.4.3 Combination equation

Values of temperature or humidity on the leaf surface are difficult or impossible to measure in practice. We can, however, apply the combination equation (3.26) to eliminate this necessity. We proceed first by specializing the energy balance equation (3.25) to the surface of a single leaf, then by combining it with (3.49a) and (3.49b) obtain

$$\lambda E_i = \frac{\epsilon_s A_i + \lambda_e \rho \Delta_i / r_{bhi}}{\epsilon_s + r_{bqi}/r_{bhi} + r_{si}/r_{bhi}}, \tag{3.50}$$

where the subscript i refers to an individual leaf, the vapor flux density E_i is positive along the outward normal to the surface, the saturation deficit Δ_i is taken just outside the leaf boundary layer, and the canopy resistance r_c has been replaced by a stomatal resistance r_{si}. All the variables are averages across the leaf surface.

Relationships among individual leaf resistances, r_{bhi}, r_{bqi}, and r_{si} and the bulk values of (3.26) can be found by replacing the variables of (3.50) by smooth averages at each level in the canopy and integrating from $0-$ to h_c+ just as in the last two sections. This operation makes it clear that, although r_{si} is truly a physiological property of the plant, r_c is strongly influenced by the distribution of temperature, humidity, and available energy in the canopy. The temperature and humidity are determined primarily by turbulent transfer, the available energy by canopy morphology and sun angle (Ross, 1975). The stomatal resistance r_{si} is an important biological variable and is often measured directly using leaf porometry. On the other hand, r_c is generally obtained by the micrometeorological methods we have described in this chapter, so that the divergence between these two quantities is of considerable interest. It has been quantified in some common situations by Finnigan and Raupach (1987) and Raupach and Finnigan (1988).

3.5 Spectra and cospectra

All of the reasons for studying turbulence spectra and cospectra that were detailed in Chapter 2 apply with equal force in plant canopies. This was apparent in the pioneering spectral measurements of Uchijima and Wright (1964) in corn, and of Allen (1968) in a plantation of larch. Unfortunately, such early studies were handicapped by the state of the art in sensor design and signal processing so the first really useful turbulence spectra date from 1974. These were obtained by Shaw et al. (1974) using servodriven, split-film anemometers in a corn crop.

3.5.1 Velocity spectra

We are particularly interested in how spectra measured in the roughness sublayer depart from the shapes and scaling relationships developed in Chapter 2. We begin, therefore, by reproducing Fig. 2.1 and indicating on it the extra physical processes that operate within the canopy. This is shown schematically in Fig. 3.11. As before, we expect shear production to produce eddies in the energy-containing range centered at wavenumbers around κ_m, where $\kappa_m \sim 1/\Lambda$ (Chapter 2). In contrast, conversion of kinetic energy of the mean flow to turbulent energy by the wake and waving production mechanisms produces eddies with a range of scales. The range is determined by the dimensions of plant elements and the thicknesses of the boundary layers upon them and the wakes behind them. As a result, this

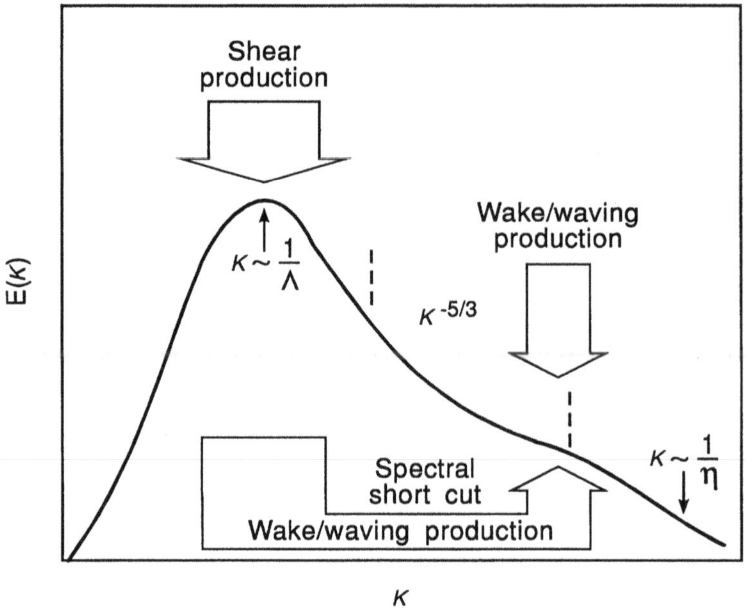

FIG. 3.11. Schematic showing energy from wake and waving production feeding directly to the bottom end of the inertial subrange of the energy spectrum, both directly through the action of the mean flow and through the spectral shortcut from large eddies that penetrate the canopy.

range brackets the bottom of the inertial subrange and the top of the viscous range. For example, the Kolmogorov microscale η is about 1 mm in canopies; this is also the typical depth of attached boundary layers on small leaves or stems.

We must also remember that the wake and waving production processes are able to extract energy from large eddies and convert it to energy at the smaller wake or waving scales. In the special case of coherent waving of cereal crops, the plant waving frequency can actually coincide with the energy-containing frequencies. Furthermore, the relative rate of viscous dissipation is much higher in the canopy than in the free boundary layer because the viscous component of canopy drag occurs through thin boundary layers, where the rate of dissipation is very high as it is in the intense shear layers in plant wakes. We saw this demonstrated clearly in the budget of turbulent kinetic energy discussed in Section 3.3.

There is considerable variation in published spectra from canopies of different types. The most striking differences are seen between measurements made in relatively open and uniform canopies (LAI $\leq$ 5) and those from forests with dense crowns (LAI $\sim$ 10) capping open trunk spaces. In Fig. 3.12 we present power spectra of u, v, and w, respectively, based on Moga forest data and typical of the more open, uniform canopies. Curves of normalized, frequency-weighted

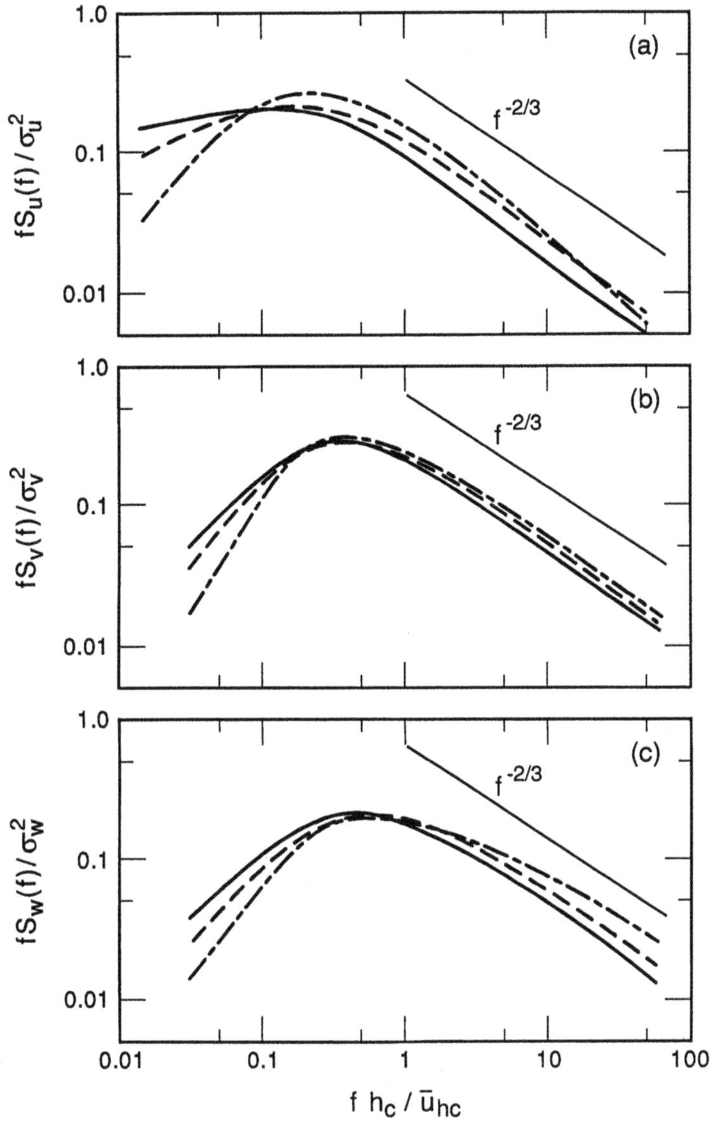

Fig. 3.12. Normalized velocity spectra from measurements above the canopy (solid), near the top of the canopy (dashed), and within the canopy (dash-dot) for the (a) u component, (b) v component, and (c) w component.

spectra $fS_\alpha(f)/\sigma_\alpha^2$ ($\alpha = u, v, w$) are plotted against the dimensionless frequency at $z/h_c = 1.5, 1.0$, and 0.5. We shall discuss the distinguishing features of these spectra, focusing on three characteristics: the position of the spectral peaks, the behavior of the various components in the inertial subrange, and the isotropy in the inertial subrange. Note first that the frequency axis is scaled as $fh_c/\bar{u}_{hc}$, where we recall that $\overline{u}_{hc}$ is, in fact, $\langle\overline{u}\rangle_{hc}$. This corresponds to the normalized frequency $n = fz/\overline{u}$, introduced in Chapter 2 for the surface layer, with the essential difference that we use fixed values of $\overline{u}$ and z appropriate to the canopy top. Note also that since λ, the eddy wavelength, is defined as $\lambda = \overline{u}/f$, then $n = fz/\overline{u} = z/\lambda$ and $h_c f/\overline{u} = h_c/\lambda$.

3.5.2 *Spectral peaks and eddy scales*

With the above choice of scaling, the positions of the peaks in $fS_u(f)$, $fS_v(f)$, and $fS_w(f)$ do not vary as we descend through the roughness sublayer to midcanopy height. Constancy in the position of the spectral peak is observed in almost all of the reliable data sets available for comparison; for example, see Shaw et al. (1974) in corn, Wilson et al. (1982) in corn, Seginer et al. (1976) and Raupach et al. (1986) in the wind tunnel, Bergström and Högström (1989) in a pine forest, and Amiro (1990) in three forest canopies (see also Table 3.2). This lack of variation also fits in well with the data of Fig. 3.4b, which (with the exception of Wilson corn) show only a weak variation of Λ_u or Λ_w with height. An important deduction from this is that the large eddies that dominate canopy transport extend through the whole depth of the foliage and into the air above. This has now been confirmed most powerfully by the three-dimensional space-time correlation maps obtained in the WT wheat canopy (Table 3.2) by Raupach et al. (1989). The position of the peak is different for the three components u, v, and w. $fS_u(f)$ peaks at a scaled frequency of $fh_c/\overline{u}_{hc} \simeq 0.15(\pm 0.05)$, whereas measured values of the $fS_w(f)$ peak cluster around 0.45 ($\pm$0.05). The peaks in $fS_v(f)$ are more variable; values range from $fh_c/\overline{u}_{hc} \simeq 0.1$ in the Shaw et al. (1974) corn to 0.35 in the Moga forest (Table 3.2). We chose the Moga value for the purpose of illustration.

This behavior is quite different from that reported for the surface layer in Fig. 2.9 of Chapter 2. In the surface layer, the peak wavelengths, $(\lambda_m)_{u,v,w}$, for the three velocity components, u, v, and w (and for the θ spectrum as well), are directly proportional to z, for any given value of $z/L > 0+$, on the stable side of neutral. The constants of proportionality are plotted in Fig. 2.9. This proportionality does not extend down to the ground but is lost through the upper part of the roughness sublayer as the curves of $(\lambda_m)_{u,v,w}$ versus z approach constant values in the canopy. Figure 3.13 illustrates this situation, where the near-neutral ($z/L = 0+$) relationships of Fig. 2.9 are projected down to the canopy top and compared with the constant values observed within the lower part of the roughness sublayer. [Note that over a tall canopy we must adjust the origin and expect $(z-d)/\lambda$ to be constant in the surface layer.]

We would reasonably expect these two relationships to merge above the vegetation, providing a smooth transition through the roughness sublayer from surface layer to canopy values. This is the kind of behavior suggested by the plots of integral length scale displayed in Fig. 3.4b, and indeed we see that this is what

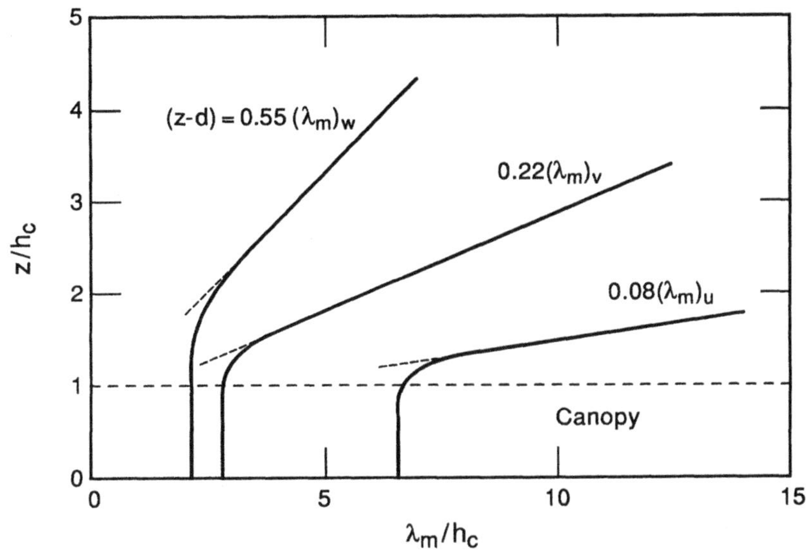

Fig. 3.13. Wavelength of the spectral peak, normalized by canopy height h_c, shown as a function of z/h_c. The peak wavelength stays invariant with height within the canopy (scaling with h_c) but quickly increases with height above the canopy top consistent with the neutral surface layer relationship in Fig. 2.9.

happens in the case of $(\lambda_m)_w$. The surface layer, horizontal peak wavelengths, $(\lambda_m)_u$ and $(\lambda_m)_v$, in contrast, shift more abruptly to their above-canopy values. The explanation for this abruptness, as well as for the success of the scaling $fh_c/\overline{u}_{hc}$ in collapsing peak wavelengths over a range of canopies from wind tunnel to forest, lies in the mechanism by which the dominant eddy structure in the upper canopy evolves.

The peaks in the surface layer u and v spectra, Figs. 2.5 and 2.6 of Chapter 2, represent eddies of such a scale that at small heights, the presence of the ground forces them to be almost horizontal motions, playing little part in the transfer of momentum. Bradshaw (1967) has called such sideways "sloshing" components, "inactive motion." These large-scale horizontal motions, however, modulate the production of canopy-scale eddies that are *not* "inactive" but crucial in the transfer of momentum and scalars, as we have seen repeatedly throughout this chapter. Raupach et al. (1989), in discussing the origin and structure of large eddies in canopies, marshalled evidence to show that the dominant eddies result from a continuous, hydrodynamic instability process linked to the inflected velocity profile in the upper canopy. As is shown in Fig. 3.3a, the maximum shear is located at $z = h_c$ and the depth of the strongly sheared region is approximately h_c.

Eddies produced by this instability process have integral length scales of order h_c, are advected downwind at speed $\overline{u}_{hc}$, and extend through the canopy and into the roughness sublayer above. The instability process is most active at times of strong shear, and these times coincide with larger gusts (the inactive motion components mentioned above) impinging on the canopy from the surface layer aloft. Large eddy production, therefore, is most active when the total instantaneous horizontal velocity $\langle u \rangle_{hc}$ exceeds $\overline{u}_{hc}$. Accordingly, the large eddies move downwind somewhat faster than $\overline{u}_{hc}$. Nevertheless, in a given canopy, the average eddy convection velocity is proportional to $\overline{u}_{hc}$ (Finnigan, 1979b) so that $\overline{u}_{hc}$ furnishes a physically based velocity scale. The link forged by the instability process between the eddy size and h_c gives a similar physical relevance to h_c as the length scale for flow in canopies.

As the canopy eddies are linked dynamically to the height scale of the vegetation in this way, there must be an abrupt transition in the peak wavelengths of u and v from the larger scale inactive motions well above the roughness sublayer to smaller, active, canopy-scale motions within the roughness sublayer. The former are represented by the asymptotically neutral values of $(\lambda_m)_u$ and $(\lambda_m)_v$ obtained from Fig. 2.9; the latter by the peak values shown in Figs. 3.12a and 3.12b.

In the discussion so far, we have assumed that Taylor's hypothesis continues to be valid within the canopy. As we saw in Chapter 2, Kolmogorov's inertial subrange relationships apply in wavenumber space and the Taylor transformation $\kappa_1 = 2\pi f/\overline{u}$ is implicit when we apply Kolmogorov's theory in frequency space. Taylor's hypothesis is, however, generally not valid when turbulent fluctuations are large relative to the mean wind, that is, when $\sigma_u/\langle \overline{u} \rangle, \sigma_v/\langle \overline{u} \rangle, \sigma_w/\langle \overline{u} \rangle \sim 1$, because, in such conditions, turbulent eddies can evolve substantially in the time taken for them to be advected past a sensor. But these are precisely the conditions that occur in plant canopies, and, not surprisingly, reservations about the use of Taylor's hypothesis are often expressed [e.g., Amiro (1990)]. A saving grace is that, as we have just pointed out, large eddies have convection velocities substantially larger than the mean wind in the canopy, particularly the mean wind in the lower canopy (Finnigan, 1979b; Raupach et al., 1989).

3.5.3 *Spectral slopes and isotropy*

Turning now to the spectral slopes of Fig. 3.12, we see that in the inertial subrange $fS_u(f)$ rolls off as $f^{-2/3}$ except within the canopy, where the roll-off rate is slightly steeper. Within the canopy $fS_v(f)$ and $fS_w(f)$, in contrast, roll off at less than $f^{-2/3}$. The only experiment where this kind of deviation from classic inertial subrange behavior is not observed is in Shaw's corn. The wind tunnel study of Seginer et al. (1976) clearly points to wake production as the source of energy in the v and w spectra, whereas the more rapid roll-off of $fS_u(f)$ can be explained in the same way. The aerodynamic drag term (3.45) is quadratic in the *total* velocity and therefore is much more efficient at extracting energy from u component fluctuations than from v and w. Unless this energy is immediately dissipated, and, of course, some of it is, the generally random orientation of eddies in plant wakes ensures that v and w fluctuations will be boosted at the expense of the u fluctuations.

A note of warning must be sounded at this point about the interpretation of wind tunnel spectra. The problems arise because of the relatively low Reynolds numbers obtainable in model canopies. It is generally agreed that to observe straight lines at $f^{-2/3}$ slope in the inertial subrange, a minimum of two decades in frequency or wavenumber must separate the energy-containing and viscous subranges. In wind tunnel models, for example, Coppin et al. (1986), this gap is generally no more than a decade. In such cases the $f^{-2/3}$ line is merely a tangent to a spectral form that, on a log-log plot, curves continuously from the energy-containing to the dissipation range. The danger lies in the possibility that this "hump" may be wrongly interpreted as a sign of wake production feeding into the inertial subrange.

In the surface layer, eddies in the inertial subrange are isotropic. As we saw in Chapter 2, isotropy implies that $S_v(f) = S_w(f) = (4/3)S_u(f)$. Within the canopy it is clear, despite a good deal of scatter in the data, that isotropy is generally violated. Spectral ratios in the inertial subrange vary from $S_w(f)/S_u(f) = 1.7$ for the Moga data to $S_w(f)/S_u(f) = 0.94$ in Shaw's corn. The data of Amiro (1990) in three different forest canopies showed $S_u(f)\colon S_v(f)\colon S_w(f) \simeq 1 \pm 0.15$, whereas the data of Baldocchi and Meyers (1988) from a dense deciduous forest fall into the same range as Moga and Shaw's corn data. We should not be too surprised at this large departure from classical isotropy, given that wake and waving production are so strongly coupled to each canopy's morphology and elastic properties.

We have described the way in which the dominant large eddies in a canopy are linked to the shear at $z = h_c$. The smaller eddies produced by the wake and waving production mechanisms display no such connection. In coherently waving cereal crops, the natural frequency of the stalks provides a further important time scale (Finnigan, 1979b), but more usually there is a broad band of length and time scales associated with these processes, ranging from the size of dominant foliage clumps, through trunk diameters, to leaf width and spanning tree swaying to leaf fluttering frequencies. In such cases, it is impossible to single out one scale as representative; we can, however, make some useful observations about the total effect of these processes on spectral shape.

We have already seen the changes in slope that occur in the inertial subrange of moderately dense canopies due to these "spectral shortcut" effects. They can also lead to overall changes in spectral shape as the wind speed increases. The change in shape follows from the quadratic dependence of aerodynamic drag and, consequently, the wake production term, on windspeed. Data of Amiro (1990) taken from a dense spruce canopy in quite strong ($\overline{u}_{hc} \simeq 3.5\ \mathrm{m\,s^{-1}}$) winds showed both $fS_u(f)$ and $fS_w(f)$ decreasing much more rapidly than $f^{-2/3}$ in the first (low frequency) two-thirds of the inertial subrange, whereas at lower windspeeds ($\overline{u}_{hc} \simeq 1.5$ m s^{-1}) the $f^{-2/3}$ law was obeyed quite well right across the range. Similar rapid rolloff of all spectral components within the canopy space was observed in the very dense forest studied by Baldocchi and Meyers (1988).

Their data exhibit another departure from the behavior we have sketched out, one to be expected in any canopy with a very dense crown and open understory. In such conditions, the depth of the region of strong shear that determines the scale of the large eddies is no longer even approximately equal to h_c. A better measure is $(h_c - d)$, which, in the case of the Baldocchi and Meyers forest, is close to $0.1h_c$. Scaling their frequency axis as $f(h_c - d)/\overline{u}_{hc}$ in fact brings the spectral peaks measured in the crown space back into the same range as all our other data, but this same scaling moves their trunk space and higher level peaks to uncomfortably low dimensionless frequencies. The reason for dwelling on this single case at some length is to emphasize that many problems in interpretation still remain, particularly when investigating natural canopies with strongly variegated foliage density, and despite the simple forms shown in Fig. 3.12, wide variations may be expected in any individual case.

3.5.4 *Spectrum of temperature*

Spectra of the most readily measured scalar, θ, have been published on only a few occasions. The measurements of Amiro (1990) show it rolling off at $f^{-2/3}$ in the inertial subrange as standard surface layer scaling predicts (Chapter 2). Its peak seems to coincide with $(\lambda_m)_u$.

3.5.5 *Cospectra*

Cospectra of u and w have been published less frequently than the individual power spectra, and it is difficult to draw any general conclusions. We expect the $fC_{uw}(f)$ slope in the inertial subrange to be $f^{-4/3}$ (Chapter 2). Amiro (1990), who presents the most extensive data set, observes it rolling off at approximately f^{-1}, as do Baldocchi and Hutchison (1987) in an almond orchard, but Shaw et al. (1974), measuring in corn, saw standard $f^{-4/3}$ behavior. Few of the published data are in a form that enables the cospectral peak to be located on the dimensionless frequency axis. The indications from Raupach et al. (1986) in the wind tunnel, in data from the Moga forest (Table 3.2), and in a wheat canopy (Finnigan, 1979a) are that it follows $(\lambda_m)_u$ and has its maximum at a dimensionless frequency about 0.15.

Cospectra of w and θ have similarly been presented on too few occasions to draw any general conclusions, but Amiro's (1990) measurements of $fC_{w\theta}(f)$ show it rolling off like $f^{-4/3}$ in the inertial subrange in accordance with surface layer theory. At this stage, the total lack of relevant published data prevents us from investigating the influence of stability on spectral shape.

3.6 Special symbols

a	leaf area per unit volume of space
A_{Li}	area of individual leaf
c	any variable
C_{di}	drag coefficient of an individual leaf
C_f, C_h, C_q	transfer coefficients for momentum, heat, and water vapor
d	displacement height
d_i	drag on the foliage
$D(z)$	aerodynamic drag, function of z
E_i	vapor flux density from an individual leaf
F_{di}	vector aerodynamic drag on individual leaf
F_h, F_q	transport rate for heat and water vapor across plant surfaces
h_c	canopy height
Pi	energy consumed by photosynthesis
q_i	humidity outside individual leaf boundary layer
q_{0i}	humidity at individual leaf surface
$\overline{q}_0$	$\overline{q}$ at height $(d + z_q)$
$q_{\text{sat}}(\overline{\theta})$	saturation specific humidity at $\overline{\theta}$
r	mixing ratio
r_a	general resistance to diffusion from plant surface
r_{ah}, r_{aq}	transfer resistance for heat and water vapor from plant surface
r_b	leaf boundary layer resistance
r_{bh}, r_{bq}	boundary layer resistance for heat and water vapor
r_{bhi}, r_{bqi}	boundary layer resistance for heat and water vapor on an individual leaf
r_c	canopy resistance
r_i	climatological resistance
r_{si}	stomatal resistance of an individual leaf
u_{hc}	wind speed at canopy top
U_i	mean wind speed outside individual leaf boundary layer
v_i'	component of foliage velocity
z_h, z_q	roughness length for temperature and humidity
Δ	specific saturation deficit
ϵ_s	dimensionless rate of change of saturated specific humidity with temperature
θ_i	temperature outside individual leaf
θ_0	θ at height $(d + z_h)$
θ_{0i}	temperature at individual leaf surface
θ_{v0}	θ_v at height $(d + z_h)$
$\phi_h^*, \phi_m^*, \phi_q^*$	similarity functions as observed above plant canopy

References

Allen, L. H., Jr., 1968: Turbulence and wind speed spectra within a Japanese larch plantation. *J. Appl. Meteor., 7*, 73–78.

Amiro, B. D., 1990: Drag coefficients and turbulence spectra within three boreal forest canopies. *Bound.-Layer Meteor., 52*, 227–246.

Baldocchi, D. D., and B. A. Hutchison, 1987: Turbulence in an almond orchard: vertical variations in turbulent statistics. *Bound.-Layer Meteor., 40*, 127–146.

Baldocchi, D. D., and T. P. Meyers, 1988: Turbulence structure in a deciduous forest. *Bound.-Layer Meteor., 43*, 345–364.

Bergström, H., and U. Högström, 1989: Turbulent exchange above a pine forest. II. Organized structures. *Bound.-Layer Meteor., 49*, 231–263.

Bradley, E. F., and J. J. Finnigan, 1973: Heat and mass transfer in the plant-air continuum. *Proc. First Australasian Heat and Mass Transfer Conference*, 23–25 May 1973, Monash Univ., Melbourne, Australia, 57–76.

Bradshaw, P., 1967: 'Inactive' motion and pressure fluctuations in turbulent boundary layers. *J. Fluid Mech., 30*, 241–258.

Brown, K. W., and W. Covey, 1966: The energy budget evaluation of the micrometeorological transfer processes within a cornfield. *Agric. Meteor., 3*, 73-96.

Cellier, P., 1986: On the validity of flux-gradient relationships above very rough surfaces (Research Note). *Bound.-Layer Meteor., 36*, 417–419.

Coppin, P. A., M. R. Raupach, and B. J. Legg, 1986: Experiments on scalar dispersion within a model plant canopy. Part II. An elevated plane source. *Bound.-Layer Meteor., 35*, 167–191.

Corrsin, S., 1974: Limitations of gradient transport models in random walks and turbulence. *Adv. in Geophys., 18A, Turbulent Diffusion in Environmental Pollution* (F. N. Frenkiel and R. E. Munn, Eds.). Academic Press, New York, 25-60.

Courant, R., 1959: *Differential and Integral Calculus, Vol. 2*. Blackie and Son Ltd., Glasgow and London, 682 pp.

Denmead, O. T., 1964: Evaporation sources and apparent diffusivities in a forest canopy. *J. Appl. Meteor., 3*, 383–389.

Denmead, O. T., 1984: Plant physiological methods for studying evapotranspiration; Problems in telling the forest from the trees. *Agric. Water Manage., 8*, 167–189.

Denmead, O. T., and E. F. Bradley, 1985: Flux-gradient relationships in a forest canopy. *The Forest-Atmosphere Interaction* (B. A. Hutchison and B. B. Hicks, Eds.). D. Reidel Publishing Co., Dordrecht, The Netherlands, 421–442.

Denmead, O. T., and E. F. Bradley, 1987: On scalar transport in plant canopies. *Irrig. Sci., 8*, 131-149.

Denmead, O. T., and I. C. McIlroy, 1970: Measurements of non-potential evaporation from wheat. *Agric. Meteor., 7* , 285–302.

Dunin, F. X., A. R. Aston, and W. Reyenga, 1978: Evaporation from a Themeda grassland. 2. Resistance model of plant evaporation. *J. Appl. Ecol., 15* , 847-858.

Dyer, A. J., 1974: A review of flux-profile relationships. *Bound.-Layer Meteor., 7* , 363–372.

Fazu, C., and P. Schwerdtfeger, 1989: Flux gradient relationships for momentum and heat over a rough natural surface. *Quart. J. Roy. Meteor. Soc., 115*, 335–352.

Ferguson, J., 1952: The rate of natural evaporation from shallow ponds. *Aust. J. Sci. Res., 5*, 315–330.

Finnigan, J. J., 1979a: Turbulence in waving wheat. I. Mean statistics and honami. *Bound.-Layer Meteor., 16*, 181–211.

Finnigan, J. J., 1979b: Turbulence in waving wheat. II. Structure of momentum transfer. *Bound.-Layer Meteor., 16*, 213–236.

Finnigan, J. J., 1985: Turbulent transport in flexible plant canopies. *The Forest-Atmosphere Interaction* (B. A. Hutchison and B. B. Hicks, Eds.). D. Reidel Publishing Co., Dordrecht, The Netherlands, 443–480.

Finnigan, J. J., and M. R. Raupach, 1987: Transfer processes in plant canopies in relation to stomatal characteristics. *Stomatal Function* (E. Zeiger, G. D. Farquar, and I. R. Cowan, Eds.). Stanford University Press, Stanford, CA, 385–429.

Gao, W., R. H. Shaw, and K. T. Paw U, 1989: Observation of organized structure in turbulent flow within and above a forest canopy. *Bound.-Layer Meteor., 47*, 349–377.

Garratt, J. R., 1978: Flux-profile relations above tall vegetation. *Quart. J. Roy. Meteor. Soc., 104*, 199–211.

Högström, U., 1988: Non-dimensional wind and temperature profiles. *Bound.-Layer Meteor., 42*, 55–78.

Inoue, E., 1963: The environment of plant surfaces. *Environment Control of Plant Growth.* Academic Press, New York, 23–32.

Jackson, P. S., 1981: On the displacement height in the logarithmic velocity profile. *J. Fluid Mech., 111*, 15–25.

Leclerc, M. Y., K. C. Beissner, R. H. Shaw, G. Den Hartog, and H. H. Neumann, 1990: The influence of atmospheric stability on the budgets of the Reynolds stress and turbulent kinetic energy within and above a deciduous forest. *J. Appl. Meteor., 29*, 916–933.

Lemon, E. R., and J. L. Wright, 1969: Photosynthesis under field conditions. XA. Assessing sources and sinks of carbon dioxide in a corn (Zea Mays L) crop using a momentum balance approach. *Agron. J., 61*, 405–411.

Maitani, T., and R. H. Shaw, 1990: Joint probability analysis of momentum and heat flux at a deciduous forest. *Bound.-Layer Meteor., 52*, 283–300.

McNaughton, K. G., and T. W. Spriggs, 1986: A mixed layer model for regional evaporation. *Bound.-Layer Meteor., 34*, 243–262.

Meyers, T. P., and D. D. Baldocchi, 1991: The budgets of turbulent kinetic energy and Reynolds stress within and above a deciduous forest. *Agric. For. Meteor., 50*, 1–16.

Monteith, J. L., 1965: Evaporation and environment. In *The States and Movement of Water in Living Organisms. Proc. XIX Symposium of the Society for Experimental Biology* (G. E. Fogg, Ed.), Cambridge University Press, 205–234.

Monteith, J. L., 1975: *Vegetation and the Atmosphere, Vol. 1. Principles.* Academic Press, New York, 278 pp.

Monteith, J. L., 1976: *Vegetation and the Atmosphere, Vol. 2. Case Studies.* Academic Press, New York, 439 pp.

Penman, H. L., 1948: Natural evaporation from open water, bare soil and grass. *Proc. Roy. Soc. London (A)., 193*, 120–146.

Penman, H. L., and I. F. Long, 1960: Weather in wheat: An essay in micrometeorology. *Quart. J. Roy. Meteor. Soc., 86*, 16–50.

Priestley, C. H. B., and R. J. Taylor, 1972: On the assessment of surface heat flux and evaporation using large scale parameters. *Month. Weather Rev., 100*, 81–92.

Raupach, M. R., 1988: Canopy transport processes. *Flow and Transport in the Natural Environment: Advances and Applications* (W. L. Steffen and O. T. Denmead, Eds.). Springer-Verlag, Berlin, 95–127.

Raupach, M. R., 1989a: Stand overstorey processes. *Phil. Trans. Roy. Soc. London, B, 324*, 175–190.

Raupach, M. R., 1989b: Applying Lagrangian fluid mechanics to infer scalar source distributions from concentration profiles in plant canopies. *Agric. For. Meteor., 47*, 85–108.

Raupach, M. R., 1991: Vegetation-atmosphere interaction in homogeneous and heterogeneous terrain: Some implications of mixed layer dynamics. *Vegetation, 91*, 105–120.

Raupach, M. R., and J. J. Finnigan, 1988: Single-layer models of evaporation from plant canopies are incorrect but useful, whereas multilayer models are correct but useless: Discuss. *Aust. J. Plant Physiol., 15* , 715–726.

Raupach, M. R., P. A. Coppin, and B. J. Legg, 1986: Experiments on scalar dispersion within a model plant canopy. Part I. The turbulence structure. *Bound.-Layer Meteor., 35*, 21–52.

Raupach, M. R., and R. H. Shaw, 1982: Averaging procedures for flow within vegetation canopies. *Bound.-Layer Meteor., 22*, 79–90.

Raupach, M. R., and A. S. Thom, 1981: Turbulence in and above plant canopies. *Ann. Rev. Fluid Mech., 13*, 97–129.

Raupach, M. R., J. J. Finnigan, and Y. Brunet, 1989: Coherent eddies in vegetation canopies. *Fourth Australasian Conference on Heat and Mass Transfer*, 9–12 May 1989, Christchurch, New Zealand, 75–90.

Ross, J., 1975: Radiative transfer in plant communities. *Vegetation and the Atmosphere* (J. L. Monteith, Ed.). Academic Press, London, 13–56.

Seginer, I., 1974: Aerodynamic roughness of vegetated surfaces. *Bound.-Layer Meteor., 5*, 383–393.

Seginer, I., P. J. Mulhearn, E. F. Bradley, and J. J. Finnigan, 1976: Turbulent flow in a model plant canopy. *Bound.-Layer Meteor., 10*, 423-453.

Shaw, R. H., G. Den Hartog, and H. H. Neumann, 1988: Influence of foliar density and thermal stability on profiles of Reynolds stress and turbulence intensity in a deciduous forest. *Bound.-Layer Meteor., 45*, 391-409.

Shaw, R. H., R. H. Silversides, and G. W. Thurtell, 1974: Some observations of turbulence and turbulent transport within and above plant canopies. *Bound.-Layer Meteor., 5*, 429–449.

Shuttleworth, W. J., 1989: Micrometeorology of temperate and tropical forest. *Phil. Trans. Roy. Soc. London B, 324*, 299–334.

Thom, A. S. 1971: Momentum absorption by vegetation. *Quart. J. Roy. Meteor. Soc., 97*, 414–428.

Thom, A. S., 1975: Momentum, mass and heat exchange of plant communities. *Vegetation and the Atmosphere* (J. L. Monteith, Ed.). Academic Press, London, 57–110.

Uchijima, Z., 1962: Studies on the microclimate within the plant communities. 1. On the turbulent transfer coefficient within plant layers. *J. Agric. Meteor. (Tokyo), 18*, 1–10.

Uchijima, Z., 1976: Maize and rice. *Vegetation and the Atmosphere*, Vol.2 (J. L. Monteith, Ed.). Academic Press, London, 34-64.

Uchijima, Z., and J. L. Wright, 1964: An experimental study of air flow in a corn plant-air layer. *Bull. Natl. Inst. Agric. Sci. (Japan), Ser. A, 11*, 19-65.

Wilson, J. D., D. P. Ward, G. W. Thurtell, and G. E. Kidd, 1982: Statistics of atmospheric turbulence within and above a corn canopy. *Bound.-Layer Meteor., 24* , 495–519.

Wright, J. L., and E. R. Lemon, 1966: Photosynthesis under field conditions. VIII. Analysis of windspeed fluctuation data to evaluate turbulent exchange within a corn crop. *Agron. J., 58*, 255–261.

Wyngaard, J. C, 1983: Boundary-layer modelling. In *Atmospheric Turbulence and Air Pollution Modelling* (F. T. M. Nieuwstadt and H. van Dop, Eds.). D. Reidel Publishing Co., Dordrecht, The Netherlands, 69–106.

Zeiger, E., G. D. Farquhar, and I. R. Cowan, Eds., 1987: *Stomatal Function*. Stanford University Press, Stanford, CA, 503 pp.

4

FLOW OVER CHANGING TERRAIN

The micrometeorologist setting out to find a field site that satisfies the requirements of horizontal homogeneity will soon be reminded that most of the earth's surface is not flat and that most of the flat bits are inconveniently heterogeneous. This is what forced the location of early pioneering experiments to remote sites such as Kansas, Minnesota, or Hay (Chapter 1), where the elusive conditions could be realized. Vital as these experiments were to the development of our understanding, they are merely the point of departure for applications to arbitrary terrain. The components of arbitrariness are two: changes in the land surface and hills. In this chapter we discuss the first of these, flow over changing surface conditions; in Chapter 5 we look at flow over hills. In the real world, the two conditions often occur together — in farmland it is the hills too steep to plow that are left covered with trees — but we separate them here to clarify the explication of phenomena and because treating them in combination would exceed the state of the art.

We simplify the problem of horizontal heterogeneity still further and discuss mainly single changes in surface conditions from one extensive uniform surface to another. Furthermore, the change will typically be at right angles to the wind direction so the resulting flow field is two-dimensional. Although multiple changes are now receiving theoretical attention (Belcher et al., 1990; Claussen, 1991), there exist as yet no experimental data for comparison.

Two types of surface change may be distinguished at the outset: change in surface roughness, which produces a change in surface momentum flux with a direct effect upon the wind field, and change in the surface availability of some scalar. Those of most interest are the active scalars, heat and moisture. (These are called active because their fluxes and concentrations affect stability and thereby turbulent mixing and momentum transfer, as we saw in Chapters 1 and 3.) We shall discover significant differences in flow behavior according to whether the wind blows from a smooth to a rough surface or a rough to a smooth surface. Similarly, flow from a cold to a hot surface produces quite a different result from a hot to a cold surface. In nature, such temperature changes are most often associated with changes in moisture availability as when the wind blows from the sea to the land. The changes in sensible and latent heat fluxes that result are coupled through the surface thermal energy balance as expressed in (3.25). In the simplest changes, such as those from open water to perfectly dry land, this coupling is easily taken into account, but over land covered with vegetation the consequences are more subtle.

The problem of surface heterogeneity can be considered on several scales. On the smallest scale the effects are confined to the surface layer. This is the classic case of local advection in micrometeorology. Local advection is of interest when choosing field sites for studies of the surface layer as we then want to know how far downwind of a change we must go to find a flow in equilibrium with the local surface. It is also important in calculating changes in temperature and humidity or rates of evaporation when the wind blows from arid to irrigated land. More fundamentally, an understanding of local advection is required to estimate surface fluxes from profile measurements once the one-dimensional conditions, upon which Monin-Obukhov similarity is predicated, are lost.

At the next larger scale we can follow the effects of the disturbance downwind until the whole depth of the ABL is affected. At this scale advective flows have received a good deal of recent attention, particularly where neutral or stable air flows onto a warm surface as in onshore breezes. Finally, at the mesoscale it is possible to track the readjustment of the whole ABL to the new surface and the reattainment of geostrophic balance. Included on this scale are thermal wind effects like the sea breeze, flows around urban "heat islands," and the question of average evaporation when wind blows from arid land to an extensive irrigation area. As we do throughout this volume, we shall avoid other than the briefest discussion of these larger scale effects and confine our interest to the ABL.

Flows over surface changes display three characteristic features:

- An internal boundary layer develops over the new surface, growing in height with downwind distance.
- Profiles of wind, temperature, and other scalars behind the change are not in equilibrium with the new surface.
- A complicated turbulent response to the change is observed with loss of the local-equilibrium character of homogeneous surface flow.

Our understanding of these effects rests mainly upon field experiments. A pervasive problem with most studies is the absence of reliable measurements of surface fluxes. The only roughness change experiment to overcome this obstacle has been that of Bradley (1968), who measured both wind profiles and surface shear stress changes in a neutral flow. Bradley studied both smooth-to-rough and rough-to-smooth changes and made unequivocal measurements of surface shear stress using a drag plate. His novel technique was to move large sheets of artificial roughness so that the fetch from the change to the drag plate and profile masts varied. Despite a variety of other less complete field studies, Bradley's data set remains the touchstone of theory.

Surface measurement of scalar fluxes is equally difficult. Dyer and Crawford (1965) used a lysimeter, the evaporation equivalent of a dragplate (see Chapter 6), to measure surface moisture flux directly at one point and to anchor an otherwise sparse data set, but the closest approach to Bradley's effort is the experiment of Itier et al. (1978) who used an array of minilysimeters to study a change in moisture over bare soil.

For studies of local advection, the wind tunnel offers opportunities. Unfortunately, one of the best known and most complete experiments on roughness change (Antonia and Luxton, 1971), which described a smooth-to-rough change, had roughness elements so large relative to the upwind inner layer depth[1] that its atmospheric equivalent would be flow to a city of skyscrapers. In the case of rough-to-smooth flow, the wind tunnel studies of Antonia and Luxton (1972) and Mulhearn (1978) offer excellent data sets of direct application to the atmosphere. The wind tunnel is also a candidate for studying scalar changes, with the added advantage of directly measurable surface fluxes if electrically generated heat is used as the scalar; unfortunately, all of the available data sets have features that compromise their application to the atmosphere. Despite the limitations of these studies, it is to the wind tunnel that we must turn for most of our information on the response of turbulence to surface changes in local advection.

Following the flow farther downwind to where the internal boundary layer is well outside the surface layer, most information comes from studies of the coastal boundary layer. Flow from sea to land with an unstable or convective internal layer has been studied at a large scale using aircraft or at a smaller scale using towers. Offshore winds usually produce stable internal boundary layers (arctic regions are the exception), and observations here are usually provided by aircraft.

The experiments of the last 25 years have been accompanied by a great deal of theoretical effort. Much of this work is discussed in the recent comprehensive review by Garratt (1990). The basic fluid mechanical principles behind the wind field changes have also been reviewed by Smits and Wood (1985). In this chapter we shall discuss only those aspects of theory that serve to unify the available data.

Our discussions revolve around two types of changes: changes in surface roughness and changes in scalar flux and concentration; in each we discuss the effects of local advection and development farther downwind. The many combinations of these features that characterize the experimental data mean that it would be presumptuous to illustrate the points we shall make with idealized examples of flow behavior. We rely instead on examples taken directly from experiments. In this respect the chapter differs substantially from the preceding ones.

[1]In a fully developed turbulent boundary layer in a wind tunnel, the inner layer is the region of constant shear stress and logarithmic velocity profile and so corresponds to the surface layer of the ABL.

4.1 Changes in surface roughness

We simplify our discussion by considering an abrupt change in surface roughness from one extensive uniform region to another. The line of surface discontinuity is perpendicular to the wind direction, which, as usual, we identify with the x direction. As the airflow encounters the new surface it either slows down because of increased surface friction (smooth-rough) or speeds up as surface friction falls (rough-smooth). The effect of this acceleration or deceleration, which is initially confined to the air layers in contact with the new surface, is diffused vertically by turbulence, and the effect of the change is felt through a steadily growing internal boundary layer.

The effects of the change are also transmitted by pressure forces, which are associated with streamline deflection at the roughness change. These effects are not confined to the internal layer. Unless there is a significant change in displacement height, such as in flow to or from a tall crop or forest, the velocity perturbations caused by this effect are negligible compared to those confined to the internal boundary layer. The only place where the pressure forces are important is in the immediate neighborhood of the discontinuity. Most theories simply ignore the pressure effect.

4.1.1 *Characterizing the strength of the roughness change*

What we require is some simple measure that will enable us to classify the magnitude of the roughness change independently of the particular airflow over it. The measure normally taken is the ratio of roughness lengths or its logarithm. We define

$$M^* = \frac{z_{01}}{z_{02}}, \tag{4.1a}$$

$$M = \ln\left(\frac{z_{01}}{z_{02}}\right) = \ln(z_{01}) - \ln(z_{02}), \tag{4.1b}$$

where subscripts 1 and 2 indicate, respectively, properties upwind ($x < 0$) or downwind ($x > 0$) of the discontinuity. [Note that some authors (e.g., Walmsley, 1989) define M^* as z_{02}/z_{01} and $M = \ln(z_{02}) - \ln(z_{01})$, reversing the order of roughness lengths.]

Although not as dependent upon flow properties as quantities like the surface stress, the roughness length z_0 is not entirely independent and only attains equilibrium values far downstream of the change. Despite this drawback it remains the best measure we have, falling as it does between a geometric description of the surface, which cannot be readily related to momentum absorption even in horizontally homogeneous conditions, and the extremely flow-dependent stress ratio. M and M^* are natural classifications for flow models that generally take the logarithmic law as a lower boundary condition (Rao et al., 1974a).

4.1.2 *Internal boundary layer depth*

Within the internal boundary layer (IBL) the flow field displays characteristics of the downwind surface. Outside it, apart from the small perturbation caused by the pressure pulse at the change, the flow field is identical to that upwind. The internal boundary layer depth δ_i is usually defined, therefore, as the height at which the downwind velocity or the shearing stress $\tau(z)$ attain fixed fractions (say 99%) of their upwind values at the same height. Note that because turbulent stresses adjust differently than mean profiles, the magnitude of δ_i depends on the definition chosen.

The height δ_i may also be identified with changes in the velocity shear $\partial\overline{u}/\partial z$. When plotted in log-linear form, a sharp discontinuity in slope is often observed at the edge of the internal layer (Bradley, 1968). A schematic diagram of the development of the internal boundary layer is shown in Fig. 4.1a. Antonia and Luxton

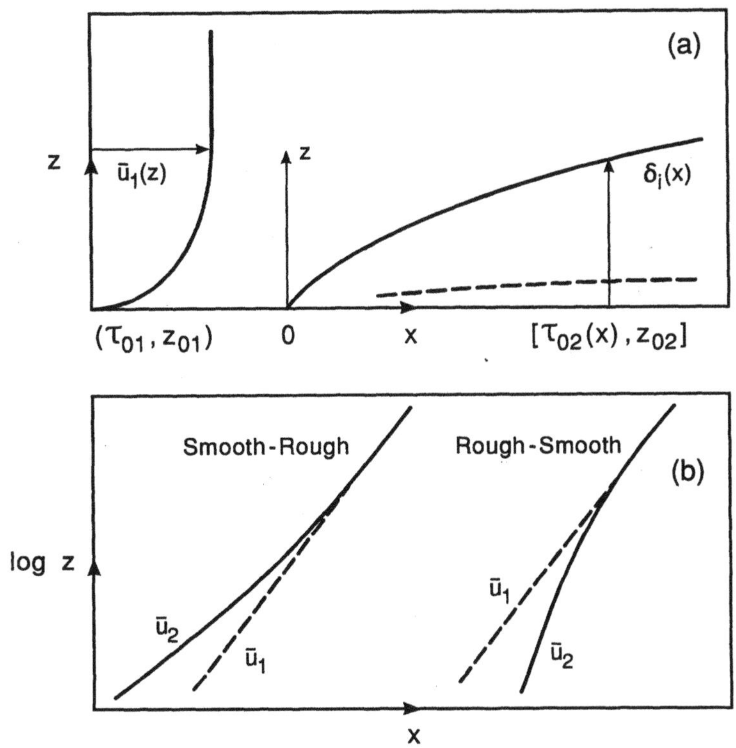

FIG. 4.1 (a) Schematic diagram of IBL growth. The hypothetical inner equilibrium region of the IBL is marked by the dashed curve. Note that this region is not expected to begin until some distance after the change in surface. (b) Logarithmic velocity profiles after a roughness change; the upwind equilibrium profile is denoted by the dashed line.

(1971, 1972) also argued on theoretical grounds and confirmed experimentally that plotting $\overline{u}$ as a function of $z^{1/2}$ would reveal sharp discontinuities in velocity gradient at $z = \delta_i$.

There are two approaches to quantifying the growth rate of the internal boundary layer: empirically based power law formulas and more complex formulas that recognize the diffusive character of the growth of the internal layer. In neutral conditions, turbulent boundary layer growth on a smooth, flat plate is proportional to $x^{0.8}$ (Schlichting, 1968). Elliott (1958) proposed an analogous empirical formula for the internal boundary layer:

$$\frac{\delta_i}{z_{02}} = A_1 \left(\frac{x}{z_{02}}\right)^n, \tag{4.2}$$

where $n = 0.8$ and the constant of proportionality A_1 has a weak dependence on the strength of the roughness change

$$A_1 = 0.75 + 0.03M. \tag{4.3}$$

Reasonable agreement has been found with expressions of this kind, although there is no agreement on the precise magnitude of the exponent or of the proportionality constant A_1. Both Bradley's (1968) smooth-rough and rough-smooth cases were well fitted by $n = 0.8$, as were Antonia and Luxton's (1971) smooth-rough wind tunnel data and Mulhearn's rough-smooth tunnel experiment in one configuration. However, Antonia and Luxton's (1972) rough-to-smooth experiment showed $n = 0.43$ and the fit with larger compendia of data (Walmsley, 1989) shows less than ideal agreement with power law expressions in general. The advantage of power law formulas nevertheless is that they give δ_i as a function of x explicitly.

The alternative approach was initiated by Miyake (see Panofsky and Dutton, 1984, p. 150). It is based on the idea that the characteristic velocity of vertical diffusion should be proportional to u_{*2}, where $u_{*2} = (\tau_{02}/\rho)^{1/2}$ is the downwind friction velocity. This is a notion commonly used in treatments of scalar diffusion from a ground source. [A good review and list of references may be found in Raupach (1983).] It follows from this assumption that

$$\frac{d\delta_i}{dx} = \frac{B_1 u_{*2}}{\overline{u}(x)}, \tag{4.4}$$

where B_1 is a constant of order one.

With the assumption that $\overline{u}(x) = (u_{*2}/k)\ln(z/z_{02})$ and the initial condition that $\delta_i = z_{02}$ at $x = 0$, equation (4.4) may be integrated to obtain an implicit formula for δ_i (Panofsky and Dutton, 1984),

$$\frac{\delta_i}{x}\left[\ln\left(\frac{\delta_i}{z_{02}}\right)-1\right]=B_1 k. \tag{4.5}$$

An immediate objection to (4.5) is that no account is taken of the influence of the upstream boundary layer and its length and velocity scales z_{01} and u_{*1} on the growth of δ_i. In an earlier version of (4.5), Jackson (1976) replaced z_{02} by $z_0' = \left[\left(z_{01}^2 + z_{02}^2\right)/2\right]^{1/2}$. This choice introduced some acknowledgment of the upstream state into the formula but in an apparently arbitrary way that lost the simple interpretation of the derivation of (4.5).

Walmsley (1989) has made a comparison of (4.2), (4.5), and Jackson's modified version of (4.5) with the available atmospheric data. Elliott's formula (4.2) was used both in M-dependent and M-independent (M set equal to 0) form. The constant B_1 in (4.5) was taken as 1.25. Overall, the Panofsky-Dutton formula (4.5) gave much better agreement with data than (4.2), which consistently overestimated δ_i regardless of whether the M-dependent or M-independent form was used.

A reliable expression for the internal boundary layer depth is a most important tool since all of the more usable descriptions we shall encounter for the development of velocity and scalar profiles behind the change rely on this parameter to furnish their vertical scale. The success of (4.5) in collapsing a range of data is, therefore, encouraging empirically as well as offering support to the concept of the diffusive growth of the internal boundary layer. One note of caution should be sounded, however. The wind tunnel data of Antonia and Luxton (1972) downwind of a rough-to-smooth change showed δ_i growing substantially more slowly than (4.5) demands. This experiment had a large positive value of M, and the downstream growth of δ_i was therefore controlled for a considerable distance by turbulence generated on the upstream rough surface, a situation that clearly contradicts the assumptions.

The lowest region, probably the lowest 10%, of the internal boundary layer is often called the equilibrium layer (Fig. 4.1a). In this region it is supposed that a new constant-stress inertial sublayer has been achieved in equilibrium with the new surface. Within this layer, local equilibrium in the sense discussed in Chapter 3 (i.e., a balance between shear production and viscous dissipation of turbulent kinetic energy with advection and transport much smaller) is assumed to exist. We shall return to this question when we consider turbulence dynamics in Section 4.3. For the moment it is sufficient to point out that the rather sparse evidence, all from wind tunnel experiments, tends to support this concept, as long as we are not too close to the roughness change.

4.1.3 *Velocity profiles*

Schematic diagrams of velocity profile changes behind a smooth-to-rough and a rough-to-smooth change are shown in Fig. 4.1b. Plotted in log-linear form, the profiles show a new logarithmic profile in the lower part of the internal boundary layer and a blending region occupying most of the internal layer across which the velocity profile adjusts to merge with the upstream logarithmic profile at $z = \delta_i$. The apparently sharp break between the log-linear profile in the internal layer and the unchanged upstream log-linear profile above δ_i is largely an artifact of the logarithmic height scale. Based on the behavior of numerical simulations some authors have suggested that a point of inflection should be observed between the inner and outer logarithmic profiles (Peterson, 1969; Rao et al., 1974a). This behavior, however, has not been observed with any certainty in experimental data.

Figures 4.2a and 4.2b display consecutive downstream profiles taken from Bradley's (1968) smooth-rough and rough-smooth cases, respectively. In each case we see the internal layer deepening and, depending on the sign of M, slowing down or speeding up as we progress downstream. The logarithmic behavior of the velocity profiles within the internal boundary layer suggests that they might be described through a modified logarithmic law,

$$\overline{u}(z) = \frac{u_{*2}}{k} \ln\left(\frac{z}{z_{02}}\right) + f(z/\delta_i), \tag{4.6}$$

where the function $f(z/\delta_i)$ has limiting values

$$f(z/\delta_i) = \frac{u_{*1}}{k} \ln\left(\frac{z}{z_{01}}\right) - \frac{u_{*2}}{k} \ln\left(\frac{z}{z_{02}}\right), \quad z/\delta_i > 1, \tag{4.7a}$$

$$f(z/\delta_i) = 0, \quad z/\delta_i \ll 1. \tag{4.7b}$$

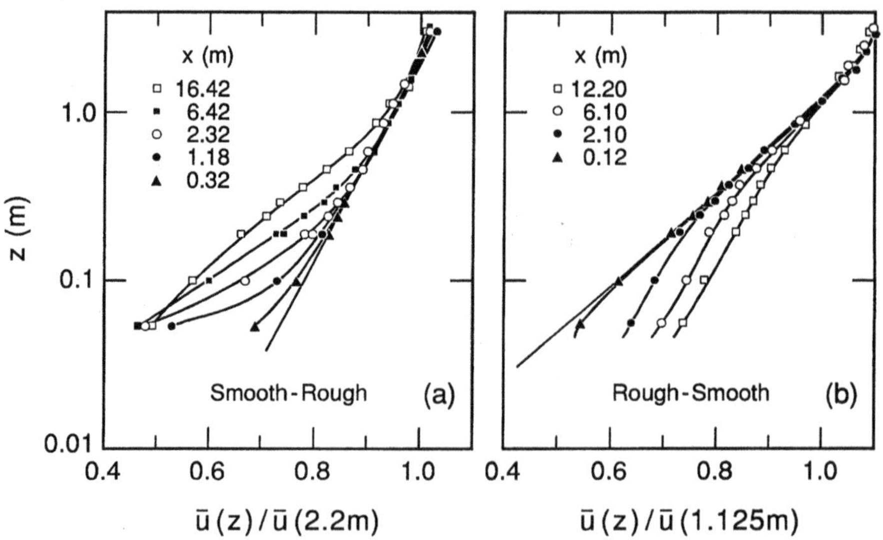

FIG. **4.2** The development of logarithmic velocity profiles after a roughness change taken from Bradley (1968). (a) Smooth-rough: $z_{01} = 0.02$ mm, $z_{02} = 2.5$ mm, $M = -4.8$; (b) Rough-smooth: $z_{01} = 2.5$ mm, $z_{02} = 0.02$ mm, $M = +4.8$.

The first limit says that outside the internal boundary layer, the upstream profile is recovered. The second limit ensures a logarithmic wall layer in equilibrium with the new surface.

A variety of forms has been proposed for f. The simplest, by Elliott (1958), has $f = 0$ for $z < \delta_i$. This choice dictates a logarithmic profile in equilibrium with the new surface all the way up to δ_i and a sharp discontinuity in slope there. Despite the contradiction between this assumption and measurements, the other predictions of Elliott's theory are, on the average, not greatly at variance with observations and, importantly, the theory is simple to apply. As a consequence it is still widely used as a first approximation and is incorporated into at least one proprietary model of airflow over complex terrain. Other assumed forms for f are reviewed by Garratt (1990).

A somewhat different approach to describing the downwind profiles was adopted by Townsend (1965, 1966) and by Mulhearn (1977). They assumed that changes in the velocity and shear stress profiles in the internal boundary layer were "self-preserving." This concept enjoys wide validity in aerodynamics, and an excellent introduction is offered by Townsend (1976). We shall encounter it in the next chapter in the context of the far wakes of hills. Self-preservation of both velocity and shear stress means that the functional forms for the vertical distributions of these quantities are invariant with fetch and that their length and velocity scales are functions of fetch only. Following Mulhearn (1977), the self-preserving forms are as follows:
For velocity,

$$\Delta u(z) = \overline{u}_2(z) - \overline{u}_1(z) = \frac{u_0}{k} g(\eta) + \overline{u}_p, \tag{4.8a}$$

and for shear stress,

$$\Delta\tau(z) = \tau_2(z) - \tau_1 = \left[\tau_2(0) - \tau_1\right] h(\eta), \tag{4.8b}$$

where τ_1 is assumed invariant with height, $\overline{u}_p$ is the velocity perturbation caused by the pressure pulse or streamline displacement, generally ignored in calculations, and u_0 is the velocity scale of the perturbation. In Mulhearn's theory, u_0 is defined as

$$u_0 = u_{*2} - u_{*1}. \tag{4.9}$$

The dimensionless height η is given by $\eta = z/l_0$, where l_0 is the vertical scale of the perturbation. Although the theory defines l_0 rigorously, for practical purposes we can take $l_0 = \delta_i$ so that $\eta = z/\delta_i$.

Forcing the forms (4.8a) and (4.8b) to be compatible with the equations of motion ensures that the unknown functions $g(\eta)$ and $h(\eta)$ obey a single ordinary differential equation that can be solved if a closure assumption is made about the relationship between $\tau_2(z)$ and $\partial\overline{u}_2/\partial z$. The assumption made by Mulhearn was that $\phi_m = \left\{kz/\left[\tau_2(z)/\rho\right]^{1/2}\right\}(\partial\overline{u}_2/\partial z) = 1$. This is a somewhat weaker assumption than that made by Elliott, which was that $\phi_m = (kz/u_{*2})(\partial\overline{u}_2/\partial z) = 1$, but it still amounts to an assumption that shear stress and velocity gradient are linked by a simple eddy-diffusivity formula in the evolving flow behind the discontinuity.

To test the validity of this assumption, independent measurements of velocity and shearing stress in the internal layer are required. These are available only from the wind tunnel experiments of Antonia and Luxton (1971, 1972) and Mulhearn (1978). Measurements of ϕ_m at a position 2 m behind Mulhearn's rough-smooth change are plotted in Fig. 4.3 and show substantial departures from unity in the central part of the internal layer. Also in Fig. 4.3 are curves of ϕ_m from the sophisticated turbulence closure model of Rao et al. (1974a). Profiles for four values of M are plotted; the one for $M = 4.8$ is the closest approximation to Mulhearn's data. Rao et al.'s model provides at least a qualitative surrogate for the nonexistent empirical results and may be taken as confirmation that the

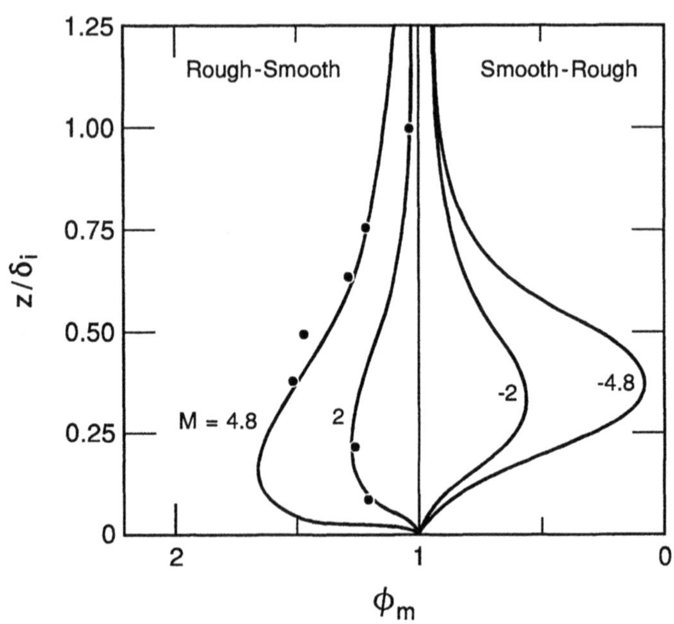

Fig. 4.3 The nondimensional wind shear $\phi_m = (kz/u_{*2})\partial\overline{u}/\partial z$ for several values of the roughness change parameter M. Solid lines are results of the model of Rao et al. (1974a). Data points come from Mulhearn (1978) and correspond most closely to $M = +4.8$.

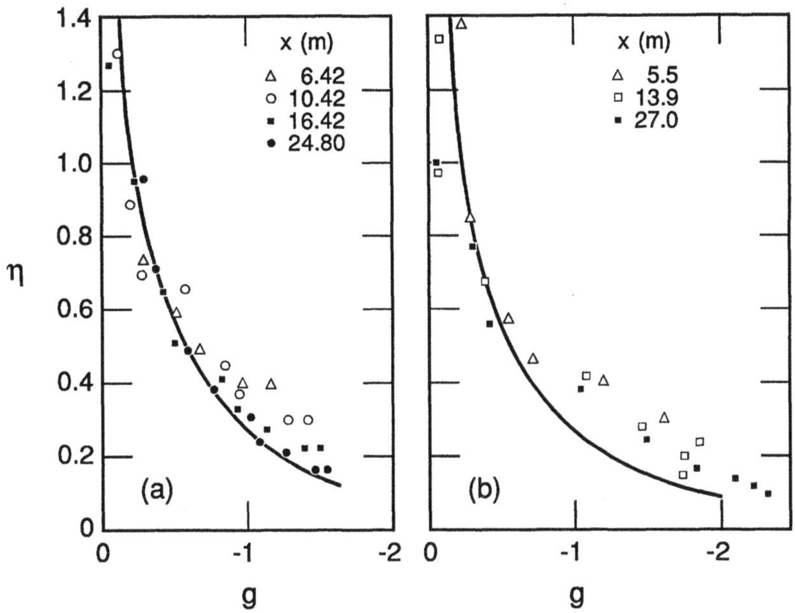

FIG. **4.4** Comparison between Mulhearn's (1977) theory and the roughness change data of Bradley (1968). (a) Smooth-rough: $z_{01} = 0.02$ mm, $z_{02} = 2.5$ mm, $M = -4.8$. (b) Rough-smooth: $z_{01} = 5$ mm, $z_{02} = 0.2$ mm, $M = +5.5$. Note that this is a different rough-smooth data set than that used in Fig. 4.2b.

relationship between shear stress and velocity gradient, at the least, takes a rather complex form. We shall return to consider this question in more detail when we discuss turbulence dynamics below. As we might expect, the departure of ϕ_m from 1 decreases as we move further from the roughness change.

Returning to Mulhearn's model, in Figs. 4.4a and 4.4b we compare his predictions with data from Bradley (1968). The quantity plotted is the dimensionless velocity perturbation $g = (k/u_0)\Delta u$ versus the dimensionless height η. Two things are noteworthy: The theory performs better for a smooth-rough than for a rough-smooth change, and it is more accurate the further we get from the discontinuity. The latter observation we should expect from the foregoing discussion; the further we proceed downwind, the better becomes the correspondence between the closure assumption $\phi_m = 1$ and observations. The fact that the theory is not grossly in error in predicting the velocity profile quite close to the change has an unfortunate corollary. It means that the surface shear stress cannot be easily inferred from velocity gradient measurements until sufficiently downwind of the discontinuity; the distance that is sufficient is substantially greater for a rough-smooth than for a smooth-rough change.

4.1.4 *Changes in surface stress*

Figures 4.5a and 4.5b reproduce data from Bradley (1968) showing changes in surface stress behind roughness changes. Figure 4.5a is from a smooth-to-rough transition, and Fig. 4.5b is from a rough-to-smooth one. The roughness of the two surfaces is characterized by the aerodynamic roughness lengths they would have in equilibrium conditions. For the rough surface this was 2.5 mm and for the smooth surface between 0.02 and 0.002 mm (Rao et al., 1974a). The corresponding M values are $M = -4.83$ (smooth to rough), and $M = 4.83$ (rough to smooth, with z_0 for the smooth surface).

Two features of these results are noteworthy: the overshoot in stress at the transition and the rapid attainment of a new equilibrium. In the smooth-rough case the downwind surface stress τ_{02} is eight times its upstream value τ_{01} immediately behind the change ($x = 0.5$ m) but has dropped to its equilibrium value of about

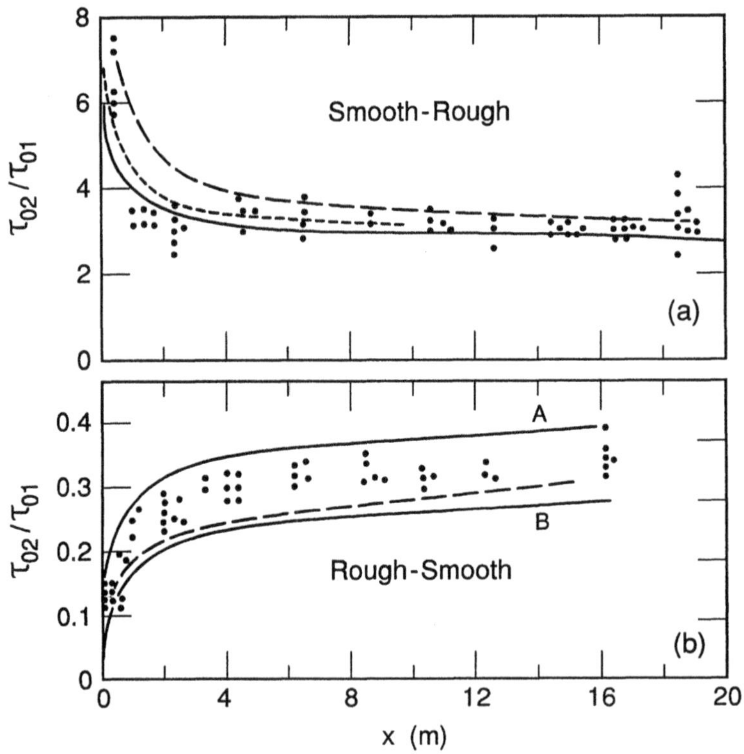

Fig. 4.5 Comparison between Bradley's (1968) measurements of surface shear stress development and various theories. (a) Smooth-rough change ($z_{01} = 0.02$ mm, $z_{02} = 2.5$ mm, $M = -4.8$): •, Bradley data; ———, Rao et al. (1974a), — — —, Panofsky and Townsend (1964); - - - - - - - - , Jensen (1978). (b) Rough-smooth change ($z_{01} = 2.5$ mm, $z_{02} = 0.02$ mm, $M = +4.8$): -A- Rao et al. (1974a); — — —, Panofsky and Townsend (1964), -B- Rao et al. (1974a), with $z_{02} = 0.002$ mm, $M = 7.1$.

$3.5\tau_{01}$ by $x = 2$ m. In the rough-smooth case the stress initially falls to $0.13\tau_{01}$ then takes somewhat longer to recover to its downstream equilibrium state of $0.3\tau_{01}$, which it does by $x = 6$ m.

The overshoot phenomenon is easily understood. In the smooth-to-rough case an airstream, traveling relatively rapidly over the smooth surface, generates a high stress on first encountering the roughness. As the new, rougher surface absorbs momentum from the air layers above it and this region of decelerated flow thickens into an internal boundary layer, the velocity of the air layer in contact with the enhanced roughness falls and so does the resulting surface stress. In the rough-to-smooth case, the air layers arriving at the new surface are initially moving slowly and the stress drops as they encounter the smoother ground. They are then able to accelerate because the smooth surface absorbs less momentum and the surface stress climbs until a new equilibrium is reached. The slower adjustment of stress after a rough-to-smooth transition is a phenomenon we shall observe in all flow properties.

The various theories we have encountered so far all make predictions about the development of surface stress after the change. We have included curves from three of them in Figs. 4.5a and 4.5b. The best agreement is provided by the second-order closure model of Rao et al. (1974a). The distinguishing feature of this class of turbulence models is that they do not link shear stress and shear by an eddy diffusivity. As a result they tend to perform much better in situations in which the turbulent flow is far from local equilibrium and eddy diffusivities are untenable (see Section 4.3 below). We see that the model of Rao et al. is the only one to reproduce the rapid stress change near the discontinuity accurately.

The model curve of Panofsky and Townsend (1964) is typical of the simpler models. Although it reproduces equilibrium conditions quite well, it completely fails to capture the sharp changes at the boundary. The simplest approach of all is that taken by Elliott (1958). His assumption of a sharp discontinuity at $z = \delta_i$ between logarithmic profiles in equilibrium with the upwind and downwind surfaces [i.e., $f = 0$ in (4.6)] leads directly to an expression for the stress ratio (Jensen, 1978):

$$\frac{\tau_{02}}{\tau_{01}} = \left[1 - \frac{M}{\ln(\delta_i/z_{02})}\right]^2. \tag{4.10}$$

This formula works surprisingly well, lying almost on top of Rao et al.'s curve. It is, however, critically dependent on the calculation of δ_i and so, in a sense, is only as good as the expression used for that quantity.

4.1.5 *Effects of roughness changes at larger scales*

The depth of the surface layer is typically 50–100 m, and the measurements and theories we have discussed so far are confined to this region. This is because the measurements used small towers, and the theories assumed a logarithmic upwind profile. We can calculate, using the formula (4.5) of Panofsky and Dutton (1984), how far downwind we can go before δ_i exceeds the height of the surface layer. If we assume a large smooth-to-rough change ($z_{01} = 0.03$ m; $z_{02} = 1.0$ m; $M = -3.5$), $\delta_i = 100$ m at $x = 700$ m. After a rough-to-smooth change ($z_{01} = 0.03$ m; $z_{02} = 0.001$ m; $M = +3.4$), $\delta_i = 100$ m at $x = 2000$ m. At fetches larger than this, we are out of the region of local advection and must consider the response of the entire ABL to the surface change.

Experimental data on the evolution of the internal layer to longer fetches in neutrally stratified ABLs are sparse but some are presented in Sempreviva et al., (1990). (As we noted in Chapter 1, neutral stratification throughout the depth of the whole ABL is an uncommon state of affairs in general.) During onshore flow, Sempreviva et al. measured wind profiles at the shoreline and at four masts up to 30-km inland. Using simple models to interpolate the data, they attempted to deduce seasonal changes in area-averaged z_0. It must be said that the results were inconclusive. The question has also been considered theoretically, however, and the results are worth recording as they emphasize that we still have a good deal to understand about this situation.

Jensen (1978), for example, used some simple arguments to point out a contradiction between the apparent early attainment of equilibrium in surface stress behind the roughness change that is exhibited in Bradley's results (Fig. 4.5) and in surface layer models and the current understanding of the magnitude of the geostrophic drag coefficient u_*/G, where G is the geostrophic wind speed. His arguments suggest that by the time the entire ABL has come into equilibrium with the new surface, the surface stress ratio τ_{02}/τ_{01} will have evolved substantially from the equilibrium observed soon after the change. To take a numerical example, Bradley's smooth-to-rough data (Fig. 4.5a) indicate an equilibrium value of $\tau_{02}/\tau_{01} = 3.5$. Jensen (1978) suggests that this will fall to $\tau_{02}/\tau_{01} \simeq 2.0$ when final equilibrium is attained.

The simple model of Taylor (1969) indicates how this equilibrium might be approached. Taylor extended a surface layer model to the whole ABL by assuming a form for an eddy diffusivity that would be valid across the whole ABL depth. For a neutrally stratified ABL, one definition of the ABL depth z_h is the height at which the mean velocity $\overline{u}(z)$ matches the geostrophic wind [see (1.1)]. With this definition, Taylor found that in both rough-smooth and smooth-rough changes the internal layer filled the whole ABL when $x/z_{01} \simeq 10^6$. (Taylor's calculations were done for $M = \pm 2.3$ and surface Rossby number $R = G/fz_{01} = 10^7$.) Up to this point there were no wind direction changes in the boundary layer. These occurred between $x/z_{01} = 10^6$ and $x/z_{01} = 10^8$, and the direction changes occurred simultaneously through the depth of the ABL. In a smooth-rough change, the angle between the surface and geostrophic wind direction (the geostrophic departure) increased, whereas in a rough-smooth change it decreased. For the parameters chosen by Taylor the magnitude of the change in geostrophic departure was about

-4° in the rough-smooth case (from 19° to 15°) and $+5^\circ$ in the smooth-rough (from 19° to 24°).

4.2 Changes in scalar flux and concentration

In this section we are concerned primarily with the active scalars, heat and moisture. Passive scalars introduced through a change in surface properties do not cause any adjustments in the wind and turbulence fields. Fluxes of the active scalars are coupled at the ground through the surface thermal energy budget, as mentioned earlier, and changes in them usually result from changes in moisture availability. Thus, if the available radiative energy remains constant as the wind blows from a dry to a wet surface, the increase in latent heat flux will be at the expense of the sensible heat flux. We will consider this coupling at greater length later in this section.

Changes in moisture flux are also usually associated with changes in z_0. The obvious example is the transition from water to land; another is between dry soil and irrigated crops, a common agrometeorological problem. The effect of a z_0 change is, as we might expect, greatest in the surface layer where it strongly modulates the effect of the scalar fluxes. (Alternatively, in local advection, we can view the buoyancy flux as strongly modulating the effect of a roughness change.) In contrast, roughness changes appear to have little effect on the development of convective internal boundary layers on larger scales.

Convective internal boundary layers develop after a cold-hot change. They have been studied most intensively in the context of onshore sea breezes. The motivation is usually problems in coastal pollution (Stunder and Sethuraman, 1985), although routing of microwave links has also been cited (Hsu, 1986). This context has meant that most studies have been relatively large scale, based on aircraft, balloon, or sodar measurements, and have ignored the surface layer. Exceptions are the tower-based experiments of Vugts and Businger (1977), Smedman and Högström (1983), and Ogawa and Ohara (1985).

In direct contrast to this, stable internal boundary layers developing after a warm-cold transition have mainly been studied in the context of local advection. This is commonly the micrometeorological situation of airflow from nonirrigated to irrigated land. The most complete experiment remains that of Rider et al. (1963), with Dyer and Crawford (1965) and Itier et al. (1978) also contributing important data sets. Stable internal layers at larger fetches occur in offshore winds and are less accessible to measurement, usually requiring aircraft (Garratt and Ryan, 1989).

4.2.1 *Internal boundary layer depth: Local advection*

We denote the scalar internal boundary layer by δ_θ for temperature and δ_q for moisture or by δ_c for an arbitrary scalar c if we do not wish to be specific. The local advection stages of internal layers following hot-cold and cold-hot transitions are drawn schematically in Figs. 4.6a and 4.6b, respectively. The development of a convective internal layer to longer fetch is depicted in Fig. 4.6c.

None of the experiments studying local advection from hot-to-cold surfaces used towers high enough to track the edge of the internal layer. Rider et al. (1963), in particular, made no temperature measurements above 1.5 m. We have slightly more information on cold-to-hot transitions, but not enough for the kind of empirical determination of the height-to-fetch relationship that was possible for the momentum boundary layer. Consequently, we look for analogies with better understood situations.

In laminar flow, the ratio of momentum and scalar boundary layer depths δ/δ_c is equal to the square root of the ratio of molecular diffusivities,

$$\frac{\delta}{\delta_c} = \left(\frac{\nu}{k_c}\right)^{1/2} = (Sc)^{1/2}, \tag{4.11}$$

ν is the kinematic viscosity and k_c the diffusivity of scalar c. Their ratio is called the Schmidt number Sc. If the scalar c is temperature, then the ratio is known as the Prandtl number Pr. (The square root is a consequence of the nature of boundary layer growth with constant molecular diffusivity.) For example, during laminar airflow over a heated flat plate, the ratio of momentum to thermal boundary layer thicknesses at any point is equal to 0.84 because the Prandtl number of air is 0.7.

As we noted earlier, the turbulent boundary layer grows approximately as $x^{0.8}$ and, in analogy with (4.11), we might expect

$$\frac{\delta}{\delta_c} = \left(\frac{K_m}{K_c}\right)^{0.8}, \tag{4.12}$$

where $K_m/K_c = Sc_t$ is the turbulent Schmidt number, K_m and K_c being eddy diffusivities for momentum and scalar c. The turbulent Prandtl number, correspondingly, is defined as $Pr_t = K_m/K_h$.

Equation (4.12) would be correct if K_m and K_c were substantive properties of the fluid such as their molecular counterparts. In reality, they are secondary characteristics of the flow field, and they, as well as their ratio Pr_t, respond to all the factors influencing the fluxes and gradients of momentum and temperature. Most of the evidence concerning the behavior of Pr_t in boundary layer flows under near-neutral conditions come from the wind tunnel. From the available data, Pr_t in the surface layer seems to be about 0.95 and varies little across the depth of the surface layer (Launder, 1976). In advective flows, however (as we shall see in Section 4.3), Pr_t may vary in the streamwise direction. Despite these caveats, when δ_i and δ_c are less than the depth of the surface layer, it is generally assumed that a relationship such as (4.12) will apply to the growth of the internal boundary layer.

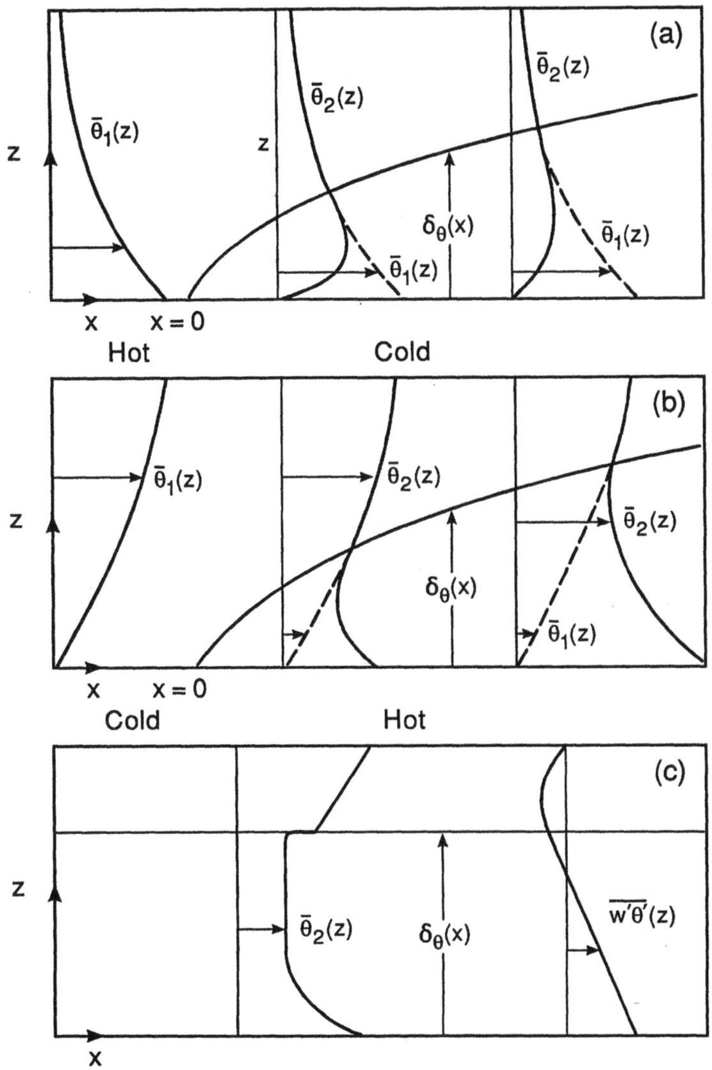

FIG. 4.6 Schematic diagrams of potential temperature profiles after changes in surface temperature or heat flux. (a) The development of an advective inversion after a hot-cold transition. (b) The growth of a CIBL after a cold-hot transition. (c) The structure of a CIBL at longer fetch. Note the jump in $\overline{\theta}_2$ at $z = \delta_\theta$, the stable temperature profile above δ_θ, and the change in sign of $\overline{w'\theta'}$ at $z \sim 0.8\delta_\theta$. Note also that the surface layer with a pronounced lapse profile may occupy a larger fraction of the total CIBL depth than in the classical convective mixed layer.

The Monin-Obukhov formulas (1.31), (1.32), and Fig. 3.2 are adjusted to ensure that $Pr_t = 1$ in neutral conditions. As we depart from neutrality, on the stable side, Pr_t remains close to 1, but on the unstable side we have, from (1.38a) and (1.40), $K_h/K_m = \phi_m/\phi_h = (1 + 16|z/L|)^{1/4}$. We should not expect strict adherence to these essentially one-dimensional formulas in advective conditions, but they give some indication of the changes to be expected.

Differences may also appear between the scalar diffusivities K_h and K_q. These are usually assumed to be equal in horizontally homogeneous flows, although K_q has not been determined nearly as reliably as K_h and K_m. In an experiment that studied a severe hot-cold, dry-wet change (flow from hot, dry ground to a flooded rice paddy), Lang et al. (1983) determined the ratio K_h/K_q directly from eddy flux and gradient measurements. They found that $K_h/K_q = 1$ at neutral stability but that the ratio decreased to 0.65 with increasing stability.

Clearly the analogy with laminar flow embodied in (4.12) and the foregoing discussion should alert us to the possibility of substantial variation among δ_i, δ_θ, and δ_q. Unfortunately, we have little direct data to test this possibility. Vugts and Businger (1977) deduced indirectly that $\delta_\theta \propto x^{0.8}$ in a cold-hot, wet-dry change, and Meroney et al. (1975) also observed $\delta_\theta \propto x^{0.8}$ in the local advection stage of a wind tunnel simulation, but the data of Ogawa and Ohara (1985) on δ_θ growth do not fit a power law. The data set of the latter authors is complicated by a sea wall at the transition that forces a momentum boundary layer growth of the form $\delta_i \propto x^{0.57}$ and probably compromised the data set for our present purposes. What their data do show is δ_θ almost twice as deep as δ_i by 160 m from the transition. At the present time under local advection, our knowledge of the scalar fetch-height relationship for both kinds of transition is seriously incomplete.

4.2.2 *Internal boundary layer depth: Longer fetches*

Cold-hot changes. The basic structure of the convective internal boundary layer (CIBL) is illustrated in Fig. 4.6c. It is usually depicted in the same way as the standard horizontally homogeneous ABL comprising an unstable surface layer surmounted by a well-mixed region and a sharp discontinuity in potential temperature separating the well-mixed from the inversion region above. This picture carries the presumption that the CIBL is growing in an initially stable boundary layer. This is usually but not always the case with onshore winds. In practice, particularly if the turbulence level in the upwind boundary layer is not negligible, a sharp temperature discontinuity is not observed at the top of the internal layer (Smedman and Högström, 1983).

Several definitions of the height of the CIBL have been used. Venkatram (1977) and Anthes (1978) use δ_θ, identifying it with the point where the potential temperature gradient above the surface layer shows a sudden change to an inversion. Others (Lyons, 1975; Raynor, 1979; Gamo et al., 1982) prefer a definition based on δ_i, the height where turbulence levels decrease sharply. Gamo et al. (1982) concluded that a δ_i based on turbulence is 1.4 times higher than a δ_θ

defined by temperature gradient. A third definition (Weisman, 1976) uses δ_H, the height at which the heat flux passes through zero. In general, $\delta_H < \delta_\theta < \delta_i$, e.g., Shao et al. (1991) in a detailed study of the CIBL observed that $\delta_i \simeq 1.2\delta_H$.

In pollution studies we are often interested in the point at which a smoke plume from a chimney stack, initially traveling horizontally and spreading slowly in the stable marine boundary layer, encounters the growing CIBL. At this point it will be rapidly mixed to ground level by the active turbulence of the CIBL, a process called fumigation. For this application it is the height based on turbulent activity that is more relevant. Formulas for CIBL height, however, invariably predict either δ_θ or δ_H.

Formulas for the CIBL height at long fetch are almost all derived by assuming the same dynamics for the CIBL as for the daytime evolution of the horizontally homogeneous mixed layer (Chapter 1). From morning to early afternoon the horizontally homogeneous mixed layer height z_i is observed to grow as $t^{1/2}$ (Tennekes and Driedonks, 1981). Models of the CIBL substitute travel time $x/\overline{u}_m$ for t, where $\overline{u}_m$ is the average mixed layer velocity. This leads to CIBL heights proportional to $x^{1/2}$. Such models implicitly assume that the CIBL has evolved to a quasi-one-dimensional state and therefore are inapplicable except at long fetches.

Smedman and Högström (1983) used the mixed layer parameterization of Tennekes and Driedonks (1981) to produce a general formula for δ_θ, which included the effects of entrainment at the inversion, mechanically (shear) produced turbulence at the ground, and convectively driven turbulence. They applied this model to results obtained at moderate fetch and found that in order to match their observations, both mechanical turbulence and entrainment were essential but convection was less important.

Stunder and Sethuraman (1985) compared seven other formulas for δ_θ with two further data sets. They found good agreement in one case with the formula of Raynor et al. (1975). This expression was based on straightforward physical reasoning and takes the form:

$$\delta_\theta = \left[C_d \frac{(T_{02} - T_{01})}{\partial \overline{T}_1 / \partial z} x \right]^{1/2}, \tag{4.13}$$

where $C_d = u_{*2}/\overline{u}_m$ is a drag coefficient, and T_{02} and T_{01} are downwind and upwind surface temperatures. Although (4.13) performed well in the particular case for which it was developed, Stunder and Sethuraman (1985) found that the formula of Weisman (1976) gave better agreement with a wider data set. Weisman's expression was

$$\delta_H = \left[\frac{2H_{02}x}{\rho c_p (\partial \overline{\theta}_1 / \partial z) \overline{u}_m} \right]^{1/2} \tag{4.14}$$

where H_{02} is the surface heat flux over land.

Venkatram (1986) has pointed out several inconsistencies in such formulas. To derive simple closed form expressions such as (4.13) and (4.14), it is necessary to assume that θ_{02}orH_{02} are constant over land, an unlikely proposition. Such formulas also predict continual growth proportional to $x^{1/2}$ as we advance inland. In fact, the growing CIBL must eventually merge with the land-based ABL, which has dynamics governed by energy availability at the surface (Tennekes and Driedonks, 1981). At this stage, it is wise to exercise care in applying any of these formulas in conditions significantly different from those for which they were developed.

Hot-cold changes. Much of the interest in the stable internal boundary layer (SIBL) at large fetch centers on offshore flow from warm land to cool sea. The growth rate of the SIBL is much smaller than the CIBL with fetches of several hundred kilometers resulting in values of δ_θ of only a few hundred meters. Mulhearn (1981) applied dimensional analysis to data obtained in offshore flow over Massachusetts Bay in the fetch range 5–100 km. His expression for δ_θ was

$$\delta_\theta = 0.0146\,\overline{u}_\delta \left(g\frac{\Delta\theta_e}{\theta_e}\right)^{1/2} x^{1/2}, \tag{4.15}$$

where $\Delta\theta_e$ is the difference in equivalent potential temperature between the sea surface and the well-mixed upwind boundary layer, and θ_e is the background equivalent potential temperature. $\overline{u}_\delta$ is a mean velocity measured near the boundary layer top (actually at $z = 300$ m).

The most detailed data set available is that of Garratt and Ryan (1989). Based on these data, Garratt (1987) and Garratt and Ryan (1989) have produced a more precise, physically based expression for SIBL growth,

$$\delta_\theta = \alpha_1 \overline{u}_{\delta_{\theta+}} \left(g\frac{\Delta\theta_e}{\theta_e}\right)^{1/2} x^{1/2}. \tag{4.16}$$

Their formula (4.16) bears a close resemblance to Mulhearn's but differs through the specification of the constant α_1, which is a function both of the geostrophic drag coefficient u_{*2}^2/G^2 and the geostrophic departure β_1, which is the angle between $\overline{u}_{\delta_{\theta+}}$, the wind just above the boundary layer, and the geostrophic wind G. At long fetches, such a dependence is to be expected. For the data set studied by Garratt and Ryan (1989), $\alpha_1 \simeq 0.024$, whereas numerical model results (Garratt, 1987) gave $\alpha_1 \simeq 0.014$. In all of these studies of the SIBL, δ_θ is defined as the point at which the profile of potential temperature (or equivalent potential temperature) merges with the upstream profile.

4.2.3 *Scalar profiles: Local advection*

Hot-cold changes. We recall our operational definition of local advection as the situation in which the growing internal boundary layer is no deeper than the upstream surface layer. Typical temperature profiles are depicted in Figs. 4.6a and 4.7 and show the development of a deepening inversion layer in an existing unstable lapse profile. This is usually referred to as an advective inversion. The curves in Fig. 4.7 are taken from Dyer and Crawford (1965) and represent the classical agrometeorological circumstance of airflow from hot dry to cool irrigated pasture.

The self-preservation hypothesis can be applied to these profiles as demonstrated by Mulhearn (1977). In analogy with (4.8), he wrote,

$$\overline{\theta}_2(z) - \overline{\theta}_1(z) = \theta_0 g_s(z/\delta_\theta), \tag{4.17}$$

where $g_s(z/\delta_\theta)$ is a profile shape function, and θ_0 a temperature scale defined as

$$\theta_0 = -\frac{H_{02}}{\rho c_p k u_{*2}} + \frac{H_{01}}{\rho c_p k u_{*1}} = T_{*1} - T_{*2}. \tag{4.18}$$

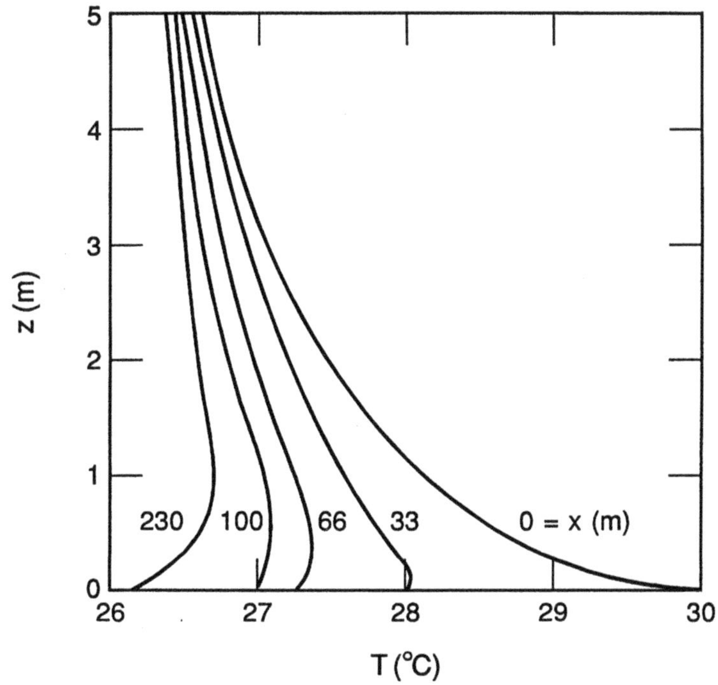

Fig. 4.7 The development of an advective inversion after flow from dry to irrigated rye grass from Dyer and Crawford (1965).

Temperature changes due to streamline displacement are ignored. The internal boundary layer height δ_θ can be determined independently or through the theory. Implicit in the treatment by Mulhearn is the identity $\delta_\theta/\delta_i = K_h/K_m = Pr_t$. Reference to the discussion earlier in this section suggests that this can only be true when $K_h \simeq K_m$, which in turn confines Mulhearn's theory to near-neutral conditions. The agreement between Mulhearn's theory and the field data of Dyer and Crawford (1965) is excellent, as Fig. 4.8 shows. Equally good agreement was achieved with the results of Rider et al. (1963), but the same caveats apply to this scalar application of the self-preserving hypothesis as to its momentum counterpart. In particular, it does not work well too close to the change of surface.

Cold-hot changes. For small departures from stability, the self-preservation theory of Mulhearn (1977) should apply to unstable conditions as it does to stable internal boundary layers, but a data set sufficiently complete to test this is not available. All that can be said is that at short fetches the available data (Ogawa and Ohara, 1985; Vugts and Businger, 1977) show the development of a lapse profile (Fig. 4.6b), whereas the data of Smedman and Högström (1983) capture the transitional situation where the CIBL has grown outside the surface layer but has not yet achieved a standard mixed-layer form. Smedman and Högström observed an inner

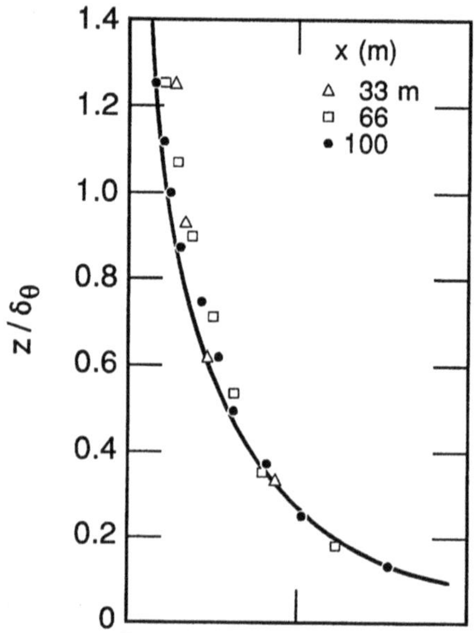

FIG. 4.8 Comparison between Dyer and Crawford's (1965) results and the theory of Mulhearn (1977).

region 50 m deep with a strong lapse profile surmounted by a well-mixed layer with weak temperature gradient extending to about 120 m above the surface. Depending on wind direction, their measurement site was 1200–1800 m downwind of the coast, and the incoming boundary layer was stable with only weak turbulence.

4.2.4 *Scalar profiles: Longer fetches*

Hot-cold changes. Mulhearn (1981) found that potential temperature profiles obtained in land-sea flow at moderately large fetch ($x > 16\,\text{km}$) could be described quite well by the expression

$$\frac{\overline{\theta}_2(z) - \theta_{02}}{\overline{\theta}_1(z) - \theta_{02}} = \left(\frac{z}{\delta_\theta}\right)^{1/4}. \tag{4.19}$$

Garratt and Ryan (1989) studied offshore flows at longer fetches ($\sim$ 100 km) and found in contrast to (4.19)

$$\frac{\overline{\theta}_2(z) - \theta_{02}}{\overline{\theta}_1(z) - \theta_{02}} = \left(\frac{z}{\delta_\theta}\right)^{2}. \tag{4.20}$$

In other words, they observed a concave rather than a convex temperature profile. They comment that the shape of their profiles was consistent with observations of the vertical divergence of temperature flux through the SIBL. The transition between the two profile shapes emphasizes the considerable distances required to achieve self-preserving forms in these weakly mixing stable internal layers.

Garratt and Ryan (1989) also measured velocity profiles and noted two characteristic features that were reinforced by Garratt's (1987) model simulation:

- The appearance of a pronounced low level maximum in the velocity profile
- A reduction in the geostrophic departure relative to the upwind overland flow

The reduction is consistent with the reduction in surface shear stress that occurs both because of the smaller surface roughness over the sea and the reduced turbulence levels in the SIBL.

Cold-hot changes. A typical well-mixed profile capped by a temperature discontinuity is sketched in Fig. 4.6c. Growing CIBLs conforming to this pattern have been documented in some detail on several occasions (Durand et al., 1989; Raynor et al., 1979; Shao and Hacker, 1990). In each of these cases, the CIBL grew into an initially stably stratified marine boundary layer. At long fetch the evidence suggests that these CIBLs behave like the horizontally homogeneous mixed layer described in Chapter 1. Their essential features are well-mixed wind and temperature profiles from the top of the surface layer to δ_θ (which is equated with z_i). Strong vertical and horizontal wind shear occurs at δ_θ, and here also a sharp gradient of potential temperature usually occurs. Models of the situation generally idealize this as a discontinuity in $\theta(z)$.

4.2.5 *Changes in surface flux: Local advection*

Surface flux changes have been studied in the wind tunnel and the field. Both approaches have their limitations. In the field there has been no scalar experiment as complete as Bradley's (1968) study of roughness change, where unequivocal measurements of the development of surface momentum flux were made. In the wind tunnel it is possible to measure the surface flux electrically if heat is used as the scalar, but it is impossible to reproduce the important coupling between the surface fluxes of radiant energy and sensible and latent heat that is so important in micrometeorology.

Flow over a step change in surface temperature has been investigated in the wind tunnel by Antonia et al. (1977), Charney et al. (1979), and Ligrani and Moffat (1985). The experiment of Ligrani and Moffat is the most relevant to micrometeorology, since they conducted their tests over a rough surface. The changes they observed in the surface flux of heat were qualitatively similar to the stress changes discussed in Section 4.1 with a large overshoot in flux just behind the change, followed by a relaxation to an equilibrium state. Of more interest to us is the situation in which the surface fluxes of the active scalars, temperature, and humidity, are coupled through the surface energy budget

$$(1-\alpha)S_d + \epsilon_\lambda L_d - \epsilon_\lambda \sigma T_0^4 - G_s = H_0 + \lambda_e E_0, \tag{4.21a}$$

$$A = R_n - G_s = H_0 + \lambda_e E_0, \tag{4.21b}$$

where α is the albedo, e_λ the emissivity, σ the Stefan-Boltzmann constant, T_0 the surface temperature, S_d the downward shortwave radiation, L_d the downward longwave radiation, and G_s the flux of heat going into storage in the ground. In (4.21a) and (4.21b), the small amount of radiant energy used in photosynthesis is neglected. Recall from Chapter 3 the shorthand terminology: R_n, the net radiation, and A, the available energy.

Following a change in surface conditions, the available energy can change because of changes in albedo or emissivity, but these are usually less important than changes in the longwave emission attendant on changes in T_0 or changes in ground heat flux as soil water content alters. All of these effects are, in turn, usually less significant than the change in energy partition between H_0 and λE_0.

If we consider first the flow from a dry surface with $H_0 \simeq A$ to a moist or wet surface, then $\lambda_e E_0$ increases rapidly at the expense of H_0. Close behind the change, the ratio $\lambda E_0/A$ can become much greater than unity, the extra energy for evaporation being supplied by a downward flux of sensible heat, which is maintained in turn by horizontal advection. We can describe this situation more formally by writing the equation for conservation of enthalpy,

$$\overline{u}\frac{\partial}{\partial x}(\rho c_p\overline{\theta}) + \overline{w}\frac{\partial}{\partial z}(\rho c_p\overline{\theta}) = -\frac{\partial}{\partial x}(\rho c_p\overline{u'\theta'}) - \frac{\partial}{\partial z}(\rho c_p\overline{w'\theta'}), \qquad (4.22)$$

and retaining terms of leading order to obtain the simplified balance

$$\overline{u}\frac{\partial}{\partial x}(\rho c_p\overline{\theta}) = -\frac{\partial H}{\partial z}. \qquad (4.23)$$

A downward flux of sensible heat at the surface, ($H_0 < 0$) means that $H(z)$ must pass through zero at some height z' so that if (4.23) is integrated from $z = 0$ to $z = z'$, it is clear that the downward flux of heat needed to sustain $(\lambda E_0/A) > 1$ is supplied by a cooling of the air in the growing internal boundary layer. Agricultural meteorologists working over green crops recognize times when the Bowen ratio β is negative [where $\beta = H_0/\lambda_e E_0 = (A/\lambda_e E_0) - 1$] as an indicator of advective conditions and as a warning to use caution in applying one-dimensional formulas.

Figure 4.9 from Dyer and Crawford (1965) illustrates the situation. The surface fluxes were obtained from mean profiles of $\overline{\theta}, \overline{u}$, and $\overline{q}$ by integrating (4.23) with the assumptions that $H = 0$, where $\partial\overline{\theta}/\partial z = 0$ and $\lambda_e E = 0$, where $\partial\overline{q}/\partial z = 0$. This is not a particularly good assumption close to $x = 0$ (as we shall see in Section 4.3), but it is reasonably accurate at longer fetches as the good agreement with the single lysimeter measurement attests. Unfortunately, it prevents us from observing whether the expected large overshoot in flux values occurs just behind the change. The data of Itier et al. (1978) obtained from minilysimeter measurements show a rapid attainment of quasi-constant moisture flux and confirm that any large overshoot in flux is confined to the region immediately behind the change.

For more detail we must turn to model simulations. Two are relevant. The second-order closure scheme of Rao et al. (1974b) uses a sophisticated description of the turbulent mixing, taking proper account of the effects of changing stability, but it uses a simplistic lower boundary condition. Over green vegetation or moist bare soil, biological and physical resistances to the movement of water vapor through the plant or through the soil ensure that the surface vapor concentration is lower than the saturated concentration at the surface temperature, namely, $\overline{q}_0 < \overline{q}_{\text{sat}}(T_0)$.

This process is parameterized by the surface or canopy resistance r_c [(3.24) in Chapter 3], which links the surface concentration to the flux:

$$\frac{E_0}{\rho} = \frac{\overline{q}_{sat}(T_0) - \overline{q}_0}{r_c}. \tag{4.24}$$

Over green vegetation, r_c is dominated by the behavior of the stomata and hence is largely a biologically determined variable, whereas over bare soil, r_c depends primarily on the soil structure and the rate at which the upper layers of soil are drying (Philip, 1967). Instead of using (4.24), Rao et al. (1974b) fixed the relative humidity $R_0 = \overline{q}_0/\overline{q}_{sat}(T_0)$ at the surface. Their boundary condition only corresponds to (4.24) in the limiting case of $R_0 = 1.0$ and $r_c = 0$, which we find over open water.

A solution of the advection problem using (4.24) as a boundary condition has been provided by Philip (1987). His treatment of the turbulent flow was much simpler than that of Rao et al. (1974b) and took no account of stability effects or roughness change, restricting its quantitative predictions to near neutral conditions and surfaces of similar roughness. Nevertheless, his analytic formulation is much more readily applied than Rao et al.'s numerical model. In practice, the predictions of both models are qualitatively similar. In Fig. 4.10 we reproduce data from Rao et al. (1974b) that shows the large jump in $\lambda_e E_0$ with $\lambda_e E_0/A \gg 1$ just behind

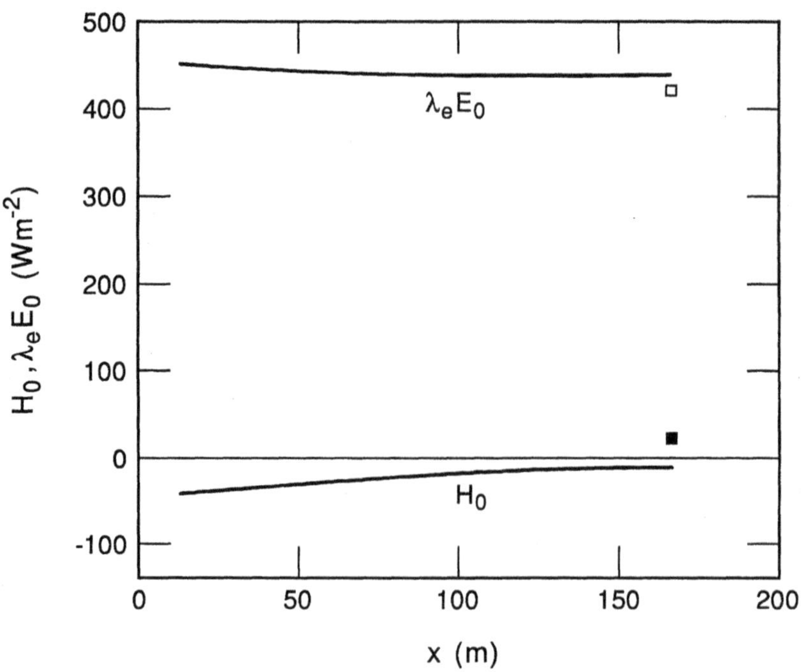

FIG. 4.9 Partition of available energy after the dry-wet transition studied by Dyer and Crawford (1965). These curves correspond to the temperature profiles of Fig. 4.7. The open and closed squares are λE_0 and H_0 values, respectively, obtained from the lysimeter.

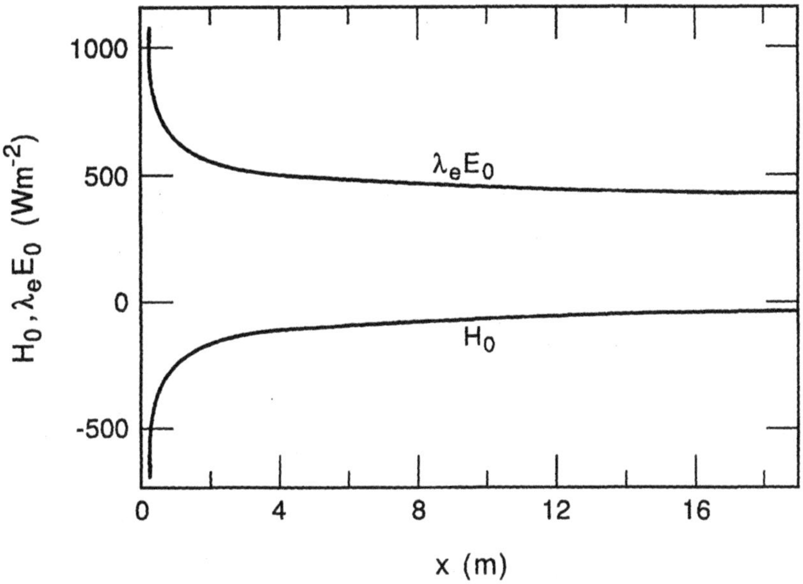

FIG. 4.10 Partition of available energy after a dry-wet transition; model results of Rao et al. (1974b).

the change followed by a rapid fall and a gentler approach to equilibrium. The behavior of $\lambda_e E_0$ is reciprocated by that of H_0.

An important deduction of Philip's (1987) analysis is that for any given A, there is a value of r_c above which an advective inversion will not form. (Philip's solution models turbulent flux by an eddy diffusivity so that negative H_0 is synonymous with a ground-based inversion.) This is illustrated in Fig. 4.11 from Philip's paper, where we see that with $A = 400$ Wm^{-2}, an inversion never forms when r_c exceeds 200 s m^{-1} and is confined to within 50 m of the change for $r_c = 100$ s m^{-1}. A typical value of r_c for green crops is 50 s m^{-1}, whereas a forest might have $r_c = 200$ s m^{-1} (Chapter 3). Behavior of this kind — the absence of an inversion — is seen in one of the cases treated by Dyer and Crawford (1965).

The wet-dry, cold-hot change does not display these complications, as a downward flux of latent heat cannot sustain an upward flux of sensible heat except in the rather special event of dew formation. Note that neither the experimental results nor the model simulations have a natural x-wise scaling, which seriously reduces their value for describing the generic case. The self-preserving model of Mulhearn (1977) has a streamwise scaling linked to the growth of the internal boundary layer, but this approach is invalid close to $x = 0$ where rapid and interesting changes in surface flux are occurring.

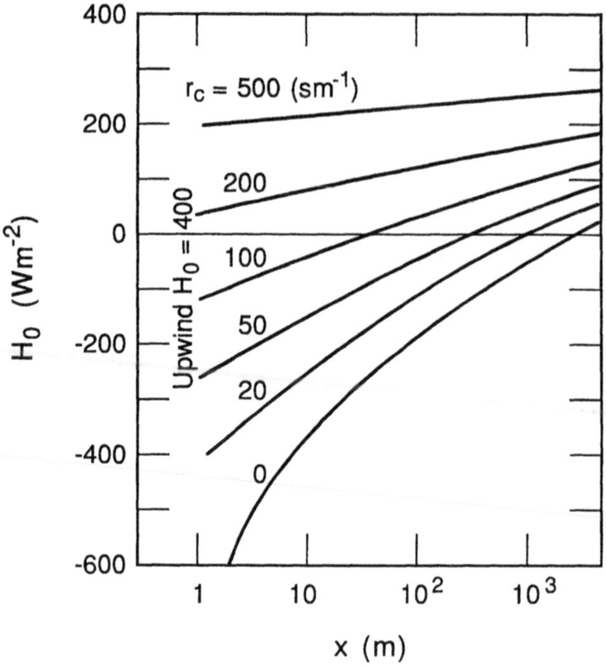

FIG. **4.11** Variation in H_0 after a dry-wet transition for various values of surface resistance, r_c; model results of Philip (1987).

4.2.6 *Change of surface flux: Longer fetches*

We saw in the last section that just behind the change, the properties of the new surface, particularly the availability of moisture as embodied in the surface resistance r_c, is of paramount importance in determining the surface flux. Recent work by McNaughton and Jarvis (1991), however, has shown how this control is relaxed as we proceed farther downstream. They considered the question of how sensitive the area-averaged evaporation is to changes in the value of r_c when we have traveled far enough from the change for the whole ABL to have adjusted to the new surface. This is an important question that arises not only in the context of estimating evaporation or other fluxes over large-scale changes in land use, but also in the topical question of parameterizing surface variability in global or regional climate models.

McNaughton and Jarvis show that environmental feedbacks can diminish (usually) or augment (occasionally) the influence of changes in surface properties. To illustrate what we mean we recall (3.21), which describes evaporation between the surface and a reference level z_r using an aerodynamic resistance r_{aq} to describe atmospheric transport of humidity,

$$\frac{E_0}{\rho} = \frac{\overline{q}_0 - \overline{q}_{z_r}}{r_{aq}} \tag{4.25}$$

and combine this relationship with (4.24) to obtain

$$\frac{E_0}{\rho} = \frac{\overline{q}_{\text{sat}}(T_0) - \overline{q}_{z_r}}{r_{aq} + r_c}. \tag{4.26}$$

Imagine now a reduction of r_c caused by, for example, irrigating a region. If other variables remained fixed, E_0 would increase; but in reality, increased evaporation would tend to increase $\overline{q}_{z_r}$ by humidifying the boundary layer while T_0 would fall, causing $\overline{q}_{\text{sat}}(T_0)$ to diminish. The formation of a stably stratified advective inversion would also reduce turbulent mixing and thereby cause an increase in r_{aq}. These three processes constitute negative feedbacks in the terminology of McNaughton and Jarvis, reducing the response of E_0 to changes in r_c.

These authors conclude that for $r_c < 50\,\text{s}\,\text{m}^{-1}$, which is the range expected of cultivated crops, changes in r_c have little effect on regional evaporation because the dominant controls on E_0 are processes on the ABL scale such as entrainment through the inversion. For r_c between 50 and 250 $\text{s}\,\text{m}^{-1}$, meteorological feedback is important but not completely dominant. This range of resistances covers most of the world's nonagricultural vegetation. McNaughton and Jarvis conclude that only for the very driest surfaces with $r_c > 250\,\text{s}\,\text{m}^{-1}$ does the surface moisture availability control evaporation on the regional scale.

This is an area of active research and the conclusions reported here may well be subject to revision; the numerical values in particular depend on the details of the ABL models they have used. Nevertheless, the basic principles of meteorological and environmental feedback are accepted by most workers in the field and are implicit in the formulation of models such as Philip's (1987) local advection treatment. We must be aware of them when estimating the downwind adjustment of area-averaged fluxes to surface change. An added refinement of this problem has been discussed by Raupach (1991), who has looked at continually changing surfaces and how processes in the whole ABL affect evaporation from them.

4.3 Turbulence

The information on turbulence response to surface change is much less complete than that on the mean fields. This means that it is very sketchy indeed. To describe local advection after roughness changes we must rely heavily on the wind tunnel even though most of the wind tunnel data are not strictly appropriate to atmospheric flows. Information about scalar changes, particularly active scalars, is even sparser except in the case of far downwind development after cold-hot changes. Unfortunately, this situation is less interesting from a fluid mechanical point of view than is local advection, since far downwind, IBLs usually develop forms that are locally similar to well-studied, horizontally homogeneous boundary layers.

4.3.1 *Changes in surface roughness: Local advection*

Smooth-rough changes. Two data sets are available: the wind tunnel data of Antonia and Luxton (1971) and the field experiment of Peterson et al. (1979), which was performed during neutral flow from sea to land. Unfortunately, the roughness elements of Antonia and Luxton's experiment were so high relative to the depth of the upwind logarithmic layer as to resemble flow from grassy fields to a city of tall buildings as pointed out earlier, whereas Peterson et al. were only able to estimate σ_u at the change and two stations downwind by combining outputs from cup and hot-wire anemometers (Chapter 6). With these caveats in mind, we can describe the changes in shear stress and variance.

Figure 4.12a shows the development of $-\overline{u'w'}$. These profiles asymptote to the kinematic surface stress u_{*2}^2 at $z = 0$. We saw in Fig. 4.5a that after a smooth-rough change this quantity showed an initial sharp rise then fell with increasing x to a smaller equilibrium value. Figure 4.12a reveals that as the near-surface value of $-\overline{u'w'}$ falls with increasing x from its initial high value, a peak in shear stress moves away from the surface. The profiles of σ_u and σ_w (Fig. 4.12b) do not display this overshooting behavior, at least down to the lowest measurement level, nor do they show a peak above the surface. No information on the lateral moment σ_v is available from these experiments. We note that the profiles of turbulence moments merge into the upstream profiles at the same height as the mean velocities. In other

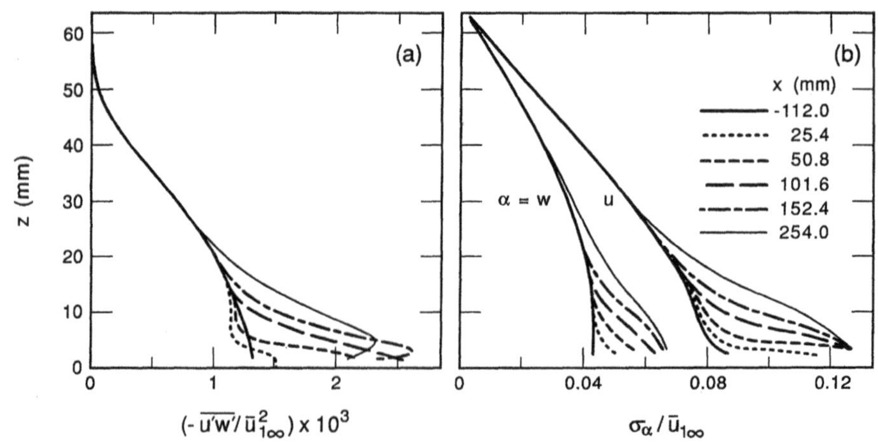

FIG. **4.12** Development of turbulent shear stress and variance after a smooth-rough change. Data of Antonia and Luxton (1971). (a) $-\overline{u'w'}(z)/\overline{u}_{1\infty}^2$, where $\overline{u}_{1\infty}$ denotes the wind-tunnel free stream velocity. (b) $\sigma_u(z)/\overline{u}_{1\infty}$ and $\sigma_w(z)/\overline{u}_{1\infty}$ with symbols as in (a).

words, δ_i defined by turbulence fluxes in this case coincides with δ_i defined by the mean flow.

We can write the equation describing the streamwise balance of mean momentum for this flow as

$$\overline{u}\frac{\partial \overline{u}}{\partial x} = -\frac{\partial}{\partial x}(\overline{u'^2}) - \frac{\partial}{\partial z}(\overline{u'w'}), \tag{4.27}$$

where we have assumed that the flow is steady, that the pressure gradient associated with streamwise displacement at the roughness change is negligible, that $\overline{w} = 0$ (these being approximations of the same order), and that viscous stresses may be ignored. We then see immediately that the positive gradients of σ_u in the streamwise direction and negative gradients of $-\overline{u'w'}$ in the vertical we observe in Figs. 4.12a and 4.12b correspond to mean flow deceleration in the internal boundary layer.

As in the horizontally homogeneous cases studied in Chapters 1 and 3, an examination of the terms in the turbulent kinetic energy equation offers important insights into the distribution of turbulence in the boundary layer. The equation for a steady, two-dimensional flow is

$$\underset{\text{I}}{\overline{u}\frac{\partial \overline{e}}{\partial x}} = -\underset{\text{II}}{(\overline{u'w'})\frac{\partial \overline{u}}{\partial z}} - \underset{\text{III}}{(\overline{u'^2} - \overline{w'^2})\frac{\partial \overline{u}}{\partial x}} - \underset{\text{IV}}{\left[\frac{\partial}{\partial x}(\overline{eu'}) + \frac{\partial}{\partial z}(\overline{ew'})\right]}$$

$$-\underset{\text{V}}{\frac{1}{\rho}\left[\frac{\partial}{\partial x}(\overline{u'p'}) + \frac{\partial}{\partial z}(\overline{w'p'})\right]} + \underset{\text{VI}}{\frac{g}{\overline{\theta}_0}(\overline{w'\theta'})} - \underset{\text{VII},}{\epsilon} \tag{4.28}$$

where $e = (1/2)\left(u'^2 + v'^2 + w'^2\right)$, ϵ represents viscous dissipation, the advection term has been simplified by assuming $\overline{w} = 0$ as in (4.27), and viscous transport has been ignored. Equation (4.28) will be recognized as a more elaborate form of (1.59) in which streamwise gradients of mean moments now appear. Note particularly an extra production term (III) corresponding to acceleration of the mean flow. Term III is positive in a decelerating flow and negative when the flow is accelerating.

In Fig. 4.13 the terms of (4.28) are plotted in budget form using data from Antonia and Luxton's (1971) experiment. Indicated on the same graph is the budget for the equilibrium boundary layer upstream. We see that the streamwise production term (III) is a gain over the internal layer, whereas advection (I) is a loss with its largest negative contribution occurring at $z \simeq \delta_i/2$. Over most of the internal layer, turbulent transport (IV) is a gain except near the surface, where it is

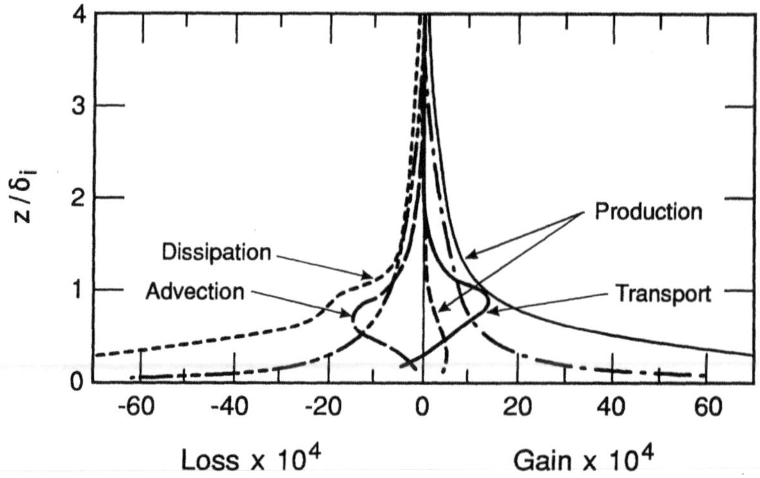

FIG. **4.13** Turbulent kinetic energy budget 111 mm behind the smooth-rough change of Antonia and Luxton (1971). Terms are normalized by the boundary layer thickness δ, where at this station $\delta \simeq 5\delta_i$, and by the free stream velocity $\overline{u}_{1\infty}$. Estimates of production and dissipation on the smooth wall upstream are included for comparison. ——, $-(\delta/\overline{u}^3_{1\infty})\overline{u'w'}(\partial\overline{u}_2/\partial z)$;——, $-(\delta/\overline{u}^3_{1\infty})\partial(\overline{w'e})/\partial z$;———, $-(\delta/\overline{u}^3_{1\infty})\overline{u}_2(\partial\overline{e}/\partial x)$; ----, $-(\delta/\overline{u}^3_{1\infty})(\overline{u'^2}-\overline{w'^2})(\partial\overline{u}_2/\partial x)$; - - - - - - - - - - -, residual; — - —, $-(\delta/\overline{u}^3_{1\infty})$ $\overline{u'w'}(\partial\overline{u}_1/\partial z)$ (upwind); — - - -—, dissipation (upwind).

a loss. [In fact, only the vertical component of turbulent transport $-(\partial/\partial z)(\overline{w'e})$ is computed. The streamwise component $-(\partial/\partial x)(\overline{u'e})$ is assumed to be much smaller and is included implicitly in the residual term. Actual measurements of this term by Mulhearn (1978) support this assumption.] Turbulent transport, therefore, represents the vertical diffusion of turbulent kinetic energy from the rough surface to the outer part of the IBL. The residual term consists of dissipation (VII) plus pressure transport (V) and the horizontal component of turbulent transport. There is evidence from some recent experiments that pressure transport tends to oppose turbulent transport but to be somewhat smaller than it. The residual therefore is primarily dissipation. Buoyant production (VI) is zero in this adiabatic flow.

Near the surface, production and dissipation are almost in balance, but over most of the IBL depth the transport terms I and IV are of the same order of magnitude as II, III, and VII. This is in sharp contrast to the situation upwind where production and dissipation are in approximate balance over most of the boundary layer depth. There are important implications here for the concept of local equilibrium, which, as we saw in Chapter 3, underpins the use of eddy diffusivities to describe turbulent transport. We return to consider this question in detail later in this section.

Rough-smooth changes. Relevant data for this case came from the wind tunnel experiments of Antonia and Luxton (1972) and Mulhearn (1978). Figure 4.14a

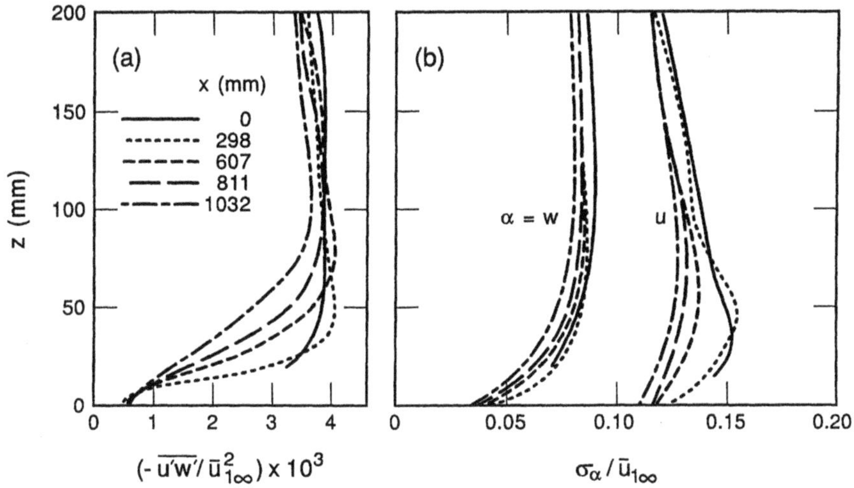

FIG. 4.14 Development of turbulent shear stress and variance after a rough-smooth change. Data of Mulhearn (1978). (a) $-\overline{u'w'}/\overline{u}_{1\infty}^2$, where $\overline{u}_{1\infty}$ denotes wind tunnel free stream velocity; (b) $\sigma_u(z)/\overline{u}_{1\infty}$ and $\sigma_w(z)/\overline{u}_{1\infty}$ symbols as in (a).

displays the development of $-\overline{u'w'}$ profiles in a similar form to (4.12a). The undershoot in $u_{*2}^2(x)$, as discussed in Section 4.1, is just apparent in the lowest values of $-\overline{u'w'}$, but this phenomenon is confined to very near the surface. The significant variations in σ_u, shown in Fig. 4.14b, follow those of $-\overline{u'w'}$, but much smaller changes, proportionately, are seen in σ_w (Fig. 4.14b); σ_v changes by a negligible amount.

The streamwise and vertical gradients of σ_u, σ_w, and $-\ \overline{u'w'}$, respectively, that we see in Fig. 4.14 correspond to mean flow acceleration in the internal boundary layer as reference to (4.27) reveals. As we noted for the smooth-to-rough case, δ_i defined by the turbulence moments corresponds to δ_i defined by the mean velocity.

The turbulent kinetic energy budget, (4.28), is displayed in Fig. 4.15 with upstream values for reference. We see that the extra terms responsible for the deviation from the upstream equilibrium form are all opposite to the smooth-rough case in Fig. 4.13. Advection (I) is now a gain and acceleration production (III) is a loss, whereas (vertical) turbulent transport (IV) is a loss in the outer part of the IBL and a gain near the surface. Shear production (II) has also fallen substantially across the IBL as has the residual term, assumed as before to be mainly dissipation. As in the smooth-rough case, the additional transport terms are of the same order as the production and dissipation in the middle of the IBL, but now this state of affairs continues all the way to the surface.

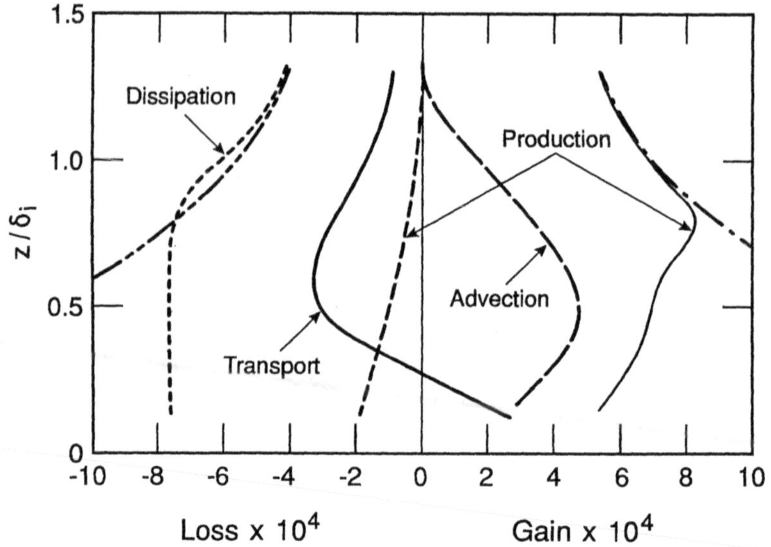

FIG. **4.15** Turbulent kinetic energy budget 607 mm behind a rough-smooth change. Data of Mulhearn (1978). Data are normalized by δ_i and $\overline{u}_{1\infty}$. Note that the normalizing length scale is different from that in Fig. 4.13. ———, $-(\delta_i/\overline{u}^3_{1\infty})\overline{u'w'}$ $(\partial\overline{u}_2/\partial z)$; ——, $-(\delta_i/\overline{u}^3_{1\infty})\partial(\overline{w'e})/\partial z$; — — —, $-(\delta_i/\overline{u}^3_{1\infty})\overline{u}_2\partial\overline{e}/\partial x$; - - - -, $-(\delta_i/\overline{u}^3_{1\infty})(\overline{u'^2} - \overline{w'^2})(\partial\overline{u}_2/\partial x)$; ----------, residual; — - —, $-(\delta_i/\overline{u}^3_{1\infty})\overline{u'w'}$ $(\partial\overline{u}_1/\partial z)$ (upwind); —- - -—, residual (upwind).

4.3.2 *Change of scalar flux and concentration: Local advection*

Here, most information comes from the wind tunnel, which means that it strictly applies to passive scalars only. (It is impossible to produce dynamically significant density gradients in most wind tunnels.) The most relevant experiment is that of Ligrani and Moffat (1985). It can be inferred from their observations following a cold-hot change on a rough surface that far enough from the change $\overline{w'\theta'}/u_{*2}T_*$ collapses to a universal shape when plotted against z/δ_i . $T_*(x)$ and $u_{*2}(x)$ were defined in terms of the local heat flux and surface stress according to the definitions (1.25a and 1.25b). Unfortunately, they do not describe profiles in the more interesting region close to the discontinuity. The results of Ogawa and Ohara (1985) closer to a cold-hot (sea-land) change show $\overline{w'\theta'}$ profiles similar to the shear stress profiles after a smooth-rough change that we saw in (4.12a), with an initial, near-surface overshoot followed by the heat flux peak moving away from the surface so that the largest streamwise gradients in $\overline{w'\theta'}$ occur about $z = \delta_\theta/4$.

Charnay et al. (1979) studied the IBL that develops over a sudden drop in surface temperature on a smooth floor. Upstream of the change, surface temperature exceeded free stream air temperature and the situation reversed downstream.

In Figs. 4.16a and 4.16b, we plot their dimensionless mean temperature and heat flux profiles, respectively. In 4.16a, we see the development of a classic advective inversion similar to the field data of Dyer and Crawford (1965) shown in Fig. 4.7. The accompanying heat flux profiles (Fig. 4.16b) correspond to the cooling of the internal layer and the transfer of heat to the surface. Except at the most downstream stations, measurements were not made close enough to the surface to explore the region where $\overline{w'\theta'}$ was negative. It is pertinent to note, however, that $\overline{w'\theta'}$ changes sign at a z value about half of that at which $\partial\overline{\theta}/\partial z = 0$.

Charnay et al. (1979) also constructed budgets of $\overline{\theta'^2}/2$, which may be thought of as a scalar equivalent of turbulent kinetic energy. The equation for $\overline{\theta'^2}/2$ to the same order of approximation as the turbulent kinetic energy budget (4.28) is

$$\underset{\text{I}}{\overline{u}\frac{\partial}{\partial x}\left(\frac{\overline{\theta'^2}}{2}\right)} = -\underset{\text{II}}{(\overline{w'\theta'})\frac{\partial\overline{\theta}}{\partial z}} - \underset{\text{III}}{\frac{1}{2}\frac{\partial}{\partial z}(\overline{w'\theta'^2})} - \underset{\text{IV}}{N_\theta}, \tag{4.29}$$

where N_θ represents the dissipation of temperature fluctuations by molecular conduction. As in (4.28), Roman numerals are used to identify the different budget

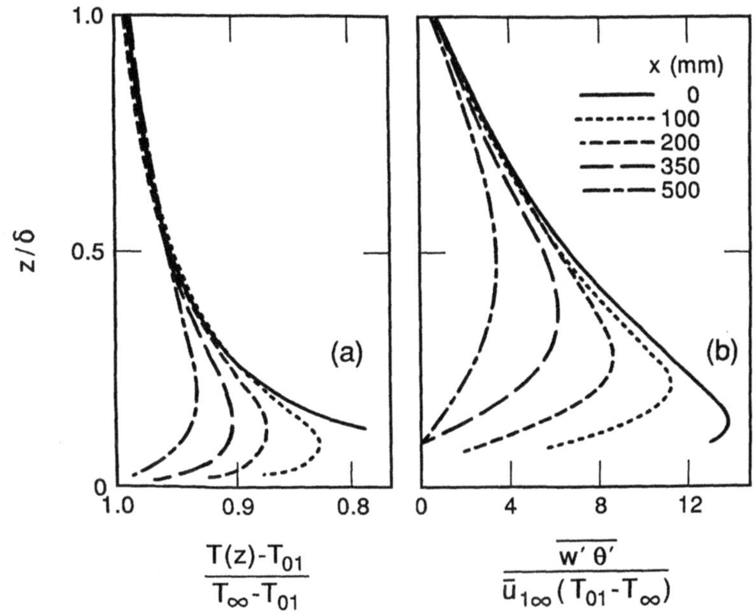

FIG. 4.16 Temperature and heat flux profiles after a hot-cold change. Data of Charnay et al. (1979). (a) Temperature profiles showing the development of an advective inversion. T_{01} is upstream wall temperature and T_∞ free stream temperature. δ is the boundary layer thickness. (b) Normalized heat flux profiles for the same experiment.

terms representing *advection* (I), *production* (II), *transport* (III), and *dissipation* (IV). The budgets before and after the change are shown in Figs. 4.17a and 4.17b. The residual term is assumed to be predominantly dissipation, N_θ. We note the same kind of structure we observed in turbulent kinetic energy budgets after roughness changes. In particular, over most of the internal boundary layer, advection and turbulent transport are of similar order to production and dissipation of $\overline{\theta'^2}/2$.

4.3.3 *Turbulent fluxes and local equilibrium: Local advection*

In Section 3.3 we discussed the problems encountered when eddy diffusivities are used to model the turbulent flux of momentum or scalars in a plant canopy. Fundamentally, the difficulties arose because eddy covariances were not simply related to the gradients of the appropriate mean quantities. For example, at some height z within the canopy a significant amount of $\overline{w'\theta'}$ was "imported" from higher levels by the turbulence itself. Linking $\overline{w'\theta'}$ and $\partial\overline{\theta}/\partial z$ by an eddy diffusivity K_h produced anomalies such as zero or negative values of K_h within the canopy.

Wyngaard (1983) has shown that a minimum prerequisite for using an eddy diffusivity to model turbulent flux in the boundary layer is that production and destruction of eddy flux should almost balance locally and be much larger than any transport term. The argument is simple and is worth reproducing here.

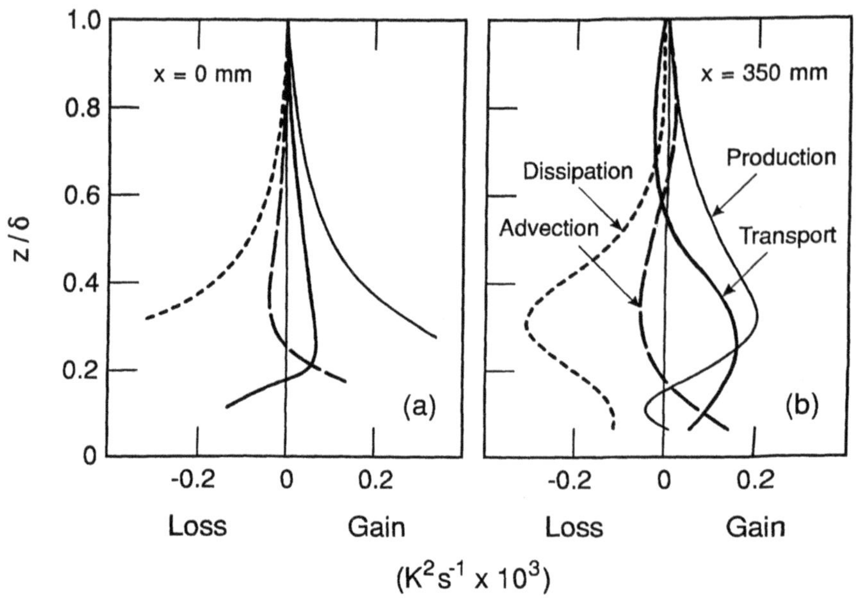

FIG. 4.17 Budget of $\overline{\theta'^2}/2$ fluctuations after a hot-cold transition. Data of Charnay et al. (1979). (a) $x = 0$ mm: ———, $-\overline{w'\theta'}(\partial\overline{\theta}/\partial z)$; — — —, $-\overline{u}_1\partial(\overline{\theta'^2}/2)/\partial x$; ———, $-(1/2)\partial(\overline{w'\theta'^2})/\partial z$; - - - - - - - -, residual; (b) $x = 350$ mm: symbols as in (a).

The equation for the vertical flux $\overline{w'c'}$ of an arbitrary scalar c in a two-dimensional steady flow is

$$\overline{u}\frac{\partial}{\partial x}(\overline{w'c'}) = -(\overline{w'^2})\frac{\partial \overline{c}}{\partial z} - \overline{c'\frac{\partial p'}{\partial z}} - \frac{\partial}{\partial z}(\overline{w'^2c'}), \tag{4.30}$$

where we have made the usual simplification of the advection term and assumed that production and turbulent transport terms involving horizontal gradients of mean quantities are much smaller than those involving vertical gradients.

In the absence of advection and turbulent transport, (4.30) reduces to

$$0 = -(\overline{w'^2})\frac{\partial \overline{c}}{\partial z} - \overline{c'\frac{\partial p'}{\partial z}}, \tag{4.31}$$

and to maintain a steady state, the pressure gradient term must be a destruction term balancing production. We expect such a destruction term to be of the general form

$$\overline{c'\frac{\partial p'}{\partial z}} \simeq \alpha_t \frac{\overline{w'c'}}{T_\epsilon}, \tag{4.32}$$

where α_t is a constant of order one, and T_ϵ an 'eddy turnover timescale' (Wyngaard, 1982). Substituting (4.32) into (4.31), we obtain

$$\overline{w'c'} = -\left(\alpha_t \overline{w'^2} T_\epsilon\right)\frac{\partial \overline{c}}{\partial z} = -K_c\frac{\partial \overline{c}}{\partial z}, \tag{4.33}$$

where the term in parentheses takes the form of an eddy diffusivity K_c. Returning to the full equation (4.30), we observe that this derivation can only proceed when advection and turbulent transport are negligible. If they are not, the essentially local connection between flux and gradient that leads to (4.33) is broken by the addition of these inherently nonlocal transport terms.

In the advective flows we have been discussing in this chapter, there have been few published budgets of eddy fluxes. An exception is Lang et al.'s (1983) measurements of some of the terms in the heat and moisture flux equations at $\delta_i/x > 10$ in an advective inversion. They found that turbulent transport was about 10% of production over a range of stability; unfortunately, advection was not measured.

Faced with the lack of this information, we turn instead to the turbulent kinetic energy and $\overline{\theta'^2}/2$ budgets. These are representative of the fundamental state of the turbulence and are taken as sensitive indicators of the state of local equilibrium. The smooth-rough turbulent kinetic energy budget displayed in Fig. 4.13 shows a clear departure from local equilibrium over most of the IBL. The only region where shear production and dissipation dominate is close to the surface. In the rough-smooth case (Fig. 4.15), the absence of local equilibrium continues all the way to the surface.

In the budget of $\overline{\theta'^2}/2$ at and after a heat flux inversion shown in Fig. 4.17, we see the important, first-order role played by advection and turbulent transport across the internal layer. Antonia et al. (1977) presented $\overline{\theta'^2}/2$ budgets after a cold-hot change and also found the transport terms to be important across most of the IBL. Only very close to the surface was local equilibrium approached.

It is clear that we should expect some problems with eddy-diffusivity formulations across the IBL. Figure 4.3 shows the wide variation in ϕ_m across the internal boundary layer for neutrally stratified smooth-rough and rough-smooth changes. We saw in Chapter 3 that $K_m = kz\, u_* \phi_m^{-1}$ in diabatic flows obeying Monin-Obukhov similarity. The large departures from $\phi_m = 1$, the value that we expect in adiabatic equilibrium flows, suggest caution in using equilibrium formulations in advective flows, but do not invalidate the use of eddy diffusivities per se.

Closer to the surface we should expect better adherence to the local equilibrium form $K_m = u_* kz$, particularly in the smooth-rough case where, as we have just seen, production and dissipation dominate at the bottom of the IBL. Figure 4.18 shows the mixing length $l_m = u_{*2}/(\partial \overline{u}_2/\partial z) = kz/\phi_m$ after smooth-rough and rough-smooth changes using data from Antonia and Luxton (1971, 1972). It is clear that close to a smooth-rough change the standard equilibrium form $l_m = kz$ with $\phi_m = 1$ does not hold; the farther downwind we proceed, the greater the depth of flow obeying this relationship, but a considerable downwind distance must elapse before the equilibrium form describes the link between momentum flux and velocity profile. The rough-smooth situation is worse with no obvious approach to $l_m = kz$ within the extent of the measurements. Referring to the discussion of the IBL in Section 4.1, we can now see that the evidence for the existence of an equilibrium layer at the foot of the IBL is far from compelling until we are well downwind. A more direct demonstration of the inapplicability of eddy diffusivities is provided by the heat flux experiment of Charnay et al. (1979). They found in an advective inversion that the height at which $\overline{w'\theta'} = 0$ was roughly half that at which $\partial\overline{\theta}/\partial z = 0$.

It is of interest to ask how far downwind of a change in surface we must go in order to use eddy diffusivity formulations and Monin-Obukhov forms for gradients and fluxes with acceptable accuracy. At this stage, no definitive answer can be provided to either of these questions. It is clear that a smooth-rough change achieves a new equilibrium more rapidly than a rough-smooth change. Similarly, the (somewhat weaker) evidence suggests that a cold-hot change comes to equilibrium faster than a hot-cold change. Physically based estimates of these distances must take account of the asymmetry of the changes. In a smooth-rough change the

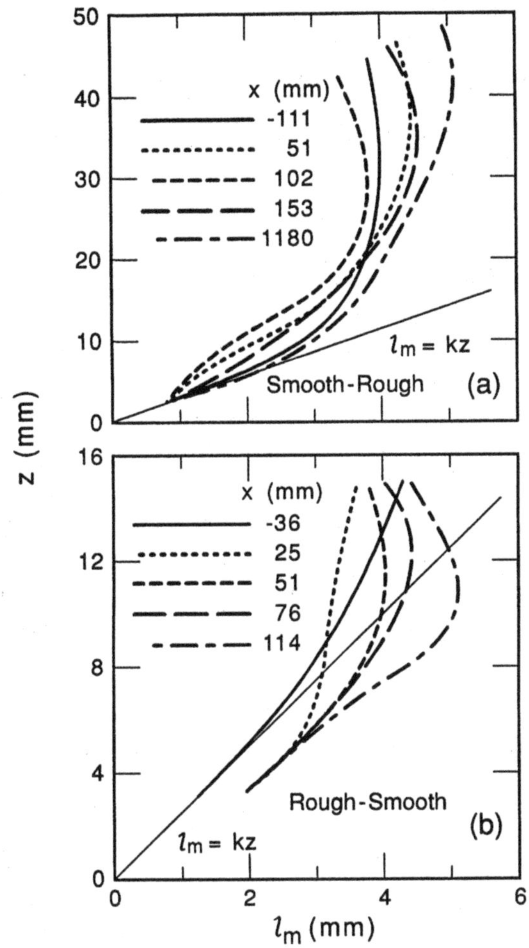

FIG. **4.18** Comparison between the near-surface value of the mixing length $l_m = (-\overline{u'w'})^{1/2} / \ (\partial\overline{u}_2/\partial z) = kz/\phi_m$ and the equilibrium adiabatic form $l_m = kz$ after roughness changes. Data of Antonia and Luxton (1972). (a) Smooth-rough. (b) Rough-smooth.

high production rate of new turbulence at the surface tends to control the growth rate of the IBL and dominate the decaying turbulence characteristic of the upwind surface. After a rough-smooth change, conversely, the decaying upwind turbulence can control the growth of δ_i for a considerable distance. Analogous arguments apply to the energetic turbulence generated downwind of a cold-hot change and its asymmetric obverse, the hot-cold transition. A survey of the experiments where the approach to local equilibrium, or more directly, the validity of the flux-gradient connection can be compared with fetch suggests that eddy diffusivities should be used with the greatest caution at fetches $x < 10\delta_i$ or $x < 10\delta_\theta$.

4.3.4 *Turbulence behavior at longer fetches*

Cold-hot changes. The study of Smedman and Högström (1983) provided data from one station at intermediate fetch, but the most detailed study to date of turbulent structure is that of Shao and Hacker (1990) and Shao et al. (1991) using aircraft in a strongly convective onshore flow. Using a local similarity scaling similar to that introduced by Nieuwstadt (1984) for the stable boundary layer, they found that velocity and temperature variances could be represented as functions of u_*, T_*, and L (the Monin-Obukhov length), as long as these scaling parameters were constructed from local rather than surface values of heat and momentum flux (Shao and Hacker, 1990). This success showed that local equilibrium or self-preserving behavior of some kind had been achieved, but because of the fetch and height dependence of the scaling parameters, it was not too helpful in comparing the global turbulent structure of different CIBLs.

Much more valuable from this viewpoint is the analysis of Shao et al. (1991). Figures 4.19a and 4.19b from their paper compare dimensionless variances of vertical velocity and temperature from their experiment and that of Smedman and Högström (1983), with the standard mixed layer data sets of Caughey and Palmer (1979), and with Lenschow et al. (1980). The variances are made dimensionless

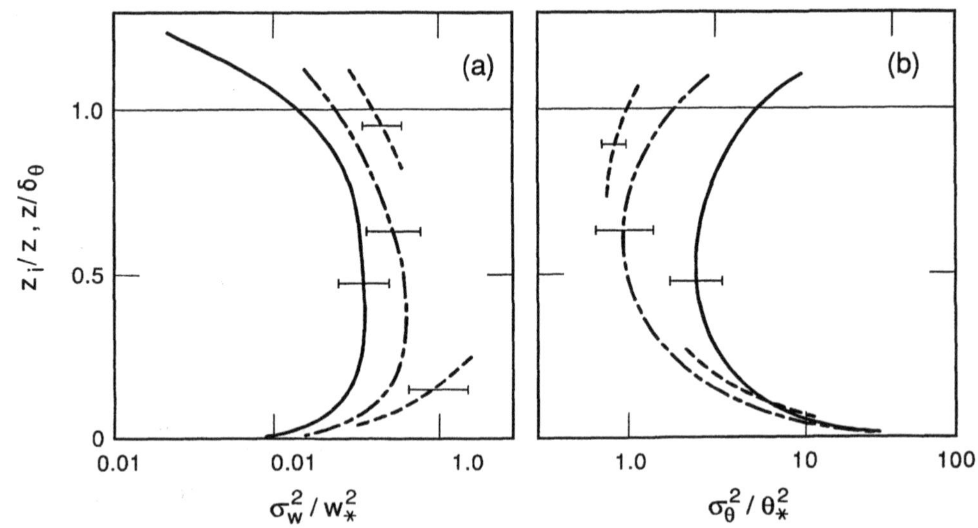

FIG. **4.19** Vertical velocity and temperature variances in two convective internal boundary layers compared with standard mixed layer data sets. Mixed layer heights are scaled with z_i and CIBLs with δ_θ. Local heat flux and δ_θ were used to form w_* for the CIBLs. Typical spread of the data points is indicated by error bars. (a) Profiles of σ_w^2/w_*^2: — - — , Shao et al. (1991); - - - , Smedman and Högström (1983), ———, Caughey and Palmer (1979) and Lenschow et al. (1980). (b) $\sigma_\theta^2/\theta_*^2$; symbols as in (a).

with the mixed layer scales w_* and θ_* introduced in Chapter 1, and the CIBL height δ_θ is used instead of z_i for the advective data.

Although the data of Shao and Hacker and Smedman and Högström are in reasonable accord, their σ_w^2/w_*^2 values tend to be larger than the horizontally homogeneous data and their $\sigma_\theta^2/\theta_*^2$ values smaller. Smedman and Högström ascribe their high σ_w^2/w_*^2 values to the increased importance of mechanically (shear) produced turbulence compared with the classical horizontally homogeneous CBL, and this explanation could also apply to the data of Shao and Hacker. The reason for the low values of $\sigma_\theta^2/\theta_*^2$ is less clear, and more research will be necessary to clear up this point; we have at our disposal after all only two data sets. Nevertheless, these rather small variations from the standard form emphasize how well the standard convective mixed layer model describes the downwind development of the CIBL and lends strong support to the formulas based on this structure that were introduced in earlier sections.

Hot-cold changes. Only Garratt and Ryan have presented relevant turbulence data. Their aircraft measurements were made in offshore flows at fetches of several hundred kilometers. Normalized variances σ_u/u_{*2}, σ_w/u_{*2}, and $(-\overline{u'w'})^{1/2}/u_{*2}$ were plotted against z/δ_θ and were seen to scatter about the curve usually taken as characteristic of nocturnal stable layers [Caughey and Palmer (1979), and Chapter 1, Fig. 1.11]. Insufficient data points were analyzed to discuss the dependence on fetch.

4.3.5 *Spectra*

The spectral measurements available are a small subset of the experiments where turbulence was recorded. Antonia and Luxton (1974) compared their smooth-rough and rough-smooth experiments and noted some success in collapsing spectra using u_{*2} and δ_i as scaling velocity and length, respectively. They also observed a major difference between the two cases. Across the smooth-rough internal layer there was a substantial increase in the low wavenumber content of $S_w(f\delta_i/\overline{u})$, but not across the rough-smooth IBL boundary. This is consistent with more energetic transport of the turbulent kinetic energy produced at the new rough surface to the edge of the IBL (where it controls the growth of δ_i) than in the contrasting, rough-smooth case.

Neither Antonia and Luxton (1974) nor Mulhearn (1978) saw significant changes in integral properties of the spectra such as the integral time scales (Chapter 3) when entering the rough-smooth internal layer. Note that it is safer to compare integral time scales in these advective situations than integral length scales formed using Taylor's hypothesis. Near the change particularly, rapid development of flow structure invalidates the use of Taylor's hypothesis.

There are no spectra available in advective inversions, but they have been measured on a couple of occasions behind cold-hot changes. Smedman and

Högström (1983) obtained data at moderately large fetch at two heights: $z/\delta_\theta = 0.1(10\,\mathrm{m})$ and $z/\delta_\theta = 0.8(75\,\mathrm{m})$. At 10 m, they found that $fS_w(f)$ followed the standard form set out in Chapter 2, but that below the spectral peak in $fS_u(f)$, data points fell squarely in the "excluded region" between $z/L = 0$ and -2.0 (see Chapter 2, Section 2.5). At 75 m, they found that $fS_w(f)$ followed standard mixed layer forms (Section 2.6), but $fS_u(f)$ exhibited more low-frequency energy than the standard forms. Both the low-level and upper-level behaviors are consistent with a greater contribution to turbulent kinetic energy from mechanically (shear) produced turbulence than we expect in a classical convective mixed layer.

Højstrup (1981) has produced a model of spectral adjustment that combines the simple model of δ_i growth and surface stress development described by Jensen (1978) (Section 4.1) with an assumption about the way a turbulent eddy will decay as it enters the internal boundary layer. The model predicts that the high wavenumber portion of the spectrum adjusts to the new surface more rapidly than the low, the adjustment time being characterized by a wavenumber-dependent timescale. The model appears to describe quite well the adjustment of both velocity and temperature spectra behind changes of roughness and heat flux in the local advection region of onshore flow (Højstrup, 1981).

The tentative conclusions that we are able to draw about the behavior of turbulence in general, highlight the venerability of the basic data sets upon which we continue to rely. New theories are invariably compared with Bradley's (1968) roughness change data or Rider et al.'s (1963) and Dyer and Crawford's (1965) studies of change in evaporation. The continued reliance on these experiments is a resounding vote of confidence in these pioneering efforts, but the many unanswered questions that have arisen during this chapter await definitive experiments for their resolution.

4.5 Special symbols

A_1, B_1	constants in the IBL wind profile equations
c	any scalar
E_0	surface moisture flux
$f(z/\delta_i)$	function modifying logarithmic law in the IBL
$g(\eta)$	velocity profile shape function
$g_s(z/\delta_\theta)$	temperature profile shape function
G	geostrophic wind
$h(\eta)$	stress profile shape function
H_0	surface heat flux
k_c	molecular diffusivity for scalar c
l_0	vertical scale of perturbation in self-preserving model
l_m	mixing length
L_d	downward longwave radiation

M	logarithm of M^*
M^*	ratio of roughness lengths
n	exponent in IBL growth equation
Pr	Prandtl number
Pr_t	turbulent Prandtl number
q_0	specific humidity at the surface
$q_{\mathrm{sat}}(T_0)$	saturation specific humidity at T_0
r_{aq}	aerodynamic resistance
R	Rossby number
R_0	relative humidity at the surface
R_n	net radiation
Sc	Schmidt number
Sc_t	turbulent Schmidt number
S_d	downward shortwave radiation
T_0	surface temperature
T_{01}, T_{02}	surface temperature upwind and downwind of temperature change
T_∞	free stream temperature in wind tunnel
$u_{1\infty}$	free stream velocity in wind tunnel
u_m	average mixed layer velocity
u_0	velocity scale in self-preserving model
$\overline{u}_\delta$	mean velocity at IBL top
$\overline{u}_{\delta_\theta +}$	mean velocity just above thermal boundary layer top
z_{01}, z_{02}	roughness lengths upwind and downwind of roughness change
α	albedo
α_1	constant in IBL growth equation
α_t	constant in scalar budget equation
β_1	geostrophic departure
δ	height of a momentum boundary layer in the wind tunnel
δ_i	height of the momentum IBL
δ_H	height at which the heat flux passes through zero
$\delta_\theta, \delta_q, \delta_c$	height of thermal, moisture, and any scalar boundary layer or IBL
ϵ_λ	emissivity
θ_e	equivalent potential temperature
θ_0	scaling temperature for the IBL
τ_{01}, τ_{02}	surface shear stress upwind and downwind of roughness change
η	ratio of z to l_0 or to δ_i

References

Anthes, R. A., 1978: The height of the PBL and the production of circulation in a sea breeze model. *J. Atmos. Sci., 35*, 1231–1239.

Antonia, R. A., and R. E. Luxton, 1971: The response of a turbulent boundary layer to a step change in surface roughness. Part 1: Smooth to rough. *J. Fluid Mech., 48* (4), 721–761.

Antonia, R. A., and R. E. Luxton, 1972: The response of a turbulent boundary layer to a step change in surface roughness. Part 2: Rough to smooth. *J. Fluid Mech., 53* (4), 737–757.

Antonia, R. A., and R. E. Luxton, 1974: Characteristics of turbulence within an internal boundary layer. *Adv. in Geophys., 18A,* In *Turbulent Diffusion in Environmental Pollution* (F. N. Frenkiel and R. E. Munn, Eds.). Academic Press, New York, 263–285.

Antonia, R. A., H. W. Danh, and A. Prabhu, 1977: Response of a turbulent boundary layer to a step change in surface heat flux. *J. Fluid Mech. 80* (1), 153–177.

Belcher, S. E., D. P. Xu, and J. C. R. Hunt, 1990: The response of a turbulent boundary layer to arbitrarily distributed two-dimensional roughness changes. *Quart. J. Roy. Meteor. Soc., 116,* 611-635.

Bradley, E. F., 1968: A micrometeorological study of velocity profiles and surface drag in the region modified by a change in surface roughness. *Quart J. Roy. Meteor. Soc., 94,* 361–379.

Caughey, S. J., and S. G. Palmer, 1979: Some aspects of turbulence structure through the depth of the convective boundary layer. *Quart. J. Roy. Meteor. Soc., 105,* 811–827.

Charnay, G., J. P. Schon, E. Alcaraz, and J. Mathieu, 1979: Thermal Characteristics of an Internal Boundary Layer with an Inversion of Wall Heat Flux, In *Turbulent Shear Flows 1* (F. Durst, B. E. Launder, R. W. Schmidt, and J. H. Whitelaw, Eds.). Springer-Verlag, Berlin, Germany, 104–118.

Claussen, M., 1991: Local advection processes in the surface layer of the marginal ice zone. *Bound.-Layer Meteor., 54,* 1–27.

Durand, P., S. Briere, and A. Druilhet, 1989: A sea-land transition observed during the COAST experiment. *J. Atmos. Sci., 46,* 96–116.

Dyer, A. J., and T. V. Crawford, 1965: Observations of climate at a leading edge. *Quart. J. Roy. Meteor. Soc., 91,* 345–348.

Elliott, W. P., 1958: The growth of the atmospheric internal boundary layer. *Trans. Amer. Geophys. Union, 39,* 1048–1054.

Gamo, M., S. Yamamoto, and O. Yokoyama, 1982: Airborne measurements of the free convective internal boundary layer during the sea breeze. *J. Meteor. Soc. Japan, 60,* 1284–1298.

Garratt, J. R., 1987: The stably stratified internal boundary layer for steady and diurnally varying offshore flow. *Bound.-Layer Meteor., 38,* 369–394.

Garratt, J. R., 1990: The internal boundary layer—a review. *Bound.-Layer Meteor., 50,* 171–203.

Garratt, J. R., and B. F. Ryan, 1989: The structure of the stably stratified internal boundary layer in offshore flow over the sea. *Bound.-Layer Meteor., 47,* 17–40.

Højstrup, J., 1981: A simple model for the adjustment of velocity spectra in unstable conditions downstream of an abrupt change in roughness and heat flux. *Bound.-Layer Meteor., 21,* 341–356.

Hsu, S. A., 1986: A note on estimating the height of the convective internal boundary layer near shore. *Bound.-Layer Meteor., 35,* 310–316.

Itier, B., A. Perrier, and G. Gosse, 1978: Présentation d'une étude analytique de l'advection III. —Vérification expérimentale du modèle. *Ann. Agron., 29* (3), 209–222.

Jackson, N. A., 1976: The propagation of modified flow downstream of a change in roughness. *Quart. J. Roy. Meteor. Soc., 102,* 924–933.

Jensen, N. O., 1978: Change of surface roughness and the planetary boundary layer. *Quart. J. Roy. Meteor. Soc.*, *104*, 351–356.

Lang, A. R. G., K. G. McNaughton, F. Chen, E. F. Bradley, and E. Ohtaki, 1983: Inequality of eddy transfer coefficients for vertical transport of sensible and latent heats during advective inversions. *Bound.-Layer Meteor.*, *25*, 25–41.

Launder, B. E., 1976: Heat and mass transport. In *Topics in Applied Physics, Vol. 12, Turbulence* (P. Bradshaw, Ed.). Springer-Verlag, Berlin, Germany, 232–287.

Lenschow, D.H., J.C. Wyngaard, and W.T. Pennell, 1980: Mean field and second-moment budgets in a baroclinic convective boundary layer. *J. Atmos. Sci.*, *45*, 3575–3587.

Ligrani, P. M., and R. J. Moffat, 1985: Thermal boundary layers on a rough surface downstream of steps in wall temperature. *Bound.-Layer Meteor.*, *31*, 127–147.

Lyons, W. A., 1975: Turbulent diffusion and pollutant transport in shoreline environments. *Lectures on Air Pollution and Environmental Impact Analyses*, American Meteorological Society, Boston, MA, 136–208.

McNaughton, K. G., and P. G. Jarvis, 1991: Effects of spatial scale on stomatal control of transpiration. *Agric. For. Meteor.*, *54*, 279–301.

Meroney, R.N., J.E. Cermak, and D.T. Yang, 1975: Modeling of atmospheric transport and fumigation of shoreline sites. *Bound.-Layer Meteor.*, *9*, 69–90.

Mulhearn, P. J., 1977: Relations between surface fluxes and mean profiles of velocity, temperature, and concentration, downwind of a change in surface roughness. *Quart. J. Roy. Meteor. Soc.*, *103* , 785–802.

Mulhearn, P. J., 1978: A wind-tunnel boundary-layer study of the effects of a surface roughness change: rough to smooth. *Bound.-Layer Meteor.*, *15*, 3–30.

Mulhearn, P. J., 1981: On the formation of a stably stratified internal boundary layer by advection of warm air over a cooler surface. *Bound.-Layer Meteor.*, *21*, 247–254.

Nieuwstadt, F. T. M., 1984: The turbulent structure of the stable, nocturnal boundary layer. *J. Atmos. Sci.*, *41*, 2202–2216.

Ogawa, Y., and T. Ohara, 1985: The turbulent structure of the internal boundary layer near the shore, Part 1: Case study. *Bound.-Layer Meteor.*, *31*, 369–384.

Panofsky, H. A., and J. A. Dutton, 1984: *Atmospheric Turbulence: Models and Methods for Engineering Applications*. Wiley-Interscience, New York, 397 pp.

Panofsky, H. A., and A. A. Townsend, 1964: Change of terrain roughness and the wind profile. *Quart. J. Roy. Meteor. Soc.*, *90*, 147–155.

Peterson, E. W., 1969: Modification of mean flow and turbulent energy by a change in surface roughness under conditions of neutral stability. *Quart. J. Roy. Meteor. Soc.*, *95*, 561–575.

Peterson, E. W., N. O. Jensen, and J. Højstrup, 1979: Observations of downwind development of wind speed and variance profiles at Bognaes and comparison with theory. *Quart. J. Roy. Meteor. Soc.*, *105*, 521–529.

Philip, J. R., 1967: The second stage of drying of soil. *J. Appl. Meteor.*, *6*, 581–582.

Philip, J. R., 1987: Advection, evaporation, and surface resistance. *Irrig. Sci.*, *8*, 101–114.

Rao, K. S., J. C. Wyngaard, and O. R. Coté, 1974a: The structure of two-dimensional internal boundary layer over a sudden change of surface roughness. *J. Atmos. Sci.*, *31*, 738–746.

Rao, K. S., J. C. Wyngaard, and O. R. Coté, 1974b: Local advection of momentum, heat, and moisture in micrometeorology. *Bound.-Layer Meteor.*, *7*, 331–348.

Raupach, M. R., 1983: Near-field dispersion from instantaneous sources in the surface layer. *Bound.-Layer Meteor.*, *27*, 105–113.

Raupach, M. R., 1991: Vegetation atmosphere interaction in homogeneous and heterogeneous terrain: some implications of mixed layer dynamics. *Vegetatio, 91*, 105–120.

Raynor, G. S., P. Michael, R. M. Brown, S. Sethuraman, 1975: Studies of atmospheric diffusion from a nearshore oceanic site. *J. Appl. Meteor., 14*, 1080–1094.

Raynor, G. S., S. Sethuraman, and R. M. Brown, 1979: Formation and characteristics of coastal internal boundary layers during onshore flows. *Bound.-Layer Meteor., 16*, 487–514.

Rider, N. E., J. R. Philip, and E. F. Bradley, 1963: The horizontal transport of heat and moisture—a micrometeorological study. *Quart. J. Roy. Meteor. Soc., 89*, 507–531.

Schlichting, H., 1968: *Boundary Layer Theory* . McGraw-Hill, New York, 747 pp.

Sempreviva, A. M., S. E. Larsen, N. G. Mortensen, and L. Troen, 1990: Response of neutral boundary layers to changes of roughness. *Bound.-Layer Meteor., 50*, 205–225.

Shao, Y., and J. M. Hacker, 1990: Local similarity relationships in a horizontally inhomogeneous boundary layer. *Bound.-Layer Meteor., 52*, 17–40.

Shao, Y., J. M. Hacker, and P. Schwerdtfeger, 1991: The structure of turbulence in a coastal atmospheric boundary layer. *Quart. J. Roy. Meteor. Soc., 117*, 1299–1324.

Smedman, A.-S., and U. Högström, 1983: Turbulent characteristics of a shallow convective internal boundary layer. *Bound.-Layer Meteor., 25*, 271–287.

Smits, A. J., and D. H. Wood, 1985: The response of turbulent boundary layers to sudden perturbations. *Ann. Rev. Fluid Mech., 17*, 321–358.

Stunder, M., and S. Sethuraman, 1985: A comparative evaluation of the coastal internal boundary-layer height equations. *Bound.-Layer Meteor., 32*, 177–204.

Taylor, P. A., 1969: The planetary boundary layer above a change in surface roughness. *J. Atmos. Sci., 26*, 432–440.

Tennekes, H., and A. G. M. Driedonks, 1981: Basic entrainment equations for the atmospheric boundary layer. *Bound.-Layer Meteor. 20*, 515–531.

Townsend, A. A., 1965: The response of a turbulent boundary layer to abrupt changes in surface conditions. *J. Fluid Mech., 22*, 799–822.

Townsend, A. A., 1966: The flow in a turbulent boundary layer after a change in surface roughness. *J. Fluid Mech., 26*, 255–266.

Townsend, A. A., 1976: *The Structure of Turbulent Shear Flow*, 2nd ed., Cambridge University Press, Cambridge, 429 pp.

Venkatram, A., 1977: A model of internal boundary-layer development. *Bound.-Layer Meteor., 11*, 419–437.

Venkatram, A., 1986: An examination of methods to estimate the height of the coastal internal boundary layer. *Bound.-Layer Meteor., 36*, 149-156.

Vugts, H. R., and J. A. Businger, 1977: Air modification due to a step change in surface temperature. *Bound.-Layer Meteor., 11*, 295–305.

Walmsley, J. L., 1989: Internal boundary-layer height formulae—a comparison with atmospheric data. *Bound.-Layer Meteor. 47*, 251–262.

Weisman, B., 1976: On the criteria for the occurrence of fumigation inland from a large lake—a reply. *Atmos. Environ., 12*, 172–173.

Wyngaard, J. C., 1983: Boundary-layer modelling. In *Atmospheric Turbulence and Air Pollution Modelling*. (F. T. M. Nieuwstadt and H. van Dop, Eds.), Reidel Publishing Co., Dordrecht, The Netherlands, 69–106.

5

FLOW OVER HILLS

We now move on to the next obstacle to understanding how the boundary layer behaves in general through the study of flow over ridges and hills. In Chapter 4 we examined simple changes in surface conditions and showed how their effects extend upwards with increasing downwind distance. The distinguishing features of the flow over those changes were a small perturbation in the pressure field and an internal boundary layer, the depth of which was controlled by turbulent diffusion from the new surface. Here, we confront not a change in surface properties but a change in surface elevation that forces large-scale changes in the pressure field. The response to this forcing is more complicated than any we have tackled so far, but the work of many scientists over the past 25 years gives us a measure of understanding of the processes involved.

In addition to extending to hillsides the kind of analyses of wind and turbulence we have already presented, there are new questions that only arise in the context of hill flows. One, with ramifications for large-scale prediction of the weather and climate, is how much drag hills exert on the atmosphere flowing over them. For large hills and mountains this problem is dominated by the behavior of the internal gravity waves initiated by hills; over lower topography, however, it involves a subtle balance between changes in the surface stress distribution and the pressure field. In questions of wind turbine siting, understanding the position and magnitude of accelerations in the mean wind becomes crucial, whereas changes to both the mean wind and turbulence are important when predicting the fate of atmospheric pollutants in hilly terrain or estimating wind loads on buildings.

The pattern of airflow around a hill is determined not only by the hill shape but also by its size. A characteristic feature of the atmosphere as a whole is its static stability, extending all the way to the ground at nighttime and down to z_i during the day. As a result, the vertical movement of air parcels that must occur as the wind flows over a hill is accompanied by a gravitational restoring force. If the hill is large enough to disturb the whole ABL, then buoyancy-driven flow patterns are important at any time of day; conversely, flow patterns around a hill on scales much smaller than the ABL depth are only affected by buoyancy when the boundary layer is stably stratified.

Hills of height $h \sim 500$ m or more occupy a significant fraction of the daytime ABL depth and always cause some disturbance of the stable atmosphere above, whereas hills of height $h \sim 100$ m are only strongly affected by buoyancy when the boundary layer is stably stratified or when the capping inversion is at a low level early in the morning.

We can quantify these conditions by comparing the time an air parcel takes to traverse a hill (and, therefore, the time during which it is displaced vertically) with the period of its vertical oscillation in the stable density gradient. This period is the inverse of the Brunt-Väisälä or buoyancy frequency N, where

$$N = \left(\frac{g}{\overline{\theta}} \frac{\partial \overline{\theta}}{\partial z} \right)^{1/2}, \tag{5.1}$$

as defined in Chapter 1.

Taking the characteristic downwind length of the hill as L_h, the half-width of the hill at half-hill height, and the characteristic velocity in the boundary layer as U, the ratio we require is the Froude number F_L,

$$F_L = \frac{U}{NL_h}. \tag{5.2}$$

For $F_L \ll 1$, the flow is affected significantly by buoyancy forces. For $F_L \gg 1$, the more familiar balance between inertial effects and turbulent friction is observed. The ratio U/N is typically 1 km so hills of kilometer length scale or less may be free of buoyancy effects for much of the daylight hours and when winds are strong, whereas hills much larger than this, say, $L_h \sim 5$ km, are always affected by stratification to some degree. Height-to-length scales of natural hills are typically in the ratio 1:10 so the corresponding hill heights are 50–500 m. (As far as we are concerned, topographical features much larger than this are mountains.) Most of the available data concerning turbulent boundary layer flow over hills have been gathered on lower hills with $h \sim 100$ m or less, and it is these data that we shall concentrate upon initially. We will return to consider buoyancy effects in Section 5.2.

Restricting our attention to hills rather than mountains, which we might define as hills with length scales greater than 10 km, brings the added convenience of allowing us to ignore explicit rotational effects. These become important when $U/fL_h \lesssim 1$, where f is the Coriolis parameter. Since $f \sim 10^4$ s^{-1} in midlatitudes, this restriction is less severe than that of thermal stratification.

To a greater degree than in the other areas we have covered, our understanding of hill flow has been shaped by mathematical modeling. This is not surprising when we realize that the expensive and delicate instruments used in one-dimensional micrometeorological experiments must be duplicated many times over to capture the two- or three-dimensional flow fields above real topography. At Askervein Hill, the site of the most complete field experiment to date, for example, 50 towers were deployed with 27 of them equipped with three-component turbulence sensors (Taylor and Teunissen, 1987).

The most influential theoretical work to emerge (Jackson and Hunt, 1975; Hunt et al., 1988a, 1988b) has relied on techniques of asymptotic matching applied to flow over low hills, where the equations of motion can be linearized. The results of this work have had a profound influence on the way we view flow over hills of much greater steepness than the linearized approach strictly accommodates. By dividing the boundary layer into several sublayers, each with essentially different flow dynamics, these linear theories provide a link with the similarity relationships that proved so useful on flat ground. We shall see that in the lowest layer, the "inner layer," whose thickness depends on the size and roughness of the hill, we can still use a modified form of Monin-Obukhov similarity, but farther from the surface the presence of advection does not allow the condition of "local equilibrium" in the turbulence that such scaling requires. On hills so steep that the tenets of linear theory are totally inapplicable, a complete description of the flow field requires a numerical model.

There is a great deal more structure to be explored in the mean flow above an arbitrary hill than in the horizontally homogeneous and advective situations we have dealt with so far. The labor involved in fully sampling the flow around a three-dimensional hill has, however, meant that almost all the available information has been gathered in winds normal to two-dimensional ridges or on the centerlines of axisymmetric hills. We must keep this restriction in mind when we generalize from the measurements we have.

A further restriction is imposed by the fact that, until recently, most published data had been gathered in near-neutral conditions. This is not entirely accidental as many of the experiments were motivated by questions of wind turbine siting and were, as a result, biased toward strong winds. At this time, we certainly lack any reliable measurements of scalar fluxes and concentrations over hills. What we think we know about scalar fluxes and gradients comes entirely from mathematical models and inference from the behavior of the velocity field.

5.1 Measurements and coordinates

We have already touched upon the logistical requirements that a hill experiment in the field can demand, but there are special measurement problems to confront as well. First, the depth of the recirculating "separation bubble" that forms if the hill is steep enough is typically equal to the hill height, which puts it out of the reach of tower-mounted instruments on all but the shallowest hills. In practice this has meant that most of the information we have about this region comes from wind tunnel simulations. This recourse also, unfortunately, has its problems. The reversing flow encountered in a separation region denies the possibility of using the standard tool of the wind tunnel, the hot-wire anemometer, because of its rectifying action. The pulsed wire anemometer does offer a viable alternative, however, and has been used in some experiments (Arya and Gadiyaram, 1986; Britter et al., 1981).

There are other problems with modeling hills in the wind tunnel that have meant it has played a minor role, at least in shaping our general understanding. To maintain a reasonable facsimile of the atmospheric surface layer in a wind tunnel, the tunnel surface roughness must exceed a certain minimum height. This ensures "aerodynamically fully rough" flow, which occurs for $u_* z_0/\nu > 5$ (Schlichting, 1968), and is the minimum requirement for accurately reproducing the atmospheric surface layer in a tunnel. Unfortunately, when a real hill is reduced in scale to fit into a wind tunnel, the surface roughness—grass, heather, stones—usually shrinks too much to satisfy the above condition. The alternatives are to work with an aerodynamically smooth model or to increase the surface roughness disproportionately. The first option allows the possibility of incorrectly modeling near-surface turbulence and phenomena such as separation that depend critically upon it; the second option means that the crucial inner layer, where changes in turbulence affect the flow dynamics, may be almost entirely occupied by the magnified roughness elements (Finnigan et al., 1990). Over a typical smooth, grass-covered hill about 100 m high, the inner layer depth is roughly 10 m and we are interested in probing its behavior in detail since it is there, as in the lower part of the surface layer on flat ground, where scalar and wind gradients are largest.

Another feature of hill flows we shall encounter is a steep gradient of shearing stress near the surface. This makes it difficult to infer the surface stress from measurements at the lowest viable level of a sonic anemometer (see Chapter 7) in the field or a hot wire anemometer in the wind tunnel. Attempts have been made to use x-configuration hot wires in the field to overcome this difficulty but with mixed success (Mason, 1986). The problem is that local surface features—the boulder, clump of heather, or surface hollow immediately upwind—may bias point measurements, and the kind of area average used in canopy studies (Chapter 3) is what is needed.

Obtaining reliable measurements at a given point above a hill is the first problem. Finding a useful conceptual framework to compare measurements at one point with those at another is the next. Rectangular Cartesian coordinates continue to be useful for locating points at which data are collected, but when it comes to interpreting the relationship between the hill shape, the turbulence, and the mean flow they are conceptually clumsy. They also present practical difficulties when interpreting measurements, particularly in the field. The conceptual problems arise because we are no longer able to identify a coordinate axis with the mean wind direction.

Over flat uniform terrain it is possible to align the x coordinate along the mean wind direction near the ground so the momentum balance in the x direction contains most of the dynamical information we require. Over hills, in contrast, we would have to take account of the momentum balances along all three Cartesian coordinate axes. Our intuition suggests, however, that there is a preferred direction, that of the local mean velocity vector, that might play the same role over hills as the Cartesian x axis does in flow over flat land.

Close to the ground, the wind vector must parallel the underlying surface for reasons of continuity, so surface-following or s-n coordinates have been used on several occasions. The equations of motion in such a system were described by Howarth (1951). In this scheme the z coordinate is the straight line normal to the surface, whereas the x and y coordinates lie on surfaces parallel to the ground. Unfortunately, s-n coordinates offer only a partial solution to our problem, because although the wind flows parallel to the ground near $z = 0$, at high levels it is approximately horizontal and there remains a mismatch between coordinate axes and the wind direction aloft.

One way out of this dilemma is to use physical streamline coordinates. In this system the x direction is along a streamline and so is always parallel to the local mean velocity vector, whereas the y and z directions are at right angles to it. Velocities, together with any other vector or tensor quantity, are referred to a *local* rectangular Cartesian frame aligned with the local streamline and are therefore the physical quantities with which we are familiar. The y and z coordinate lines are *orthogonal trajectories* to the streamlines, the three sets of coordinate lines forming an orthogonal, curvilinear coordinate system.

These coordinates are sketched in Fig. 5.1, which represents the case of flow over a three-dimensional hill. The z lines intersect the surface at right angles whereas the x and y lines lie on stream surfaces of which the ground is the lowest one. In these coordinates $\overline{v} = \overline{w} = 0$ by definition, so the total velocity vector has components $(\overline{u} + u'), v', w'$, corresponding to the x, y, and z directions, respectively.

As we should expect in a curvilinear coordinate system, the equations of motion have extra terms to account for spatial variation in the coordinate directions. We see this in the momentum equation for two-dimensional, steady flow in the x direction (as over a ridge extending infinitely in the y direction),

$$\overline{u}\,\frac{d\overline{u}}{dx} = -\frac{1}{\rho}\frac{d\overline{p}}{dx} - \frac{d\overline{u'^2}}{dx} + \left(\frac{\overline{u'^2} - \overline{w'^2}}{L_a}\right) - \frac{d}{dz}\left(\overline{u'w'}\right) + 2\left(\frac{\overline{u'w'}}{R}\right) + g_x\left(\frac{\overline{\theta} - \overline{\theta}_0(x)}{\overline{\theta}_0(x)}\right), \tag{5.3}$$

where g_x is the component of the acceleration-due-to-gravity vector in the x direction and viscous terms have been ignored; they are almost always negligible in

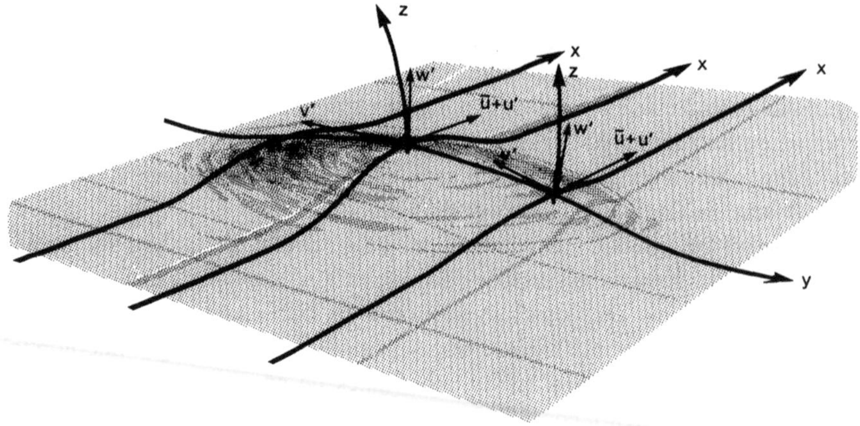

FIG. 5.1 Orthogonal curvilinear coordinate system for describing flow around curved surfaces. The x axis is aligned with the local streamline, whereas the y and z axes are normal to it as shown in the figure.

turbulent atmospheric flows. Care must be taken with the form of the buoyancy term, the last term on the right-hand side. Problems arise because the origin of z coordinates does not lie on a geopotential. This is not a difficulty of streamline coordinates alone but with any surface-following system. The reference temperature $\overline{\theta}_0$ is the value of potential temperature in the undisturbed upwind flow at the same height as the (x, y, z) point under consideration. Since this height changes as we traverse a hill along a line of constant z, $\overline{\theta}_0$ must be a function of x.

R is the local radius of curvature of x coordinate lines and L_a an acceleration length scale. In two-dimensional flow, L_a is equal to $\hat{R}$, the radius of curvature of the z lines as shown in Fig. 5.2. L_a and R are related to the divergence and vorticity of the flow in the following way:

$$\frac{1}{L_a} = \frac{1}{\overline{u}}\frac{d\overline{u}}{dx}, \tag{5.4}$$

$$\frac{1}{R} = \frac{1}{\overline{u}}\left(\Omega_y + \frac{d\overline{u}}{dz}\right), \tag{5.5}$$

where Ω_y is the component of the mean vorticity in the y direction. L_a and R are simultaneously geometric parameters of the coordinates and properties of the flow field. Their appearance in (5.3) emphasizes the essential difference between streamline coordinates and more conventional choices: the fact that the flow field itself determines the coordinate frame. We shall see below that $1/L_a$ and $1/R$ are important parameters for describing flow dynamics.

Moving from two dimensions to flow over the centerline of a three-dimensional hill, the streamwise momentum equation becomes

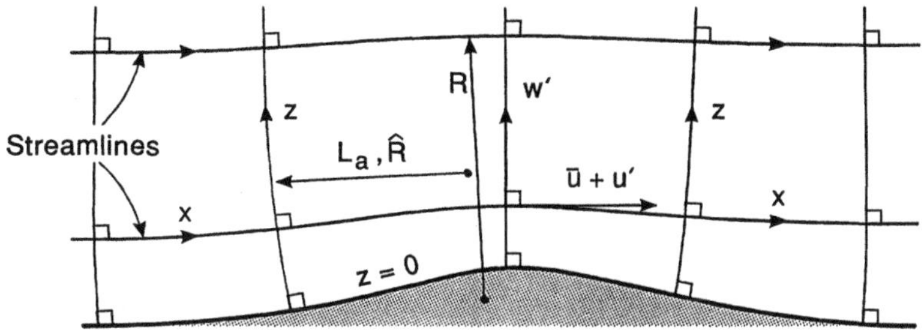

FIG. 5.2 Local radius of streamline curvature R and acceleration length scale L_a defined for flow over a two-dimensional ridge.

$$\overline{u}\,\frac{d\overline{u}}{dx} = -\frac{1}{\rho}\frac{d\overline{p}}{dx} - \frac{d\overline{u'^2}}{dx} + \left(\frac{\overline{u'^2} - \overline{w'^2}}{L_a}\right) - \frac{d}{dz}\left(\overline{u'w'}\right)$$
$$+ 2\left(\frac{\overline{u'w'}}{R}\right) + \frac{\overline{u'w'}}{L_b} + \left(\frac{\overline{w'^2} - \overline{v'^2}}{L_c}\right) + g_x\left(\frac{\overline{\theta} - \overline{\theta}_0}{\overline{\theta}_0}\right), \quad (5.6)$$

where we have ignored viscous terms as before, as well as second-order terms in $1/R, 1/L_a, 1/L_b$, and $1/L_c$, which only become appreciable on very steep hills.

The length scales L_b and L_c characterize the departure of the flow from two dimensionality as we now encounter curvature in the y lines. They are related to r, the local radius of curvature of the y coordinate lines by the following formulas:

$$\frac{1}{L_b} = \frac{1}{r}\frac{dr}{dz}, \quad (5.7a)$$

$$\frac{1}{L_c} = \frac{1}{r}\frac{dr}{dx}. \quad (5.7b)$$

L_a is no longer equal to $\hat{R}$, the radius of curvature of the z lines. It now satisfies

$$\frac{1}{L_a} = \frac{1}{\hat{R}} + \frac{1}{L_c}. \quad (5.8)$$

Equation (5.6) contains all the terms necessary to characterize flow on the centerline of an axisymmetric hill. But it does not describe completely the general three-dimensional flow over topography where another parameter, the torsion or twist of the streamlines, becomes important. Since we do not have useful experimental information about the general situation, however, it is unnecessary to introduce this complication.

In both (5.3) and (5.6), the extra parameters of dynamical significance are $1/L_a$ and $1/R$. $1/L_b$ and $1/L_c$ can be represented in terms of $1/R$ and $1/L_a$ by integral formulas, but it is much more revealing to write them in terms of r as in

(5.7) because r, like R, can often be approximated in the near-ground region by adding z to the surface values of streamwise and lateral radii of curvature.

It is clear from (5.3) and (5.6) that we have simplified the advection term on the left-hand side of the equations at the expense of adding extra terms to the right-hand side. These reflect the influence of curvature and acceleration on the turbulent stress divergence. Note also that (5.3) and (5.6) contain no partial derivatives $\partial/\partial x$, only ordinary, or "directional," derivatives d/dx along the coordinate lines. This is a necessary consequence of attaching a Cartesian reference frame to curvilinear coordinates. It does not affect the interpretation of the equations in any way but must be accounted for when calculating in streamline coordinates. This point, together with the derivation of (5.3) and (5.6) and the related equations up to second order, can be followed in Finnigan (1983, 1990).

The use of these coordinates avoids a practical problem in field and wind tunnel measurements—that of aligning instruments with, or rotating data into, an arbitrary Cartesian or surface-following coordinate frame. To do this the alignment of instruments relative to the chosen coordinate frame must always be accurately known, an exceedingly difficult requirement in practice. To use streamline coordinates, instruments need only be pointed roughly into the mean wind and the data rotated ex post facto so that $\overline{v} = \overline{w} = 0$. A final rotation about the x axis is then performed to fix the y and z directions. Formally, the z direction is the direction of the principal normal to the streamline, that is, the direction in which the streamline appears to have its maximum curvature. Practical ways of finding this direction are discussed in Chapter 6.

No single reference frame works in all situations, and streamline coordinates are no exception: They are difficult to use in regions of reversed mean flow. A practical way out of this difficulty is to use the idealized streamlines of inviscid, potential flow about the hill as a bridge between the real streamlines of the attached, nonreversing flow upwind of the separation bubble and the real streamlines of the wake flow behind the reattachment point. An even simpler procedure, requiring no calculations, is to use surface-following coordinates to bridge the separation region although a smooth fit to the real streamlines may be harder to achieve with this choice. The procedure is illustrated in Fig. 5.3. It provides continuity of the coordinate system across the separation region. A disadvantage is that we can no longer attach physical meaning to terms in the equations over this region. We shall see when we discuss the turbulence in the separation bubble in Section 5.3, however, that there, the conventional Reynolds separation of flow into mean and turbulent components has little relevance or explanatory power so the disadvantage is a minor one.

A comparison of streamline and surface-following coordinates, as used, for example, by Finnigan et al. (1990) and Bradshaw (1973), respectively, shows that all of the terms appearing in streamline equations have counterparts in the surface-following equations. The latter also has a set of extra terms accounting

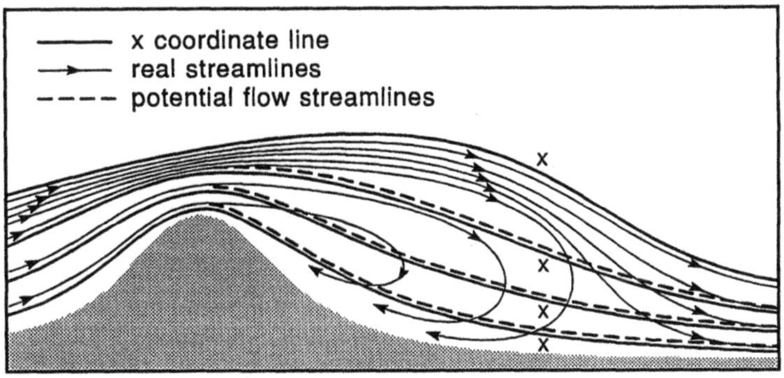

Fig. 5.3 Reversed mean flow in the wake of a hill when separation occurs compared to idealized streamlines for inviscid potential flow that can be used to define the x coordinates.

for the local mismatch between coordinate and flow direction. Near the surface, where this mismatch is usually small, these terms are of second order (and are often ignored in applications) so the streamline equations may be viewed as a first approximation to the surface-following equations or vice versa. This is a point of view we find convenient to adopt in this chapter as it allows us to present and interpret data gathered in surface-following coordinates as if they were really in the local streamline reference frame. The error involved in doing this is less than typical error bounds attached to field measurements. We shall also revert to standard rectangular Cartesian coordinates where necessary. Such occasions will be clearly labeled; otherwise, streamline coordinates should be assumed.

5.2 Mean flow

We begin by examining the mean flow under neutral conditions and postpone a discussion of the influence of stratification until later in this section. Imagine we are recording the velocity along streamlines that approach an isolated two-dimensional (2-d) ridge roughly at right angles. Close to the surface, we see the flow decelerate slightly at the upwind foot of the hill before accelerating to the hilltop. If the hill is steep enough, this deceleration may cause a small separation bubble to form at the foot of the hill. The wind reaches its maximum velocity above the hilltop, then decelerates behind the hill. If the hill is steep enough downwind, a separation bubble forms in which the flow direction at the lowest level is opposite to that above (Fig. 5.4). Whether or not a separation bubble forms, a wake region develops behind the hill with a marked velocity deficit extending for many hill heights downwind. If we follow the higher streamlines, we observe monotonic acceleration to the crest followed by deceleration behind the hill.

Following the streamlines over the crest of an axisymmetric hill, we observe the same features with one difference: The upwind deceleration is absent. It is

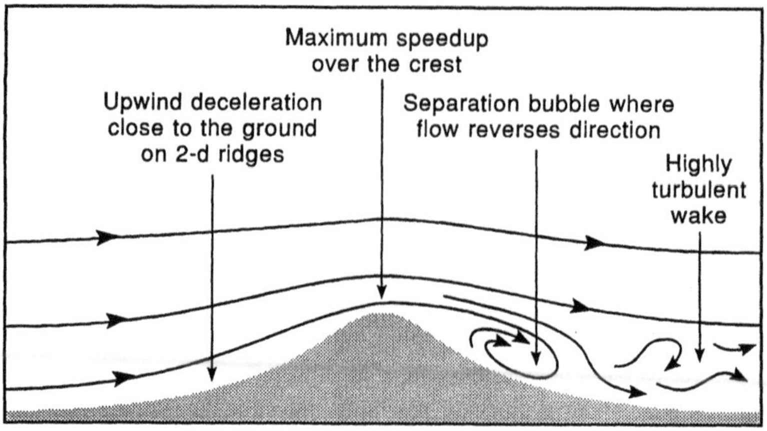

FIG. 5.4 Flow over a two-dimensional ridge showing formation of a separation bubble when the downwind slope is steep enough.

replaced by a region of lateral flow divergence as the streamlines divide to pass around the hill. This lateral divergence decreases as we move up the hill centerline, disappearing completely at the crest.

This evolution of the mean wind is displayed in two different ways in Figs. 5.5 and 5.6. In Fig. 5.5 we present typical vertical profiles upwind, on the crest and behind the hill. The heights, distances, and velocities are normalized by parameters we will define later. The height scale is logarithmic to display the near-surface structure in the profiles. Note the large increase in velocity near the ground on the hilltop and the velocity deficit in the hill wake. The same features can be seen in Fig. 5.6 as we follow two streamlines, one close to the ground, starting, let us say, at a height of $0.05h$ upwind, and a higher streamline starting at $0.5h$. Along the lower streamline, in the two-dimensional case, both the upwind and downwind decelerations are marked as is the speedup at the crest. On the higher streamline we see no upwind deceleration and the speed variation in general is less pronounced.

To make the information displayed in Figs. 5.5 and 5.6 more concrete, we can imagine the profiles developing over a ridge or axisymmetric hill of height $h = 100$ m, length scale $L_h = 200$ m, and surface roughness $z_0 = 0.02$ m, which is typical of grazed pasture. We shall soon define the horizontal scale L_h more precisely.

Before we attempt to quantify the behavior we have just described, it is helpful to introduce a division of the flow field into three separate regions: an *inner layer* extending from the surface to a height l, an *outer layer*, and a *wake region*. These are sketched in Fig. 5.7. The justification for this division is the essentially different dynamical processes that dominate in each region. In the outer layer, mean flow patterns are determined by inertial forces; turbulent friction is relatively unimportant and we can treat the mean flow as essentially inviscid. In

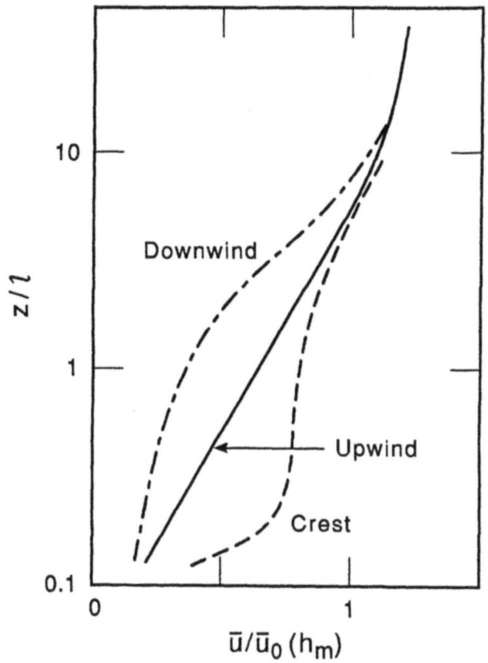

FIG. 5.5 Normalized wind profiles observed upwind, at the crest, and at the downwind foot of a two-dimensional ridge.

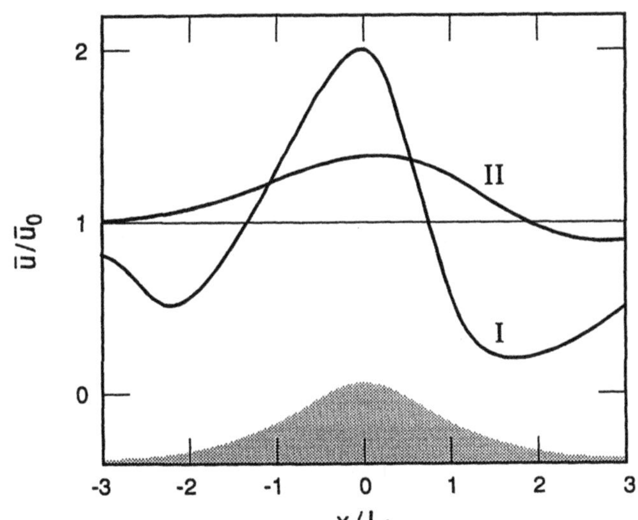

FIG. 5.6 Normalized mean wind in the inner (I) and outer (II) layers plotted as a function of dimensionless distance from the hill crest.

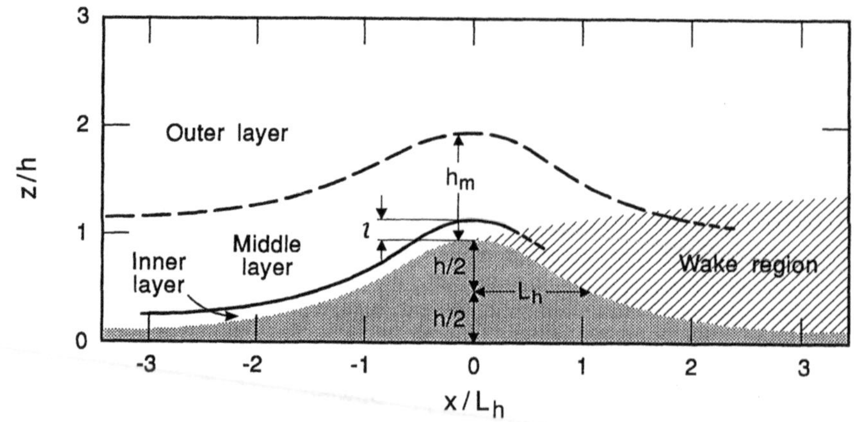

FIG. 5.7 Schematic of flow over a two-dimensional ridge showing different regions of the flow: inner, middle, outer, and wake regions.

the inner layer, turbulent friction plays a major role not least in causing separation to occur, whereas in the separation bubble and in the wake region turbulence dominates.

5.2.1 *Speedup*

The magnitude of the speedup on the hilltop holds obvious interest both for wind-power enthusiasts and for engineers interested in predicting wind loads on buildings. According to convention, the fractional speedup Δs is defined as

$$\Delta s = \frac{\overline{u}(x,z) - \overline{u}_0(z)}{\overline{u}_0(z)} = \frac{\Delta\overline{u}(x,z)}{\overline{u}_0(z)}, \tag{5.9}$$

where $\overline{u}_0(z)$ is the reference wind profile sufficiently far upwind to be removed from any influence of the hill. We further define Δs_{max} as the value of Δs where $\Delta\overline{u}$ is maximum; it occurs at some height above the hill crest. On hills low enough to avoid separation, which in practical terms means that the downwind hill slope should be less than about 10°, the speedup can be understood in a qualitative way using ideas borrowed from the linearized theories.

Flow accelerations are driven by the pressure field that develops around the hill. If the hill were embedded in a uniform, unsheared flow with upstream velocity U, the pressure field could be calculated using classical potential flow theory. The pressure perturbation that developed over the hill would have a characteristic magnitude $\sim \rho U^2$. This would act upon the flow, whose characteristic velocity was U, to produce a speedup Δs. The spatial distribution and absolute magnitude of Δs would depend on the hill shape.

In reality, hills are embedded in shear flows with different characteristic velocities in their inner and outer layers. In the outer layer we take $U \sim \overline{u}_0(h_m)$, the reference velocity at h_m, the height at which the shear in the upwind profile ceases to be important in the flow dynamics. It is defined by the formula (Hunt et al., 1988a):

$$h_m = L_h \left[\ln\left(\frac{L_h}{z_0}\right)\right]^{-1/2} \tag{5.10}$$

This definition allows a useful subdivision of the outer layer by specifying a *middle layer* ($h_m > z > l$), the inviscid but rotational part of the outer layer where turbulent stresses have negligible effect on the mean flow but where the role of the mean vorticity cannot be ignored in calculations. L_h may be taken as the distance from the crest to the half-height point (see Fig. 5.7).

In the inner layer we take $U \sim \overline{u}_0(l)$, the reference velocity at height l, the inner layer depth. Typically, $l/h_m \sim 0.1$ and, if the approach flow $\overline{u}_0(z)$ is logarithmic with realistic values of z_0, then $\overline{u}_0(h_m) \gg \overline{u}_0(l)$. The pressure field is, however, still determined primarily by the outer flow field so the pressure perturbation in both layers has characteristic magnitude $\rho\overline{u}_0^2(h_m)$. In the outer layer this produces a speedup Δs similar to that predicted by potential theory but in the inner region, where the background velocity is so much smaller it produces Δs values much larger than those in the outer regions.

We loosely defined the inner layer as the region where turbulent stresses affect the changes in the mean flow. Before we can propose quantitative expressions for speedup we need to be more specific. We define l, therefore, as the height at which mean flow advection, streamwise pressure gradient, and cross-stream divergence of shearing stress [the first, second, and fifth terms in (5.3)] are of comparable magnitude. For $z \gg l$ the effect of the turbulence stress becomes negligible, whereas for $z < l$ it becomes the dominant term. To quantify this definition requires some assumption to be made about the form of the mean velocity and shear stress profiles over the hill. The simplest assumption is that the hill constitutes only a small perturbation to an existing logarithmic velocity profile. This assumption is appropriate over low hills where linearized theories apply and leads to the following definition for l:

$$\frac{l}{L_h} \ln\left(\frac{l}{z_0}\right) = c, \tag{5.11}$$

where c is a constant of order 1. The value taken by c in the two-layer theory of Jackson and Hunt (1975) and in the more comprehensive theory of Hunt et al. (1988a) that has superseded it is $c = 2k^2$, where k is von Karman's constant. We shall encounter other choices for c later in this chapter. Over the hypothetical hill used as the basis of Figs. 5.5 and 5.6, with $h = 100$ m, $L_h = 200$ m, and $z_0 = 0.02$ m, we find that $l = 10$ m and $h_m = 66$ m.

The linear theory of Hunt et al. (1988a) produces an expression for speedup in the inner layer ($z < l$)

$$\Delta s(x, z) = \frac{\Delta \overline{u}(x, z)}{\overline{u}_0(z)} = \frac{h}{L_h} \left(\frac{\overline{u}_0^2(h_m)}{\overline{u}_0(l)\overline{u}_0(z)} \right) \zeta(x, z_0). \tag{5.12}$$

ζ is a function of order 1 that factors the precise shape of the hill into the expression for Δs. The hill shape is important because it determines the details of the pressure field around the hill. ζ is also responsive to the surface roughness, z_0.

A useful general form for the hill shape is the so-called Witch of Agnesi profile. In two dimensions, it is given by

$$z = h \left[1 + \left(\frac{x}{L_h} \right)^2 \right]^{-1}. \tag{5.13}$$

[The hill shape defined by (5.13) is often used in mathematical models of hill flow because it has a particularly simple Fourier transform; it also shows that the definition of L_h as the distance to the half-height point is not accidental.] With this choice for the hill shape we can compute $\zeta(x, z_0)$. Specializing still further to the case of maximum Δu, which occurs for $x = 0$ and $z \simeq l/3$, we have

$$\Delta s_{\text{max}} = \frac{\Delta \overline{u}_{\text{max}}}{\overline{u}_0(l/3)} = \frac{h}{L} \left(\frac{\overline{u}_0^2(h_m)}{\overline{u}_0(l)\overline{u}_0(l/3)} \right) (1 + 1.8\delta), \tag{5.14}$$

where $\delta = [\ln(l/z_0)]^{-1}$. The value of Δs_{max} given by (5.14) and its location at $z = l/3$, both predictions of the Hunt et al. (1988a) theory, match observations on Askervein Hill (Taylor and Teunissen, 1987) and Cooper's Ridge (Coppin et al., 1986) quite closely. $l/3$ seems to be a reasonable estimate of the height of Δs_{max} on a wide range of hill shapes, although it must be said that the absence of sharp peaks in the Δu profile ensures a good deal of scatter in the experimental data. Certainly, at this stage, no clear difference can be detected between the location of the peak in $\Delta \overline{u}$ on two-dimensional and axisymmetric hills.

In the middle layer, ($h_m > z > l$), the same theory predicts a different expression for speedup. This now becomes

$$\Delta s = \frac{\Delta \overline{u}(x, z)}{\overline{u}_0(z)} = \frac{h}{L_h} \left(\frac{\overline{u}_0^2(h_m)}{\overline{u}_0^2(z)} \right) \xi(x), \tag{5.15}$$

where ξ is a factor of similar magnitude to ζ, which plays the same role above the inner layer. The crucial parameters of Hunt et al.'s (1988a) linearized theory are summarized in Table 5.1. The theory of Hunt et al. (1988a), although comprehensive, is not simple to apply to arbitrary hill shapes. Taylor and Lee (1987), basing their calculations on the earlier theory of Jackson and Hunt (1975), formulated

Table 5.1 Parameters Used in the Linearized Theory of Hunt et al. (1988a) to Calculate Speedup

Parameter	Symbol	Formula
Inner layer height	l	$(l/L_h)\ln(l/z_0) = 2k^2$
Middle layer height	h_m	$h_m = L_h[\ln(L_h/z_0)]^{-1/2}$
Scale size	δ	$\delta = [\ln(l/z_0)]^{-1}$

simple guidelines for estimating wind speed changes over small-scale topographic features without separation. We can summarize their recommendations as follows:

$$\Delta s_{\max} \simeq 1.6h/L_h, \quad \text{for axisymmetric hills,} \tag{5.16a}$$

$$\Delta s_{\max} \simeq 0.8h/L_h, \quad \text{for two-dimensional escarpments,} \tag{5.16b}$$

$$\Delta s_{\max} \simeq 2.0h/L_h, \quad \text{for two-dimensional ridges.} \tag{5.16c}$$

A good deal of experimental evidence both from the wind tunnel and field measurements suggests that (5.16a, 5.16b, 5.16c) are generally accurate within $\pm 15\%$ (Finnigan, 1988), whereas on low hills that fit the assumptions of linear theory more closely, the agreement is even better.

When hills become steep enough to establish steady separation regions, many of the basic tenets of linearized theory break down. Instead of the pressure field being approximated by potential flow around the hill shape, it is the flow around the hill shape plus separation bubble that is important and that must be used to calculate L_h. The appearance of the separation bubble and its streamwise growth with increasing hill steepness, a phenomenon we shall discuss more fully in the next section, put an upper limit on $\Delta s_{\max}$ of about 1.25 and this value is attained on axisymmetric hills where the separation bubble size is minimal.

5.2.2 *Separation*

When separation occurs, it changes the wind field not only in the separation region itself but over the entire hill. It does this because the large-scale pressure field that develops around the hill and drives the flow perturbations is dominated by flow in the outer layer, and this in turn reacts as if the hill and its attached separation bubble formed a single obstacle. It follows that quantifying the onset and extent of separation is necessary before we can apply the speedup formulas (5.12)–(5.16).

We define the separation point operationally as the x position where the near-surface flow reverses. We can write this condition precisely as

$$\left.\frac{d\overline{u}}{dz}\right|_{z=0} < 0. \tag{5.17}$$

The flow can only reach condition (5.17) by passing through a point where $d\overline{u}/dz$ at the surface is zero; in laminar flow this coincides with the point where the surface shear stress vanishes. In turbulent separation, however, we find that eddy diffusivity parameterizations of the stress fail near the separation point so this coincidence need not occur. In fact, at the present time we lack any coherent predictive theory of turbulent separation so we shall confine ourselves to qualitative remarks and a review of the empirical evidence.

At the surface $d\overline{u}/dz = -\Omega_y$, where Ω_y is the cross stream component of mean vorticity, the only nonzero component of vorticity at the surface. In an attached boundary layer, Ω_y is negative ($d\overline{u}/dz > 0$). The rate of generation of Ω_y at the surface is equal to $-(1/\rho)\, d\overline{p}/dz$; in other words, an adverse (positive) streamwise pressure gradient at the wall generates a flux of positive Ω_y (negative $d\overline{u}/dz$) into the flow (Morton, 1984). The resultant vorticity (or shear) at the wall and the consequent possibility of separation [through (5.17)] are determined by the balance between this flux, the advection of vorticity from upstream, and the turbulent diffusion of vorticity to the wall from higher in the boundary layer. An illuminating discussion of this balance may be found in Lighthill (1989) and a more quantitative treatment in Finnigan et al. (1990).

In fully turbulent atmospheric boundary layers, any agency that reduces near-wall advection of negative Ω_y from upstream or reduces diffusion of negative Ω_y toward the surface in a region where the pressure gradient is positive will tend to encourage separation. Increasing momentum absorption at the surface by an increase in surface roughness is one way of reducing near-surface advection while a particularly pervasive reducer of turbulent diffusion is the convex streamline curvature that occurs over a hilltop. We will treat this mechanism in greater detail when we consider the turbulent structure; for the moment, we point out that the extremely high curvature and accompanying adverse pressure gradient at a sharp edge ensures the separation of flow there. This is the one situation where the separation point is easy to predict. At a sharp edge also, all of the deceleration necessary for separation occurs at the edge itself, and the measurable surface flow may appear to accelerate right up to the point where it separates.

If the mechanism of flow separation is poorly understood, the recirculating flow in the bubble region between separation and reattachment is even less well known. What data we do have come from wind tunnel studies because the bubble depth is characteristically of order h, too high to be explored by the meteorological towers of field experiments. Unfortunately, many of these model studies have used smooth hills in rough wall boundary layers, which allowed measurements to be made close to the surface but introduced the complications of rough-smooth transition discussed in Chapter 4. Nevertheless, they have provided answers to the basic question of how steep a hill must be before separation occurs.

The available data on two- and three-dimensional hills were summarized by Finnigan (1988). On smooth two-dimensional ridges, whether smoothly curved like natural hills or of triangular cross section, the critical downwind hill slope angle for steady separation is about 18°, but this angle reduces with increasing surface roughness. The limit for a very rough two-dimensional ridge, such as that studied by Finnigan et al. (1990), which had $h/z_0 \simeq 400$, seems to be about 10°.

The streamwise extent of the separation bubble depends to a significant degree on the angle the separating boundary layer makes with the mean flow direction. For example, on a smooth two-dimensional ridge with triangular cross section and 30° slope, Snyder and Britter (1987) detected a separation bubble about $3h$–$4h$ in horizontal extent, whereas on a triangular ridge of 60° slope, Arya and Shipman (1981) measured a bubble extending for $13h$ downstream and attaining a maximum depth of $2.5h$. The bubble extent also increases with surface roughness so that on a very rough $(h/z_0 = 120)$ 30° triangular ridge, Castro and Snyder (1982) observed a bubble of depth $\sim h$ that extended downwind about $11h$ from the hill crest. On more naturally curved hills the separating shear layer is roughly parallel to the mean flow and the downwind extent of the separation bubble is usually only $2h$ or $3h$.

As the hill's horizontal aspect ratio decreases from two-dimensional to axisymmetric, the critical angle for separation increases whereas the extent and width of the separation region shrinks. On a smooth cone, 30° seems to be the minimum angle to ensure steady separation, but adding surface roughness reduces this angle to about 20°. The extent and width of the separation bubble is now so dependent on the precise hill geometry that the available data do not allow us to discern any meaningful trends.

One striking conclusion is the sensitivity of separation to surface roughness even on very abrupt hills where we would expect topographical effects to dominate totally. The search for a useful theory of rough wall turbulent separation has not yet produced simple applicable rules, although it is an area of active research. Consequently, we have avoided including results from mathematical hill flow models in this brief survey, as those that do predict separation have yet to be matched confidently with experiment.

The structure of the mean flow within a separation bubble is complicated. Separated shear layers are unstable so that even the separation bubbles behind two-dimensional ridges may spontaneously break up into cells with three-dimensional circulation, whereas the bubbles behind truly three-dimensional hills display complex topologies closely connected to the shape of the hill. One important distinction can be made, however, between two- and three-dimensional separation regions following the work of Hunt et al. (1978) and Tobak and Peake (1982). This is that three-dimensional separation bubbles are not closed; there is a mean inflow and an outflow to and from the separation region. Two-dimensional bubbles, in contrast, are bounded by a closed stream surface that joins the separation and reattachment lines. It follows that transfer of scalars from a two-dimensional separation region can only occur by turbulent diffusion across this stream surface, but in the three-dimensional case mean flow advection can play a role. This is an important result in the context of atmospheric pollution dispersal.

Many of the points we have covered in this section can be followed in much more detail in Finnigan (1988), Taylor (1988), and Taylor et al. (1987).

5.2.3 *The wake region*

Even in the absence of a true separation bubble, the near wake region just behind the hill has a structure that is still quite specific to a particular hill shape. The far wakes of isolated hills, however, display some universal features in common with those behind other surface obstacles such as wind breaks and buildings. Their properties have been comprehensively reviewed by Taylor (1988). In Fig. 5.8 we define characteristic parameters for the wake behind a two-dimensional ridge. The wake depth l_w is defined as the height at which the velocity attains a fixed fraction, say 95%, of the velocity of the upstream undisturbed profile at the same height, whereas the wake velocity deficit, $\overline{u}_w$ is the difference between wake and upstream velocities at the same z value. The position of the virtual origin x_0 varies from case to case depending on conditions in the near wake.

A variety of theories has proposed that the wake will display a "self-preserving" behavior expressed in the following power laws:

$$l_w(x - x_0) = A(x - x_0)^a, \tag{5.18a}$$

$$\overline{u}_w = B(x - x_0)^b f(z/l_w), \tag{5.18b}$$

where the velocity deficit $\overline{u}_w$ is defined in Fig. 5.8 and the constants A and B as well as the shape function $f(z/l_w)$ depend on the character of the near wake. In the two-dimensional case there is general agreement that $b = -1$, whereas a mixture of theory, wind tunnel tests, and field experiments has not decided between two possibilities for a: $a = 0.5$ or $a = 1$.

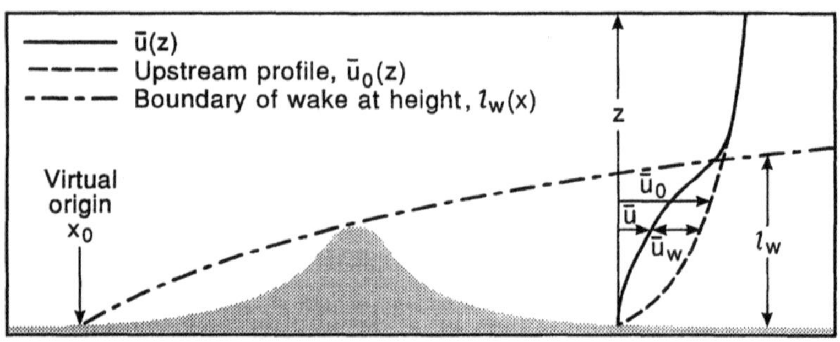

FIG. 5.8 Characteristic parameters for the wake region behind a two-dimensional ridge.

In Taylor's (1988) words, the information on three-dimensional wakes "has a distinct signal-to-noise problem." We are now required to consider growth in the y as well as the z direction and define corresponding exponents a_z and a_y in the obvious way, while the shape function becomes $f(y/l_{wy}, z/l_{wz})$. There seems to be agreement that $a_z = 0.5$, while Lemberg (1973) suggests that $a_y = 0.5$ also. There is more consensus on the value of b with wind tunnel measurements and theory clustering close to $b = -1.5$.

The relatively slow decay of the velocity deficit indicated by these formulas tells us that the hill wake may be detectable many tens of hill heights downwind. Equation (5.18b) relates the strength of the wake at any point to the initial velocity deficit and this, in turn, is much larger behind hills with well-established separation bubbles than behind those without separation. The far-field behavior of the wake makes an important contribution to the estimation of the total drag force on a hill, and we will return to it when we consider distributions of turbulence moments over the hill.

5.2.4 *Effect of stability on the mean flow*

So far we have only discussed the neutral case where $N \simeq 0$. Turning now to diabatic flows we can conveniently group the effects of stability into classes based on the value of the Froude number we introduced early in this chapter and a new number $F_h(= U/Nh)$ defined using the hill height h. These classes are

1. Weakly stable: $u_*/NL_h \leq 1$; $F_L > 1$; $F_h \gg 1$.
2. Moderately stable: $F_L \leq 1$; $F_h > 1$.
3. Strongly stable: $1 \geq F_h > 0$.
4. Weakly unstable: $1 \geq -l/L > 0$, where L is the upwind Monin-Obukhov length.
5. Moderately unstable: $-l/L \simeq 1$.
6. Strongly unstable: $-l/L \gg 1$.

Note that $U = \overline{u}_0(h_m)$ is the scaling velocity used for calculating F_L and F_h.

The effects of weak and moderate stability and of weak and moderate instability can be understood using the same ideas we used to explain speedup earlier in this section, that is, through the way the perturbation pressure field responds to changes in the outer layer flow. In weakly stable through moderately unstable conditions, the pressure field changes primarily because its scaling velocity $\overline{u}_0(h_m)$ varies according to the predictions of Monin-Obukhov similarity, increasing relative to $\overline{u}_0(l)$ in stable conditions and decreasing in unstable. In moderately stable flows, the effect of buoyancy is more severe as the dynamics of the outer layer flow respond directly to buoyancy forces.

The perturbation pressure field that develops around a hill is intimately connected to the vertical velocity the hill induces. In the outer layer over low hills, this is well described by the linearized inviscid equation for $\overline{w}$, the mean vertical velocity in Cartesian coordinates. Even over steeper hills, the outer layer responds to a hill shape smoothed out by the presence of separation bubbles so the linearized equation may still be appropriate. In two dimensions, it takes the form

$$\frac{\partial^2 \overline{w}}{\partial x^2} + \frac{\partial^2 \overline{w}}{\partial z^2} + \left[\frac{N^2}{\overline{u}_0^2} - \frac{1}{\overline{u}_0} \left(\frac{\partial^2 \overline{u}_0}{\partial z^2} \right) \right] \overline{w} = 0, \qquad (5.19)$$

where we have reverted to Cartesian coordinates to write this equation in standard form. Equation (5.19) is derived in the Boussinesq approximation, which assumes that, whereas variations in density are responsible for the buoyancy term, they may be neglected in the inertial terms of momentum equations. For the amplitudes of vertical displacements encountered in boundary layer flows, the Boussinesq approximation is always valid. Its assumptions are set out clearly by Smith (1979). A slightly more complicated version of (5.19) applies in three dimensions (Smith, 1979), but the points we intend to make are not affected.

The first two terms in (5.19) are the inertial or acceleration terms, whereas the terms within the square brackets describe, respectively, the influence of buoyancy and shear (strictly, mean flow vorticity) on $\overline{w}$ and, consequently, on the pressure field. Choosing a height $z = L_h$ to place us firmly in the outer layer, noting that with a moderately stable (or unstable) logarithmic wind profile $U = \overline{u}_0(h_m) \simeq \overline{u}_0(L_h)$ and representing $d^2\overline{u}/dz^2$ by U/L_h^2, we see that the ratio of the shear to the buoyancy term in (5.19) is F_L^2.

Strong stability and instability involve more profound changes to the total flow around the hill. It is now no longer profitable to think of a pressure field, determined primarily by the outer flow, driving accelerations in an inner layer. At the same time, even over low hills, buoyancy forces produce such large perturbations in the outer layer that a linearized equation such as (5.19) is not appropriate. This will be clear as we consider the various stability classes in more detail.

Weak stability ($u_*/NL_h \leq 1$; $F_L > 1$; $F_h \gg 1$). Under weak stability conditions the perturbation pressure field changes in magnitude in response to changes in the velocity field in the outer layer. These changes can be quantified by inserting Monin-Obukhov diabatic forms into (5.12), (5.14), and (5.15), the equations specifying speedup in the inner and middle layers. In doing this it is necessary to modify not only the scaling velocity $\overline{u}_0$ but also the scaling heights l and h_m, since the formulas given for these quantities are derived with the assumption of neutral logarithmic wind profiles. The required formulas for l and h_m are given in Table 5.2.

Figure 5.9 compares speedup modified in this way with data measured at a height of 8 m on the crest of Cooper's Ridge (Coppin et al., 1986). The unpublished data in Fig. 5.9 are reproduced with their kind permission. Cooper's Ridge is one

Table 5.2 Diabatically Modified Formulas for l and h_m

Parameter	Diabatic Form[a,b]
l	$(l/L_h)[\ln(l/z_0) + (\psi_m)_{z=l}] = 2\,k^2$
h_m $(z/L > 0)$	$h_m/L_h = [\ln(h_m/z_0) + 5\,h_m/L]^{-1/2}$
h_m $(z/L < 0)$	$h_m/L_h \simeq \{(1 + 15h_m/\lvert L\rvert)^{-1/4}/[\ln(h_m/z_0) + (\psi_m)_{z=h_m}]\}^{1/2}$

[a]In determining these expressions, the values for ψ_m given in Chapter 1 have been used.
[b]L is measured upwind, but both l and h_m must be determined iteratively.

of the few data sets where boundary layer measurements have been made over a range of stability. On Cooper's Ridge, l, the inner layer depth, was 17 m, so Fig. 5.9 describes conditions in the middle of the inner layer. The Obukhov length L was measured in the undisturbed upwind profile.

On the stable side of the graph the modified formula (5.12) works reasonably well down to a value of $L \simeq 40$ m. By using the Monin-Obukhov forms for wind and temperature profiles, it is possible to express F_L as a function of z/L. When we do so we conclude that a value of $L = 40$ m corresponds to $F_L \simeq 1.5$. In stratification stronger than this the limits of weak stability are clearly exceeded, and we should not expect these simple modifications to Δs formulas to be successful.

Comparisons between the Cooper's Ridge data at the top of the inner layer and the middle layer formula (5.15) show agreement similar to that in Fig. 5.9 up to $L = 40$ m, but the formula seriously overestimates the speedup at greater stabilities. The reason is that simply modifying the scaling velocities $\overline{u}_0(l)$ and $\overline{u}_0(h_m)$, leaving the pressure distribution functions ζ and ξ unchanged, takes no account of the changes buoyancy is effecting in the outer layer flow and, consequently, in the pressure gradient that drives the speedup. We shall see what these changes are in the next section.

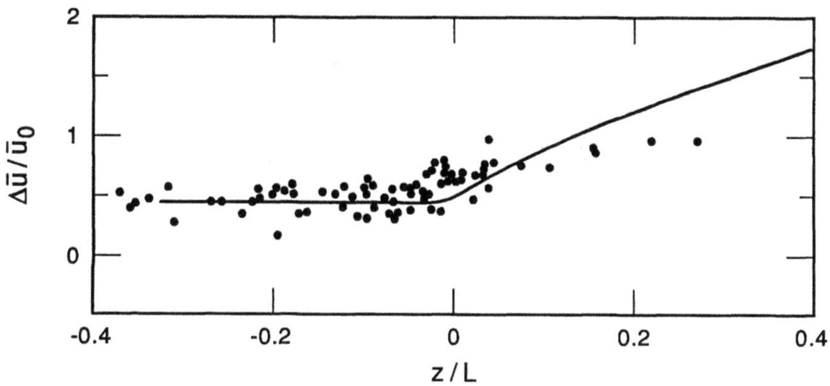

FIG. 5.9 Normalized speedup measured at $z/l = 0.5$ on the crest of Cooper's Ridge compared to (5.12) with Monin-Obukhov adjustment of the scaling velocities.

Moderate stability ($F_L \leq 1$; $F_h > 1$). We now enter a regime where few experimental data are available for comparison, but a good appreciation of the changes to be expected is provided by the analytical solution of (5.19). This can be obtained by taking $\overline{u}_0$ and N^2 to be constant in the outer layer. To match the solution to surface parameters such as z/L, we assume the constant values of $\overline{u}_0$ and N^2 are those that hold at $z = h_m$ and that below this height, Monin-Obukhov profiles apply. With these assumptions we are able to use the theory of Hunt et al. (1988a, 1988b) to calculate speedup in the inner and middle layers.

The result of doing this for a two-dimensional ridge with the Witch of Agnesi profile (5.13) and L_h, h, and z_0 values matching Cooper's Ridge are displayed in Figs. 5.10a and 5.10b. In Fig. 5.10a we plot $\Delta\overline{u}(l/3)/\overline{u}_0(h_m)$, the normalized

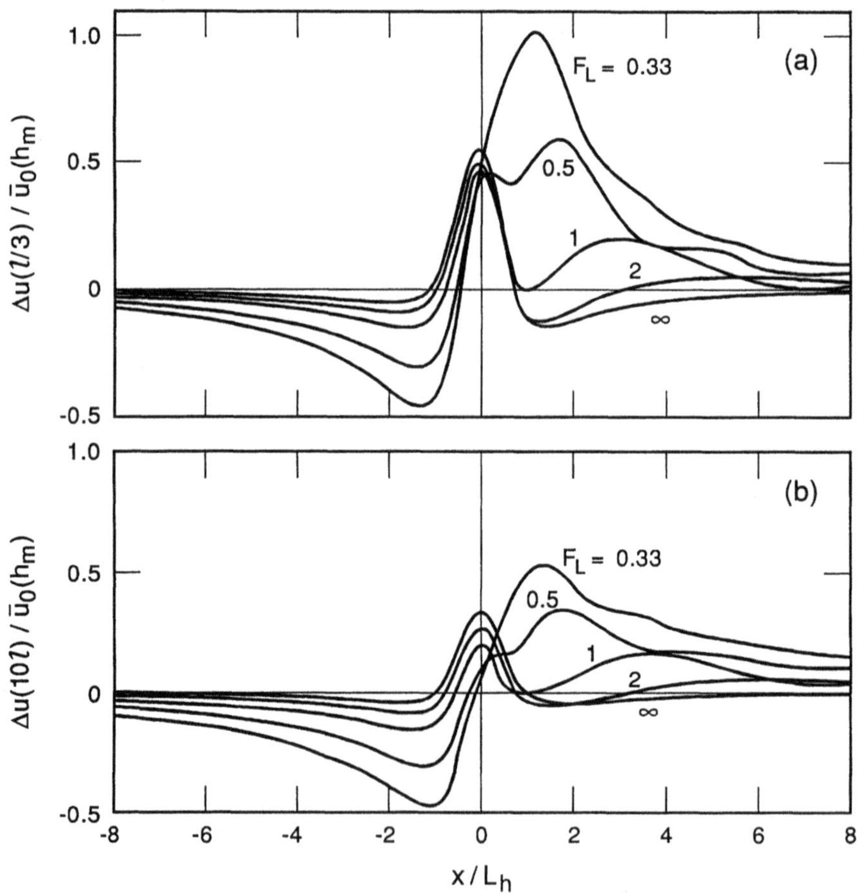

FIG. 5.10 (a) Inner layer speedup shown as a function of Froude number F_L for a two-dimensional hill with a Witch of Agnesi profile. (b) Outer layer speedup over the same hill.

velocity perturbation at a height of $z = l/3$ for a series of F_L values. This reference height was chosen so that under neutral conditions ($F_L = \infty$) on the crest ($x/L_h = 0$), $\Delta\overline{u}$ would equal $\Delta\overline{u}_{\max}$. Between $F_L = \infty$ and $F_L = 2$ little change is observed in the $\Delta\overline{u}$ profile but by $F_L = 1$, the deceleration region at the upwind foot of the hill has started to deepen, the speedup at the crest is markedly reduced, and a secondary maximum has appeared behind the hill.

With further increases in stability to $F_L = 0.5$ and $F_L = 0.33$, these features come to dominate the flow with a substantial upwind lull, much reduced speedup on the crest, and strong wind speed maxima on the downwind side of the hill. It appears that, when stability is strong, the most sheltered place to stand is in front of a hill and the windiest spot is behind it. By $F_L = 1$, also, the periodic velocity perturbations that denote lee waves have started to appear behind the hill.

Moving to the outer layer in Fig. 5.10b, essentially the same features are observed with quantitative rather than qualitative differences. The maximum in $\Delta\overline{u}$ at the crest is lost sooner, the lee waves behind the hill are more pronounced, and by $F_L = 0.33$ the upstream lull and downstream speedup are almost symmetrical in magnitude and position about the crest. Complete symmetry of those features about the hilltop is achieved in the "hydrostatic limit," which occurs when the pressure field is determined entirely by the density distribution in the outer layer, the dynamical effects of velocity variations being completely overwhelmed. It is interesting to compare the near symmetry in $\Delta\overline{u}$ in the outer layer at $F_L = 0.33$ with the strong asymmetry in the inner layer, where shear and turbulent stress exert a strong influence.

Although setting N^2 and $\overline{u}_0(z)$ constant enables us to generate analytic solutions to (5.19), these are not very realistic choices for the atmosphere. Much more common is for N^2 to be a strong function of height, either reducing sharply above a ground-based inversion at nighttime or increasing sharply through an elevated inversion, then dropping to a smaller value above the inversion during the day. It is also common to observe strong wind shear at an elevated inversion.

These situations have been investigated analytically by Hunt et al. (1988b) and numerically by Carruthers and Choularton (1982), whereas the classical approach to the structure of the outer layer flow is presented in detail by Smith (1979). These studies reveal the extreme sensitivity of the details of the flow to the vertical distribution of $l_s^2(z)$, where $l_s^2(z) = [(N^2/\overline{u}_0^2) - (1/\overline{u}_0)d^2\overline{u}_0/dz^2]$, the coefficient of $\overline{w}$ in (5.19).

For example, the existence of trapped lee waves depends on the so-called Scorer condition being fulfilled. For a Witch of Agnesi profile the condition is $l_s^2 < \pi^2/L_h^2$ aloft but $l_s^2 > \pi^2/L_h^2$ at some lower level. In practice this requires a strongly stable layer near the ground and stability decreasing or $\overline{u}_0(z)$ increasing strongly aloft. Trapped lee waves can be very energetic phenomena, producing destructive downslope winds in the lee of hills or regions of recirculating flow called rotors, which are hazardous to aircraft.

The sensitivity of the upper layer flow and pressure fields to elevated inversions is the reason why hills that occupy a significant fraction of the ABL depth are never unaffected by stability, even though the air flowing directly over them may be neutral or unstably stratified. Unfortunately, the resulting flow depends not only on the inversion strength and height but also critically on the magnitude of the stability above and below the inversion. For a detailed treatment of many of the combinations we can encounter in practice, refer to Hunt et al. (1988b).

Strong stability $(1 > F_h > 0)$. In strong stability linear approximations are no longer acceptable, even over gentle topography. Indeed, it is clear that in the linear calculations used in the last section and illustrated in Fig. 5.10, buoyancy produces large amplitude perturbations in the outer layer flow, which of themselves invalidate the linearization assumptions. We turn for guidance instead to consideration of the energy balance and to hydrostatic approximations for the pressure field.

Over small hills $(h \sim 100\,\mathrm{m}), F_h \ll 1$ usually corresponds to a strong, ground-based inversion; the limit may also be attained with weaker stability but on a much larger scale over mountain ranges. In such circumstances, the energy required to lift fluid parcels over the hill may exceed the kinetic energy available in the approach flow. These ideas led Sheppard (1956) to enunciate the dividing streamline concept. The dividing streamline separates streamlines that pass over the hill from those that go around it as we illustrate in Fig. 5.11. The height of the dividing streamline far upwind, h_s, is of great importance in pollution studies, since a contaminant plume that originates above h_s can pass over the hill, whereas a plume starting below h_s may impinge on the hill with consequent high concentrations at the surface.

Sheppard's (1956) work produced an integral formula for h_s that can be written (in Cartesian coordinates) as

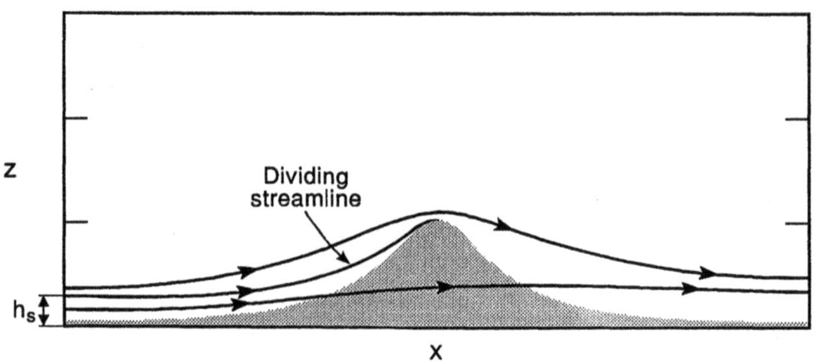

Fig. 5.11 Illustration of the dividing streamline concept. With three-dimensional hills, in the presence of a strong ground-based inversion, the flow below this streamline will go around the hill rather than over the top.

$$\frac{1}{2}\,\overline{u}_0^2(h_s) = \int_{h_s}^{h} (h - z)N^2(z)dz, \tag{5.20}$$

where the left-hand side can be interpreted as the kinetic energy of a fluid parcel located at a height h_s far upstream and the right-hand side as the potential energy the parcel gains in being lifted to a height h through the stable density gradient ρ_0/gN^2. Equation (5.20) is an implicit equation that must be solved numerically, but by making the simplifying assumptions of constant $N^2(z)$ and constant $\overline{u}_0(z)$ we obtain the explicit formula:

$$\frac{h_s}{h} = (1 - F_h). \tag{5.21}$$

Equations (5.20) and (5.21) are based on energy balance considerations only, neglecting continuity and departures from hydrostatic behavior. Smith (1990), in presenting an alternative derivation, has commented that Sheppard's emphasis on potential energy and neglect of pressure changes in determining h_s invalidates his derivation. Nevertheless, both the simple (5.21) and more complete formula (5.20) match surprisingly well to towing tank measurements on a range of hill shapes (Snyder et al., 1985) and the limited field data available (Hanna and Strimaitis, 1990). A complete discussion of the dividing streamline idea, together with a description of extensive laboratory experiments designed to test it, is provided by Snyder et al. (1985). They comment that in times of strong upwind shear (5.20) should be regarded as a lower limit on h_s.

If the fluid below a given streamline is unable to flow over the hill, it must flow around it. If the hill is a two-dimensional ridge, this cannot occur and the upwind flow is blocked below h_s. Before this situation is reached, however, a variety of interesting phenomena occur. These are inherently nonlinear but can be analyzed analytically with the simplifying assumptions of two dimensionality, constant velocity in any vertical plane, and hydrostatic pressure; the nonlinearity in the advection terms is retained. Under these circumstances the well-known shallow water flow equations (Turner, 1973) apply, where the state of flow over the hill is determined by two parameters: the upstream Froude number F_h and the ratio of hill height to inversion height. The basic categories of supercritical flow (where the inversion rises over the hilltop) and subcritical flow (where it drops) as well as the possibility of hydraulic jumps, the abrupt, turbulent transitions from one state to another, are discussed in detail by Lamb and Britter (1984). They also investigated the three-dimensional case where even the draconian assumptions of constant $\overline{u}_0(z)$ and hydrostatic pressure are insufficient to avoid a numerical treatment.

Experimental measurement of these phenomena has been confined to studies in towing tanks or flumes. They have been documented near mountain ranges such as the Sierra Nevada (Turner, 1973). But over the lower boundary layer hills we are concerned with, the shallow water equations are usually too far from reality to have more than qualitative explanatory power; in this restricted role they have proved quite useful.

Weak and moderate instabilities $(1 \gtrsim -l/L > 0)$. For this case the speedup can be calculated by inserting the Monin-Obukhov expressions for $\overline{u}_0$ into (5.12), (5.14), and (5.15) in the same way as we did for weakly stable flows. The results can be seen in Fig. 5.9, the comparison with data from Cooper's Ridge. The changes in Δs are much less for negative, unstable values of z/L than for stable stratification. This occurs because, in unstable flows, the wind shear and, therefore, the ratio of $\overline{u}_0(h_m)$ to $\overline{u}_0(l)$ decreases with increasingly negative z/L, whereas the opposite occurs in stable flow. Above the surface layer in times of unstable stratification, we approach the constant velocity (zero shear) state of the mixed layer.

Strong instability $(-l/L \gg 1)$. The very small, negative values of F_h^2 required to meet this condition are necessary before unstable buoyancy forces can play a role in the dynamics of the surface layer mean flow over the hill. With realistic values of surface heat flux, they can only occur at times of almost zero mean wind speed. Then, anabatic, upslope flows may develop on hillsides. Although this phenomenon has certainly been observed, there are at present no reliable quantitative data available.

In this hurried journey through the various stability categories we have done little more than point to the vast range of fascinating phenomena to be studied. We have concentrated on near-surface effects and on the relatively few useful formulas at hand. For most of these we have relied on the analytical, linear, asymptotic theories of Hunt et al. (1988a, 1988b). Numerical models of boundary layer flow that incorporate stability and might be appropriate to steeper hills are not yet available. We have also touched on the analytic solutions for outer layer flow as well as the shallow water approximations. These have a more venerable history than boundary layer studies. Classic papers by Queney (1948) and Scorer (1949) were among the first in a long series of distinguished contributions on lee waves [see Smith (1979) for a summary], whereas studies of the shallow water equations date back to Lord Rayleigh and Lord Kelvin.

5.3 Turbulence

The linear asymptotic models that provided such a valuable guide to the behavior of the mean flow can offer no comparable help when we consider the turbulence in the flow. To predict quantitatively the behavior of individual turbulence moments requires a much more sophisticated approach: the employment of second-order closure models (Wyngaard, 1983). As yet, there have been few applications of such models to hills; however, we shall find that the qualitative ideas they embody are indispensable in interpreting the data.

The main concepts we need are *local equilibrium, rapid distortion,* and *turbulence memory*. We have already encountered local equilibrium in several contexts in earlier chapters. This is the state of affairs where production and dissipation of turbulence are in balance, with advection and turbulent transport playing little part in the budget of turbulent stress or kinetic energy. Local equilibrium leads naturally to the representation of turbulent fluxes by eddy diffusivities. For example, in this region we should expect the momentum flux, $-\rho\overline{u'w'}$, to be proportional to the mean shear, $d\overline{u}/dz$, as it is in the surface layer over homogeneous flat ground.

Rapid distortion is the antithesis of local equilibrium. It occurs when the mean flow, from which the turbulence derives its energy, is changing too quickly for the turbulence to come into any kind of equilibrium with it. The aspects of the mean velocity field that are important in rapid distortion are the rates of strain since it is these that appear in the production terms of second-moment equations such as (1.59) or (3.40). These rates of strain are expressed conveniently in streamline coordinates as shear, $d\overline{u}/dz$, streamwise acceleration, $d\overline{u}/dx$ or $\overline{u}/L_a$, and curvature or centrifugal strain, $\overline{u}/R$.

The time it would take for turbulence to attain an equilibrium state, if the strain rates were held constant, defines the turbulence memory. It can be estimated by comparing the kinetic energy of the turbulence with the rate it is being destroyed by dissipation. As we examine the behavior of turbulence moments above a typical hill we shall encounter a region of local equilibrium at the bottom of the inner layer while rapid distortion prevails above the inner layer. We shall also find the concept of turbulence memory invaluable in interpreting the observations. These concepts will be discussed in greater detail when we consider turbulence structure later in this section. As we did for the mean flow, we begin by concentrating on neutral conditions and postpone consideration of the effects of stratification until later.

5.3.1 *Turbulent stresses*

To demonstrate the changes in turbulent second moments or Reynolds stresses we form typical ratios of the various standard deviations to their values far upwind. These ratios will be presented in two ways: as z-wise profiles at the hill crest and as x-wise profiles along three streamlines located at different heights. The vertical profiles draw on the largest body of data because upwind and hilltop measurements have been made in all experiments. The streamwise variations rely on a more limited data set; fortunately, this includes some of the most reliable measurements. Normalized vertical profiles are presented in Fig. 5.12: streamwise and lateral moments σ_u and σ_v in Fig. 5.12a, and σ_w and $-\overline{u'w'}$ in Fig. 5.12b. The

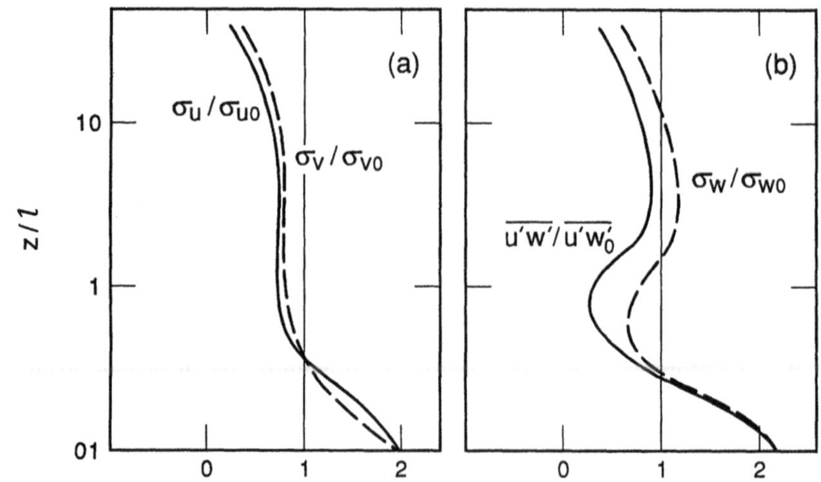

FIG. 5.12 Typical profiles of standard deviations of wind components and of shear stress at hill crest.

height scale is logarithmic to display the near-surface structure. Although these are representative profiles, they may be thought of as developing over the same model hill we introduced in Section 5.2 as the basis for Figs. 5.5 and 5.6.

We are struck immediately by the differences between the behavior of the streamwise and lateral moments and that of σ_w and $-\overline{u'w'}$. Both σ_u and σ_v, after exceeding their upwind values at the surface, fall rapidly as z increases, cross parity with upwind about $z/l = 0.3$ and remain at a fairly constant, reduced level thereafter. σ_w and $-\overline{u'w'}$, in contrast, after also exceeding their upwind levels near the surface, crossing parity at a slightly lower height than σ_u, attain a minimum value near $z/l = 1$, then recover to a secondary maximum. In the case of σ_w this once again exceeds upwind values, but $-\overline{u'w'}$ does not recover to the same extent. Sufficiently far above the surface all the moments must approach their upwind values. The precise form of the variation of the moment ratios with z/l, particularly the heights at which they cross parity with upwind ratios, depends on the detailed shape of the hill as well as its surface roughness. Nevertheless, the generic features presented here are observed in all cases with relatively minor variations.

Upwind, the moment ratios satisfy $\sigma_u : \sigma_v : \sigma_w : u_* = 2.4 : 1.9 : 1.25 : 1$ as we discussed in Chapter 1. Close to the bottom of the inner layer we can expect local equilibrium conditions, but this does not mean these ratios are preserved. In fact, the available experimental data suggest that although the ratio σ_w/u_* is essentially unchanged from upwind to hill top, σ_u/u_* and σ_v/u_* both fall. Note, however, that the very steep gradients exhibited by all the moments near the ground (a feature that is deemphasized by the logarithmic height scale of Fig. 5.12) makes unequivocal measurements of the limiting surface values a difficult undertaking. A reason for this change in the near-surface moment ratios can be found in the influence that turbulence structure aloft exerts on the horizontal moments near the surface.

We saw in Chapter 1 that σ_u and σ_v do not obey surface layer scaling but reflect instead some of the properties of the mixed layer well above the surface. This is possible because horizontal motions are not restricted by the proximity of the ground. These "extra" contributions to horizontal variance have been called inactive motions (Bradshaw, 1967) since they do not contribute to $-\overline{u'w'}$ and, hence, to momentum transport. In an analogous way, the near surface values of σ_u and σ_v reflect the reduced values of σ_u and σ_v at higher levels, and the ratios σ_u/u_* and σ_v/u_* fall in consequence.

As a final comment on these changes in the moment profiles, we observe that $u_*/u_{*0} \simeq 1.6$ whereas, for the same hill, we should expect that $\Delta s_{\max} = 2.0$. In other words, the hilltop surface friction or drag coefficient $u_*/\overline{u}(l/3)$ has fallen to 80% of its value on flat land. We return to this important point in Section 5.4.

Next, in Figs. 5.13 and 5.14, we follow the ratios along three streamlines. We choose streamlines originating at $z/l = 0.05, 0.5$, and 3.0 far upstream and denote them as I, II, and III, respectively. Streamline I is in the local equilibrium region at the bottom of the inner layer, and along I the ratios of σ_w/σ_{w0} (Fig. 5.14a) and $\overline{u'w'}/(\overline{u'w'})_0$ (Fig. 5.14b) conform closely to each other. Their evolution reflects the changes in near-surface mean shear that correspond to the upwind deceleration close to the surface, the speedup at the crest, and the deceleration in the wake. These were demonstrated in Fig. 5.6. The speedup over the crest leads to greatly increased shear below $l/3$ and a corresponding increase in σ_w and $-\overline{u'w'}$, whereas the velocity reduction in the wake leads to a decrease in the moments there. σ_u/σ_{u0} and σ_v/σ_{v0} (Figs. 5.13a and 5.13b) follow much the same pattern with differences corresponding to the inactive motion component and so reflecting their behavior on the higher streamlines.

When we move to streamlines II and III we see variations that are strikingly different from those near the surface. Beginning with σ_u along streamline II (Fig. 5.13a), we now observe a significant rise in this moment at and just behind the upwind deceleration region, a fall on the hill crest to well below upstream values, then a large increase downwind as the streamline enters the wake. This pattern is repeated along III apart from the disappearance of most of the upwind rise in the moment. We recall from Fig. 5.6 that the upwind deceleration region does not extend out to this streamline.

In Fig. 5.14a, σ_w displays even more variation. Along II this moment drops at the upwind deceleration point then climbs rapidly in the region of concave surface (and streamline) curvature on the upwind face of the hill. This is followed by a rapid fall over the hill crest, which is a region of convex surface and streamline curvature. Finally, σ_w increases again when the wake region is encountered just as

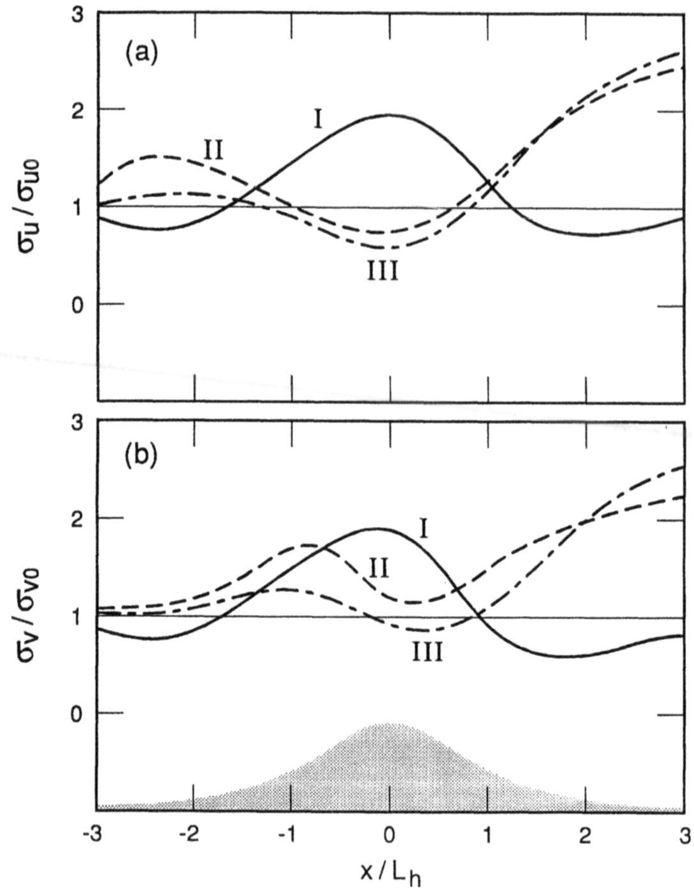

FIG. 5.13 Standard deviations of (a) the u component and (b) the v component (normalized by their reference values in the undisturbed flow) shown as they evolve along streamlines originating at three heights far upstream: I $(z/l = 0.05)$; II $(z/l = 0.5)$; III $(z/l = 3.0)$.

did σ_u. Generally, similar behavior can be seen on the higher streamline III except that the response to the deceleration region has disappeared and the reduction over the crest is now severe enough to drop the moment ratio below 1.

The behavior of $-\overline{u'w'}$ on these higher streamlines mirrors that of σ_w with decreases upwind on II, a rise on both II and III in the upslope region of concave curvature, a decrease on the crest, and a large increase in the wake (Fig. 5.14b). Changes in the σ_v ratio on streamlines II and III are shown in Fig. 5.13b. They show a neutral response to upwind deceleration, a rise in the same position as σ_w and $-\overline{u'w'}$, and a corresponding fall over the crest. The steep rise in the wake follows all the other moments.

All of the moments show some differences in their behavior over an axisym-

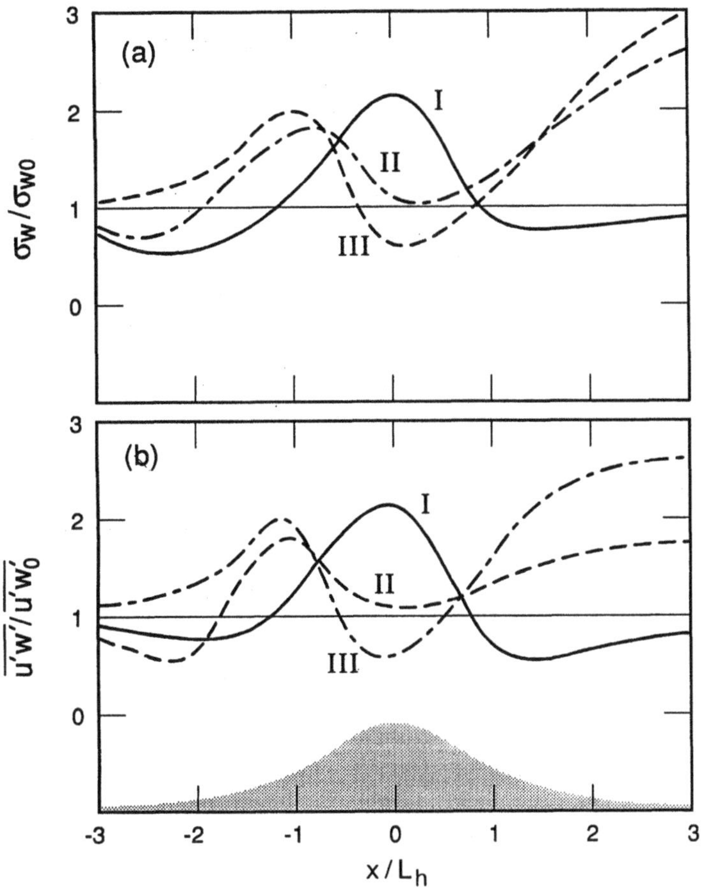

Fig. 5.14 As in Fig. 5.13 but for (a) the standard deviation of w and (b) the shear stress.

metric as opposed to a two-dimensional hill. To begin with, the upwind deceleration is absent in the axisymmetric case so all the responses we have connected with this feature disappear also. Furthermore, the lateral flow divergence that replaces this deceleration and continues on the upwind hill centerline leads to enhanced increases in σ_v and, to a smaller extent, $-\overline{u'w'}$ in the region of concave surface curvature. Behind the hill there is some evidence that σ_v values on streamlines II and III on the centerline would be much less than in the two-dimensional case (Gong and Ibbetson, 1989). This, presumably, is a consequence of the bilateral symmetry of eddies behind an axisymmetric hill as well as lateral flow convergence there; it is unlikely to be present behind less regular, three-dimensional hills.

The positions of the three streamlines were carefully chosen to lie in the near-surface, local equilibrium region, I, the middle of the inner layer, II, where we should not have strayed too far from local equilibrium, and the rapid distortion region III. Changes in the moments along these representative paths enable us to draw some general conclusions about the changes in turbulent stresses. To begin with, we have observed a near-surface region where moments respond to changes in shear, changes that may be estimated from the variation in $\overline{u}$ along the streamline through $z/l = 0.3$, the position of maximum speedup. In this region the stress ratios do not vary much as we pass over the hill. The response to flow distortion is approximately the same for each moment.

At higher levels we observed striking changes, with σ_u apparently increasing in the region of flow deceleration and decreasing over the hilltop, whereas σ_w and $-\overline{u'w'}$ increased strongly when surface and streamline curvature was concave then decreased greatly on the hilltop, possibly to well below their upwind values. The behavior of σ_v is supported by fewer experimental data than the other moments but is somewhere between σ_u and σ_w. All the moments increased strongly in the wake. Away from the surface, evidently, the moments all respond differently to flow distortion. This is the main reason why simple models are inadequate to predict turbulence over hills.

We must take care not to extend these divisions of the flow field dynamics naively into the wake region. Although it is true that local equilibrium must prevail very close to the surface behind the hill, the dynamics of the separation and wake region in general are very different from those of the attached flow upwind of the crest. A primary difference is the region of elevated shear that develops above the strong velocity deficit in the wake (Fig. 5.5). This shear has its maximum at about the hill height h and results in an inflected velocity profile for typically 10 or more hill heights downwind. The result is a pattern of mean and turbulent velocities that have many of the characteristics of a mixing layer (Wygnanski and Fiedler, 1970). For example, all the moment ratios peak at the level of maximum shear, $z = h$, attaining numerical values of 2 or 3 for the first 10 or more hill heights downwind [see, e.g., Arya and Shipman (1981), Arya and Gadiyaram (1986), Finnigan et al. (1990)]. Along with this enhancement in turbulence magnitude, turbulence length scales also increase greatly, being set by the width of the "free shear layer" rather than distance to the ground. Most of the data on the wake come from the wind tunnel for precisely this reason; the depth of the region of strong turbulence as well as the large length and time scale of the eddies place them outside the reach of tower-based field instruments.

We have not yet offered any justification for ascribing the variations in moments to deceleration in the case of σ_u and curvature and lateral divergence in the case of σ_v, σ_w, and $-\overline{u'w'}$. Nor have we explained why we expect local equilibrium near the surface and rapid distortion above. This will be the concern of the next section, where we shall introduce some different interpretations for the inner layer depth l and show how the concepts of turbulence memory and rapid distortion lead to qualitative estimates of turbulence response over arbitrary topography.

5.3.2 *Turbulence structure*

When we dealt with flow over flat, uniform terrain in Chapter 1, we were able to appeal to well-established similarity relations backed up by extensive experimental data to bring order to the turbulence moments. Over hills, no equivalent similarity laws exist except in restricted regions of the flow. For an alternative guide to the structure of the second moments we turn to their governing equations. We shall confine ourselves to a simplified treatment of the properties of these equations. You can find them discussed in much greater detail in Wyngaard (1983) and Tennekes and Lumley (1972), as well as in many other texts.

The equations for turbulent second moments describe the relationship between the mean flow and the Reynolds stresses that act upon it. Adopting the standard viewpoint, turbulent stresses or moments result from the interaction of existing turbulence with the mean flow strain rates. The stresses so produced are then advected along mean streamlines, diffused by the turbulence itself, and destroyed by the isotropizing effect of pressure fluctuations (in the case of shear stresses) or by the dissipative effect of viscosity (in the case of normal stresses). The level of turbulent stress at any point in the flow is the outcome of the balance between these four processes.

The energy-containing eddies (Chapter 2) comprise the scales of motion that interact most strongly with the mean flow. Their size, intensity, and structure, particularly their anisotropy, are determined primarily by the characteristics of the mean strains that produced them. These eddies are important because they are responsible for most of the observed stress.

Production. The mean strain field encountered over a hill can be conveniently decomposed into a superposition of basic strains. Those important in the generation of stresses appear in the production terms of the appropriate equations. In streamline coordinates on the centerline of an axisymmetric hill, these equations reduce to the following:

Turbulent kinetic energy budget:

$$\overline{u}\frac{d\overline{e}}{dx} = -\overline{u'^2}\left(\frac{\overline{u}}{L_a}\right) + \overline{v'^2}\left(\frac{\overline{u}}{L_c}\right) + \overline{w'^2}\left(\frac{\overline{u}}{L_a} + \frac{\overline{u}}{L_c}\right) - (\overline{u'w'})\left(\frac{\overline{u}}{R} + \frac{d\overline{u}}{dz}\right) + \frac{g_x}{\overline{\theta}_0}(\overline{u'\theta'}) + \frac{g_z}{\overline{\theta}_0}(\overline{w'\theta'}) + \begin{pmatrix}\text{Turbulent}\\ \text{diffusion}\end{pmatrix} + \epsilon, \tag{5.22}$$

σ_u^2 Budget:

$$\overline{u}\frac{d\overline{u'^2}}{dx} = -2\overline{u'^2}\left(\frac{\overline{u}}{L_a}\right) - 2(\overline{u'w'})\frac{d\overline{u}}{dz} + 2(\overline{u'w'})\left(\frac{\overline{u}}{R}\right) + 2\frac{g_x}{\overline{\theta}_0}(\overline{u'\theta'}) + \begin{pmatrix}\text{Turbulent}\\ \text{diffusion}\end{pmatrix} + \begin{pmatrix}\text{Pressure strain}\\ \text{interaction}\end{pmatrix} + \epsilon_u, \tag{5.23}$$

σ_v^2 Budget:

$$\overline{u}\frac{d\overline{v'^2}}{dx} = +2\overline{v'^2}\left(\frac{\overline{u}}{L_c}\right) + \begin{pmatrix}\text{Turbulent}\\ \text{diffusion}\end{pmatrix} + \begin{pmatrix}\text{Pressure strain}\\ \text{interaction}\end{pmatrix} = \epsilon_v, \tag{5.24}$$

σ_w^2 Budget:

$$\begin{aligned}\overline{u}\frac{d\overline{w'^2}}{dx} &= -4(\overline{u'w'})\left(\frac{\overline{u}}{R}\right) + 2\overline{w'^2}\left(\frac{\overline{u}}{L_a} + \frac{\overline{u}}{L_c}\right) + 2\frac{g_z}{\overline{\theta}_0}(\overline{w'\theta'}) \\ &\quad + \begin{pmatrix}\text{Turbulent}\\ \text{diffusion}\end{pmatrix} + \begin{pmatrix}\text{Pressure strain}\\ \text{interaction}\end{pmatrix} + \epsilon_w,\end{aligned} \tag{5.25}$$

$\overline{u'w'}$ Budget:

$$\begin{aligned}\overline{u}\frac{d}{dx}(\overline{u'w'}) &= -2\overline{u'^2}\left(\frac{\overline{u}}{R}\right) + \overline{w'^2}\left(\frac{\overline{u}}{R}\right) - \overline{w'^2}\frac{d\overline{u}}{dz} - (\overline{u'w'})\left(\frac{\overline{u}}{L_c}\right) \\ &\quad + \frac{g_x}{\overline{\theta}_0}(\overline{w'\theta'}) + \frac{g_z}{\overline{\theta}_0}(\overline{u'\theta'}) \\ &\quad + \begin{pmatrix}\text{Turbulent}\\ \text{diffusion}\end{pmatrix} + \begin{pmatrix}\text{Pressure strain}\\ \text{interaction}\end{pmatrix},\end{aligned} \tag{5.26}$$

where only the advection and production terms have been written out in full. g_x and g_z are the x and z components, respectively, of the acceleration due to gravity vector, and ϵ_u, ϵ_v, and ϵ_w are the viscous dissipation rates of $\overline{u'^2}$, $\overline{v'^2}$, and $\overline{w'^2}$, respectively. Pressure strain interaction describes the isotropizing effect of turbulent pressure fluctuations. This is the main destruction process for shear stresses and is responsible for transferring energy among the three normal stresses $\overline{u'^2}$, $\overline{v'^2}$, and $\overline{w'^2}$. As such, it disappears from the turbulent kinetic energy equation, where the normal stresses are summed. Turbulent diffusion includes both turbulent and pressure transport. Equations for flow over a two-dimensional ridge are recovered by setting $1/L_c$ equal to 0.

The important strain rates are shear $(d\overline{u}/dz)$, streamwise acceleration $(\overline{u}/L_a = d\overline{u}/dx)$, centrifugal acceleration or rotation $(\overline{u}/R)$, and lateral divergence $(\overline{u}/L_c)$. It is worth repeating that $1/L_c$ is not a fundamental property of the flow field as it can always be written in terms of integrals of $1/L_a$ and $1/R$, but it is more convenient to work with the simpler form. In completely general three-dimensional flows other strain rates also play important roles. The equations above demonstrate one of the most useful features of streamline coordinates: The strain rates are given relative to the mean streamlines along which the turbulence is advected. This greatly simplifies the interpretation of the production terms.

It is apparent from (5.23) through (5.26) that production of each stress depends on a different mix of strain rates and, furthermore, that the influence of an individual production rate is conditioned by the magnitude of the stress that multiplies the strain rate. For example, $\overline{u'^2}$ production responds to shear $d\overline{u}/dz$, acceleration $\overline{u}/L_a$ and curvature $\overline{u}/R$, but it is most sensitive to acceleration because $\overline{u'^2}$ is typically $-5\overline{u'w'}$. In contrast, $\overline{w'^2}$ responds to curvature, acceleration, and lateral divergence with a twofold weighting to curvature, whereas $-\overline{u'w'}$ is dominated by curvature.

Because all of the strain rates except $d\overline{u}/dz$ vanish at the surface, shear dominates at the bottom of the inner layer. The other strain rates become relatively more important as we move away from the surface and their distributions about the hill are strikingly dissimilar. This is illustrated in Fig. 5.15, where we have plotted strain rate changes along two streamlines over the archetypal hill of Fig. 5.4, the first at $z/l = 0.2$, the second at $z/l = 3.0$. All the strains are normalized with $d\overline{u}_0/dz$ on the same streamline.

Along streamline I (Fig. 5.15a), which is within the inner layer below the height of maximum speedup, we observe changes in $d\overline{u}/dz$ closely following the velocity changes shown in Fig. 5.6, whereas the other strain rates are an order of magnitude smaller. Upwind, only $d\overline{u}/dz$ is nonzero. On streamline II shown in Fig. 5.15b, the picture is quite different. Here $d\overline{u}/dz$ drops to a low value on the crest and both $\overline{u}/R$ and $\overline{u}/L_a$ are of comparable magnitude. The distribution of these extra strains about the hill is also markedly different; $\overline{u}/R$ is symmetric about the hill crest, whereas $\overline{u}/L_a$ is skew symmetric. Included in both figures is an estimate of $\overline{u}/L_c$ for the axisymmetric hill. This strain rate also is skew symmetric about the crest. One feature of the strain field not illustrated in Fig. 5.15 is the region of strong shear forming the upper boundary of the wake. A streamline starting at $z \simeq h_m$ upwind would encounter this. As the turbulent fluid is advected over the hill the strain distribution is impressed upon the production terms of (5.22) through (5.26). The anisotropy of the strains combined with the anisotropy of the production terms is a first clue to the observed distribution of stresses.

Dissipation. Viscous dissipation occurs at scales too small for changes in the mean flow to affect it, but the rate of dissipation is controlled by the rate at which energy can be passed from larger to smaller eddies in the eddy cascade process (Chapter 2). This process, which is mediated by nonlinear interaction between eddies, is not indifferent to changes in the mean strain field. The time scale characterizing both dissipation rate and the pressure-strain destruction of shear stress is the eddy lifetime or turnover time T_ϵ. We can define T_ϵ in various ways, the most appropriate one in nonequilibrium flow being

$$T_\epsilon = \frac{e}{\epsilon}. \tag{5.27}$$

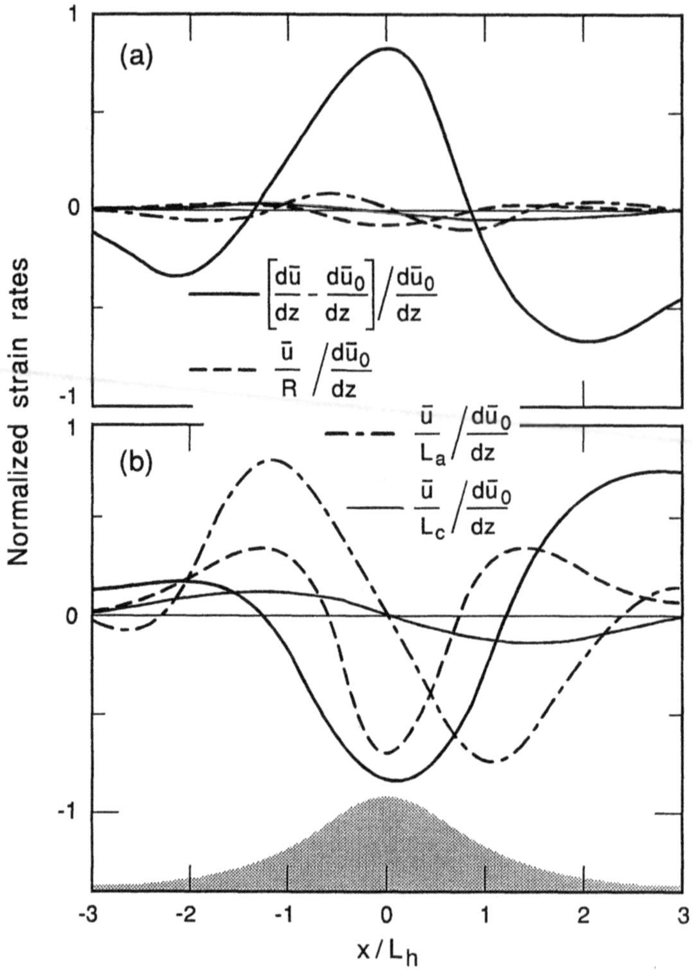

Fig. 5.15 Changes in strain rate relative to upwind shear $d\overline{u}_0/dz$. (a) Along streamline I ($z/l = 0.05$). (b) Along streamline II ($z/l = 0.5$).

If production of the turbulent kinetic energy e were switched off, T_ϵ would represent the "e-folding" time of turbulence decay. Physically it can be thought of as the time that elapses between an eddy appearing as a result of interaction with the mean flow and losing its identity through nonlinear interaction with its neighbors. Small values of T_ϵ tell us that dissipation is proceeding rapidly relative to the local level of turbulent kinetic energy.

Turbulence memory. Combining T_ϵ with the mean velocity defines l_ϵ, the average distance an eddy is advected in its lifetime:

$$l_\epsilon = \overline{u}T_\epsilon. \tag{5.28}$$

Typical values of l_ϵ on streamlines I, II, and III of Figs. 5.13 and 5.14 are plotted in Fig. 5.16. l_ϵ is normalized with L_h. L_h is a reasonable measure of the distance over which strain rates change significantly. We see that on the lower streamline I, where $z/l \simeq 0.05, l_\epsilon/L_h$ is substantially smaller than 1 so that eddies encounter essentially constant strain rates during their lifetimes; they have a short memory. On streamline II at $z/l \simeq 0.5, l_\epsilon/L_h$ hovers around 0.5, but on streamline III at $z/l \simeq 3.0, l_\epsilon/L_h$ never drops far below 2 so eddies at this level reflect a range of strains in their structure; they have a long memory.

If we make the simplifying assumption that flow over the hill is a small perturbation on a logarithmic background flow, we can use the Monin-Obukhov forms of Chapter 1 to estimate l_ϵ. We take $e \simeq 3u_*^2$ and $\epsilon = u_*^3/kz$, whence $T_\epsilon \simeq z/u_*$, and

$$\frac{z}{l_\epsilon} \ln\left(\frac{z}{z_0}\right) = k. \tag{5.29}$$

Comparing (5.29) with the expression for l, the inner layer depth given in Table 5.2, we obtain another interpretation for l as the height at which the average

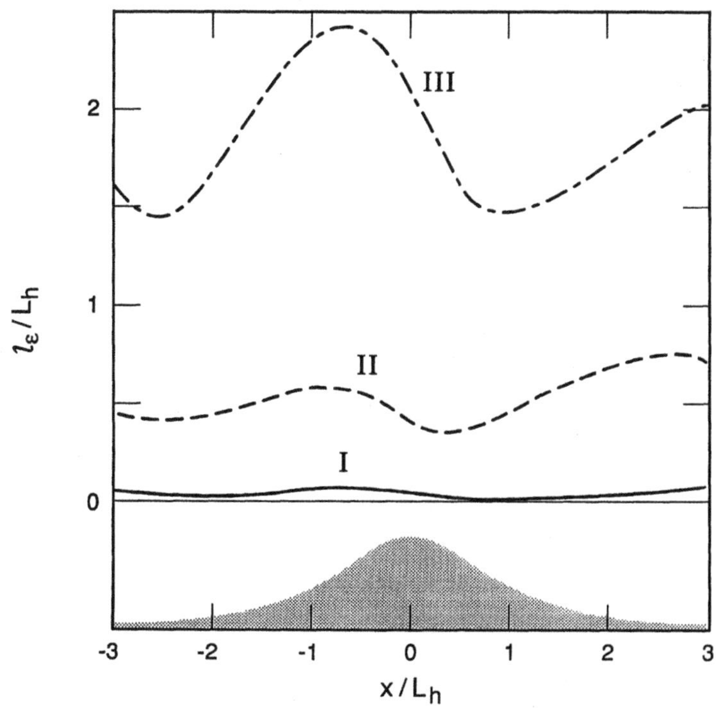

Fig. 5.16 Changes in turbulence memory l_ϵ relative to hill halflength L_h along streamlines: I ($z/l = 0.05$); II ($z/l = 0.5$); III ($z/l = 3.0$).

distance an eddy is advected in its lifetime approximately equals L_h. At heights much less than l, eddies encounter little change in the strains they are subjected to through their lifetime; when z is much larger than l, the opposite is true.

In turbulent shear flows the eddy size, as described by integral length scales (see Chapter 2), is observed to be closely proportional to l_ϵ; larger eddies are longer lived than smaller ones and are advected farther in their lifetimes. Finnigan (1988) demonstrated that this correspondence continues to hold true in the strongly advective flow over hills. If eddy size is small compared with the scale over which mean strain rates change, the turbulence will exhibit local homogeneity. Homogeneity in the x direction means that advection will be small compared to production or dissipation; homogeneity in the z direction leads to small relative values of turbulent transport. In Chapter 3 we saw that transport in plant canopies was significant precisely because this condition was violated.

Local equilibrium. We can now see why we should expect local equilibrium at the bottom of the inner layer. Eddy length scales are small there, the turbulence is relatively homogeneous so that advection and transport are small, and production must be balanced by dissipation. Conversely, well above the inner layer all these conditions are reversed. The turbulence is strongly inhomogeneous and advection is strong. In fact, a third interpretation of the top of the inner layer defines it as the height at which advection and production are in balance (Finnigan, 1988).

We saw earlier in Figs. 5.5 and 5.12 that shear and shear stress, the components of the leading production term of the turbulent kinetic energy equation (5.22), both peak at the bottom of the inner layer so this region satisfies all the requirements for local equilibrium (Townsend, 1976): Large production rates, negligible advection or transport, and dissipation necessarily balancing production.

Rapid distortion. Well above the inner layer there is no balance between production and dissipation. The large values of l_ϵ/L_h there mean that dissipation is unable to come into equilibrium with rapidly changing production rates. This is known as the rapid distortion region for two reasons. First, because the mean strain rates change rapidly compared to the eddy turnover time T_ϵ; second, because the mean flow in this region is essentially unaffected by turbulence and so turbulence moments may be calculated by the kinematic, linearized turbulence approach of rapid distortion theory (Hunt and Carruthers, 1990).

In the rapid distortion region, however, the turbulence is strongly inhomogeneous and all the basic strain rates exert their influence on production with comparable weight, producing a situation that cannot be handled easily with current, analytic rapid-distortion methods. Nevertheless, we shall find that the observed stress distribution in this region can be described qualitatively in terms of production rates, advection, and eddy lifetime. Not surprisingly, the most difficult region in which to predict the stresses is around $z = l$, where production, advection, and the nonlinear processes of pressure strain interaction and dissipation are all of comparable magnitude.

Wake turbulence and the separation region. Far downwind the wake flow relaxes to a self-preserving form and both shear stresses and mean velocity can be handled by similarity methods (Taylor, 1988). Much less tractable is the near-wake region, including the separation bubble. Despite some detailed wind tunnel investigations of this region (Arya and Gadiyaram, 1986; Arya and Shipman, 1981; Snyder and Britter, 1987), there has been little attempt to link the turbulent structure to the mean flow in a consistent way.

A unique feature of the flow field here is the point of inflection that develops in the mean velocity profile just behind the hill. If there is a separation bubble this feature begins as a free shear layer bounding the separation region and continues for at least 10 hill heights downwind. This whole region, as a result, displays turbulence characteristics similar to that of a mixing layer (Wygnanski and Fiedler, 1970) rather than a boundary layer. These include a peak in all turbulence moments at the height of maximum shear, which is at $z \simeq h$ just behind the hill and increases slowly downwind; large turbulence intensities compared with boundary layers; and turbulence length scales of the order of the mixing layer depth, that is, of order h. The large size of the turbulent eddies relative to the scale over which the strain rates vary ensures that the turbulence is strongly inhomogeneous so that turbulent transport is significant and local equilibrium does not exist. A similar situation is observed at the top of a plant canopy (Raupach et al., 1989).

The near wake behind a three-dimensional hill is more complicated. Turbulent stress levels display a marked bimodal distribution in the y direction with minima on the centerline and peaks in intensity to either side. This results from the pair of trailing, counterrotating vortices that develop behind an axisymmetric hill (Arya and Gadiyaram, 1986). Less symmetrical three-dimensional hills have even more complex vortex and turbulence distributions behind them.

If the turbulence structure of the near wake is poorly understood, that of the separation bubble is virtually terra incognita. In three dimensions in particular, the flow is so complex and unsteady and turbulence scales so large, that the instantaneous velocity field never approaches the time-mean field. In such a situation it is doubtful if the Reynolds decomposition of the flow into mean and turbulent components has other than a purely formal meaning. It seems more likely that understanding of this feature of the flow will require quite different methods to those we have used thus far.

We can now return to see how well these general considerations fit the observed changes in the stresses. We begin with streamline III of Figs. 5.13 and 5.14, which is firmly in the rapid distortion region. Here, we expect changes to be dominated by changes in production terms. Because l_ϵ/L_h exceeds 1 on this streamline, the turbulence structure should reflect a range of upwind strains weighted by their proximity to the measurement point. This is indeed the case. Referring to equation (5.23) and Fig. 5.13a, we see σ_u fall as the flow accelerates over the hill crest, because $\overline{u}/L_a$ is positive and $-2\overline{u'^2}(\overline{u}/L_a)$ is a loss term for σ_u^2, and we see σ_u recover in the wake, where the flow decelerates again.

In contrast, $-\overline{u'w'}$ along III (Fig. 5.14b) responds mainly to curvature. It rises steeply through the upwind concave region, where $\overline{u}/R$ is positive and $-2\overline{u'^2}(\overline{u}/R)$ is a gain for $-\overline{u'w'}$, then falls over the crest, where this term changes sign. A simple rule for working out the sign of R is that R is positive if the center of curvature of the streamline lies in the positive z direction. Hence R is negative over the hilltop and positive on either side as we saw in Fig. 5.15b. There is significant 'cross talk' between production terms in their response to the various strain rates. For example, the rise in $-\overline{u'w'}$ through the upwind concave region feeds into $\overline{w'^2}$ through $-4\overline{u'w'}(\overline{u}/R)$ of (5.25) and augments the response of $\overline{w'^2}$ there. A similar process occurs in reverse over the hilltop.

Variations along streamline I present few surprises. The dominant strain rate is shear, which varies as shown in Fig. 5.15. The ratios σ_u/u_*, σ_v/u_* and σ_w/u_* do not vary much (see comments made earlier in this section), and the behavior is what we might expect in a local equilibrium region from our experiments in the neutral surface layer over flat land (Chapter 1).

Much more surprising is the behavior on streamline II. This streamline is at a height of $z/l = 0.5$, and we might not have expected large departures from local equilibrium behavior. In fact, the stresses behave in essentially the same way as in the rapid distortion region. Indeed, we now see a rise in σ_u^2 following the upwind deceleration on this streamline. This rise is linked to the term $-2\overline{u'^2}(\overline{u}/L_a)$ of (5.23). Although shear is not small except just over the crest (we are above the height of $\Delta\overline{u}_{\text{max}}$), the effect of the additional strains is overwhelming.

It is clear from the above that response to curvature dominates the behavior of $-\overline{u'w'}$, whereas σ_u^2 responds predominantly to acceleration and σ_w^2 to both strains with curvature exerting the greater influence. This is true not only in the outer region but also at $z = 0.5l$. In fact, one of the most unexpected features of the data is that production terms provide a reliable qualitative guide to changes in the moments even within the inner layer, where nonlinear processes are important. Clearly pressure-strain interaction and dissipation modify but do not negate changes due to production.

It is well known that shear flows are extremely sensitive to the imposition of "extra" strain rates, particularly acceleration and curvature, the relative change in Reynolds stress typically being an order of magnitude greater than the relative change in strain (Bradshaw, 1973; Smits and Wood, 1985; Sreenivasan, 1985; Townsend, 1980). This magnified response to additional strain dictates that only at the bottom of the inner layer, well below $z/l = 0.5$, can the changes in moments be linked to changes in shear alone.

As a quick guide to the changes in second moments over arbitrary hills, we have summarized these responses in Table 5.3. This sets out the primary reaction of stresses to strain rate changes on various parts of the hill. It does not take account of the cross-talk effects we alluded to earlier, whether caused by changes in one stress feeding into another's production term or by pressure-strain transfer of energy between stresses. Despite this, it is a reliable if qualitative guide to much of the available data as long as two points are kept in mind:

1. At any point on a hill, there will be a combination of strains to consider and these must be accorded their proper weight.
2. The turbulence memory must be taken into account when deciding what strain rates may be influencing stress at a given location.

Buoyancy curvature analogy. Since shearing stress is the most important agent of cross streamline momentum transfer and thereby of the overall flow structure, it is proper to look more closely at curvature. We might reasonably expect that a numerical model of turbulence would be required to do this in such a complex flow regime. Indeed, the second-order closure model of Zeman and Jensen (1987)

Table 5.3 Reaction of Stresses to Strain Rates on Various Parts of a Hill

Strain Rate	**Formula**	**Position**	**Primary Stress Response**
Acceleration	$\frac{u}{L_a} > 0$	Upwind of crest	$\overline{u'^2}$ falls $\overline{w'^2}$ rises
Deceleration	$\frac{u}{L_a} < 0$	Upwind foot of hill on 2-d ridges; behind the hill	$\overline{u'^2}$ rises $\overline{w'^2}$ falls
Shear increases	$\left(\frac{d\bar{u}}{dz} - \frac{du_0}{dz}\right) > 0$	Upwind of crest for $z < l/3$ (excluding upwind foot of hill on 2-d ridges); outer boundary of the wake	$\overline{u'^2}$ rises $-\overline{u'w'}$ rises
Shear decreases	$\left(\frac{d\bar{u}}{dz} - \frac{du_0}{dz}\right) < 0$	Upwind hill foot near ground; above $z = l/3$ over hill crest; bottom of the wake	$\overline{u'^2}$ falls $-\overline{u'w'}$ falls
Positive curvature	$\frac{\bar{u}}{R} > 0$	Upwind slope of hill; behind hill, if no separation	$\overline{w'^2}$ rises $-\overline{u'w'}$ rises
Negative curvature	$\frac{\bar{u}}{R} < 0$	Over the hilltop	$\overline{w'^2}$ falls $-\overline{u'w'}$ falls
Lateral divergence	$\frac{\bar{u}}{L_c} > 0$	Upwind foot of axisymmetric hills	$\overline{v'^2}$ rises $-\overline{u'w'}$ increases
Lateral convergence	$\frac{\bar{u}}{L_c} < 0$	Downwind of axisymmetric hill without separation	$\overline{v'^2}$ falls

gave a convincing demonstration of the importance of including curvature in such models. Fortunately, there exists a well-developed heuristic approach to the influence of curvature on shear flows, the "buoyancy curvature analogy," that has more explanatory power.

Recognition of the analogy between centrifugal and diabatic effects in a shear flow goes back at least as far as Lord Rayleigh (Chandrasekhar, 1961), but the analogy was formalized in a useful way by Bradshaw (1969). In this approach the net effects of curvature are summarized in the curvature Richardson number R_c, which is assumed to influence the flow in the same way as the gradient Richardson number Ri of Chapter 1. In streamline coordinates R_c is defined as

$$R_c = \frac{2\overline{u}}{R\Omega_y}, \tag{5.30}$$

where $\Omega_y = (\overline{u}/R - d\overline{u}/dz)$ is the y component of vorticity. In analogy with its diabatic namesake, R_c is the ratio of centrifugal to shear forces on the fluid. R_c is also the "flux curvature Richardson number" as we see by comparing shear and curvature production terms in the turbulent kinetic energy equation (5.22). At the bottom of the hill, where streamlines are concave, R_c is negative and the flow behaves as if it were unstably stratified; over the hilltop R_c is positive and the flow appears stably stratified.

In Fig. 5.17 we show R_c on the three streamlines of Figs. 5.13 and 5.14. We recall that these streamlines had z/l values of 0.05, 0.5, and 3.0, respectively. Along streamline I, R_c never exceeds 0.1 but on the higher streamlines, particularly over the hill crest, where du/dz is small (above $z/l \simeq 0.3$) R_c reaches large positive values of 2 or 3. We can clearly expect substantial effects on flow structure if the analogy holds.

In turbulent flows the response to positive and negative R_c is qualitatively quite different (Hoffmann et al., 1985; Muck et al., 1985). In a concave, unstable region an inviscid instability mechanism promotes the growth of streamwise vortices similar to the Görtler vortices of laminar flow and cross-stream momentum, and scalar transfer is greatly increased. The effect of convex, stabilizing curvature (positive R_c) is much closer to true diabatic stability. Its immediate manifestation is a reduction in the extent of fluid particle movement in the z direction. This damping of cross-stream movement is reflected immediately in integral length scales of cross-stream fluctuations (Chapter 2) as is the earlier increased motion in the unstable concave region (Finnigan, 1988).

The buoyancy curvature analogy is a useful guide to turbulence response throughout the flow regime; an equivalent heuristic picture of the effect of acceleration is only available in the local equilibrium region. There it combines with the buoyancy curvature analogy to modify Monin-Obukhov similarity in a consistent way so we can apply it over topography. This will be the subject of Section 5.4.

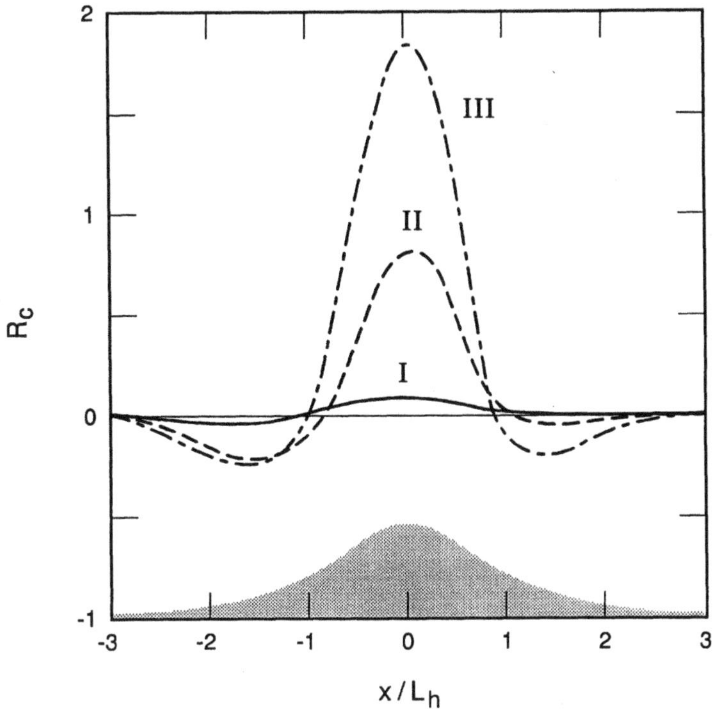

FIG. 5.17 Curvature Richardson number R_c along three streamlines: I ($z/l = 0.05$); II ($z/l = 0.5$); III ($z/l = 3.0$).

More complete accounts of the effects of curvature and acceleration and their relationship to rapid distortion can be found in Finnigan (1988) and Finnigan et al. (1990).

5.3.3 *Spectra and integral scales*

In the last section we made passing mention of the behavior of various integral scales over the hill. Unfortunately, very few spectra have been published to date and it is difficult to draw firm conclusions about the way the spectral forms of Chapter 2 might be modified. Hilltop and upwind spectra have been presented by Mason and Sykes (1979) and Mason (1986), whereas information about changes in various integral scales over a wind tunnel model and some real hills can be found in Finnigan (1988).

5.3.4 *Effect of stability*

A natural classification arises when we compare turbulence generation by buoyancy with that caused by strain rates induced by the presence of the hill. We can characterize the first by the flux Richardson number, R_f, and the second by the curvature Richardson number, R_c. R_f is not easily expressed in terms of simple scales but we noted in Chapter 1 that

$$R_f = \frac{K_h}{K_m} Ri, \tag{5.31}$$

and for $z/L > 0$; $K_h/K_m = 1$

$z/L < 0$; K_h/K_m is an increasing function of $-z/L$ with $K_h/K_m = 2.4$ at $z/L = -2$.

These relationships are only exact where Monin-Obukhov theory holds, but they are indicative of the relative values of R_f and Ri over a much wider range.

If we now take h as a characteristic length scale for shear outside the inner layer, we can write

$$\frac{1}{F_h^2} = \frac{(g/\overline{\theta})(\partial\overline{\theta}/\partial z)}{(U/h)^2} \simeq Ri. \tag{5.32}$$

Outside the inner layer, we saw in Fig. 5.17 that $R_c \sim 1$ and we can ask, under what conditions can $|R_f|$ attain comparable values? The answer can be summarized as follows:

$R_f = 1$ corresponds to $F_h \simeq 1$, which implies strong stability.
$R_f = -1$ corresponds to $F_h^2 \simeq -2$, which implies moderate instability.

(Note that $Ri = 1$ for $R_f = 1$ and $Ri \approx -0.5$ for $R_f = -1$ in Fig. 1.8, Chapter 1.) Here we have used the classifications introduced in Section 5.2 to provide a link with the influence of stability on the mean flow.

Within the inner layer the appropriate scale for shear is l and we should write $R_f \simeq 1/F_l^2$, where $F_l = U/Nl$ is still another Froude number, in this case based on l. There, the typical value of R_c is 0.1 (Fig. 5.17). However, F_l is roughly equal to $0.1F_h$ so that within the inner layer too, strong stability or instability is necessary for diabatic effects to overwhelm hill-induced flow distortions in the production of turbulence.

We have little empirical information on the behavior of turbulence in times of strong stability. On a large scale, energetic turbulence may be generated in hydraulic jumps, although these are uncommon features of boundary layer hill flows. Another possibility is generation of turbulence by lee waves, particularly if the lee waves break. Such extreme occurrences are sometimes reported by aviators in the lee of substantial hills. On the unstable side we enter the realm of anabatic flow and the approach to free convection. No quantitative data appear to describe this limit. Most interest, therefore, centers on the regime of weak stability and instability where turbulence moments respond to changes in the mean strain rates; these in turn result from changes in upwind profiles as Monin-Obukhov theory predicts and as we saw in Section 5.2.

It is natural to ask whether the Obukhov length continues to play a useful role in characterizing local diabatic effects on a hill. First, we recognize that in streamline (or surface-following) coordinates, the hill slope naturally induces two Obukhov lengths: L_x and L_z, corresponding to the x and z components of the gravity vector. Similarly, the reference potential temperature $\overline{\theta}_0$ of (1.24) must be replaced by the position-dependent $\overline{\theta}_0(x)$, as we pointed out in Section 5.1. The biggest obstacle to the usefulness of z/L is, however, placed by the assumptions implicit in the derivation of Monin-Obukhov similarity itself. These are the use of u_* as the velocity scale characterizing cross-stream momentum transfer and the adoption of "moving equilibrium" as the foundation of the asymptotic matching process that leads to the logarithmic law. Moving equilibrium states that only the local values of the governing length scales, z_0 and L, are important and not their upstream histories (Yaglom, 1979). Finnigan (1992) has shown that over hills, moving equilibrium can hold only at the bottom of the inner layer, that is, in the local equilibrium region. The steep gradients of shearing stress near the surface (Fig. 5.12b) confine the use of u_* as a velocity scale to the same thin layer.

This restricts application of Monin-Obukhov scaling and the use of z/L to $z \lesssim l/3$. The relationship between l and L_h (Table 5.2) ensures that this region is only of consequential depth on long shallow hills. On such hills also, $g_z \gg g_x$ so that, if L is used, we may ignore L_x and set $L = L_z$. A way of combining the use of L with L_a and R_c to provide a complete characterization of turbulence in the inner layer is suggested in the next section.

5.4 Local similarity over hills

In a local equilibrium flow we expect to describe fluxes by eddy diffusivities such as those implicit in the Monin-Obukhov formulas of Chapters 1 and 3. The failure of flux-gradient relationships, when local equilibrium does not hold, was seen in plant canopies in Chapter 3. Monin-Obukhov or local similarity theory lies at the root of most descriptions of the surface layer as we saw in Chapters 1 and 2. The logarithmic law, its most famous prediction, links surface fluxes of momentum, heat, and other conserved scalars to their mean vertical gradients. The ability to deduce these fluxes from simpler measurements of the mean quantities is an experimental tool whose power is difficult to overestimate. It would be exceedingly valuable if we could apply this technique on a hill, because it would enable surface fluxes to be inferred without making turbulence measurements in the steep, near-surface gradients of eddy flux that are encountered there. (See, e.g., Fig. 5.12b.)

A method of extending local similarity to topography has been proposed by Finnigan (1992). It relies upon the natural characterization of flow distortion provided by the parameters R and L_a of streamline coordinates. In this scheme the Monin-Obukhov relationship in Chapter 1 for momentum (1.26) becomes

$$\frac{d\overline{u}}{dz} = \left(\frac{u_*}{kz}\right)\phi_m(z/L, z/L_a, R_c). \tag{5.33}$$

The dependence on R_c is the formal expression of the buoyancy curvature analogy, and the inclusion of z/L_a, another dimensionless parameter, introduces a buoyancy acceleration analogy. We discussed briefly the interpretation of z/L in Section 5.3. Equation (5.33) shows how diabatic effects combine with flow distortion in this formulation; for the rest of this section, however, we will restrict ourselves to the neutral case for clarity. Hence, (5.33) becomes

$$\frac{d\overline{u}}{dz} = \left(\frac{u_*}{kz}\right)\phi_m(z/L_a, R_c). \tag{5.34}$$

The local equilibrium region is restricted to $z/l \lesssim 0.3$. Both L_a and R are always much larger than l and, in the near wall region, $R_c \simeq (\overline{u}/u_*)(z/R)$ so that the numerical values of the arguments of (5.34) in its domain of applicability are much less than 1. As a result, the unknown function ϕ_m may be replaced by the leading terms in its Taylor series expansion about $\phi_m(0,0)$:

$$\frac{d\overline{u}}{dz} = \frac{u_*}{kz}\left(1 + \alpha\frac{z}{L_a} + \beta R_c\right), \tag{5.35}$$

where

$$\alpha = \left(\frac{d\phi_m}{d(z/L_a)}\right)_{z/L_a=0} \qquad \text{and} \qquad \beta = \left(\frac{d\phi_m}{dR_c}\right)_{R_c=0}$$

are constants.

Comparison with an extensive body of wind tunnel data (Finnigan, 1992) has produced values for β:

$$\begin{aligned} \beta &= 4.0 \quad \text{for} \quad R_c < 0, \\ \beta &= 5.0 \quad \text{for} \quad R_c \geq 0. \end{aligned} \tag{5.36a}$$

The values above closely match those given in the Monin-Obukhov formula (1.31) (with R_c substituted for z/L), confirming the buoyancy curvature analogy in a quantitative way.

Comparison with a much smaller body of data has produced a value for α in the accelerated flow

$$\alpha = -5.0 \qquad \text{for} \qquad z/L_a > 0. \tag{5.36b}$$

The problem with fixing α is that strongly decelerated boundary layers separate so that substantial negative values of z/L_a are unobtainable. On the other hand, strongly accelerated, smooth wall boundary layers tend to relaminarize and data from accelerated rough walls are sparse indeed. Separating boundary layers, however, have no local equilibrium regions, so it is a reasonable step to take α to be continuous through $z/L_a = 0$ and use the value given in (5.36b) for the small region of the decelerated domain where the theory is applicable.

Equation (5.35) may be integrated to obtain a modified logarithmic law:

$$\overline{u}(z) = \frac{u_*}{k}\left(\ln\frac{z}{z_0} + \alpha\frac{z}{L_a} + \hat{\beta}R_c\right), \tag{5.37}$$

where $\hat{\beta}$ is the leading coefficient in the expansion of $\int \phi_m dz$ about $R_c = 0$. In this form it has been applied to velocity profiles on the top of Cooper's Ridge (Coppin et al. 1986), Nyland's Hill (Mason, 1986), and on the upwind face of Askervein (Bowen and Teunissen, 1986). The results obtained are encouragingly consistent, but it must be emphasized that a good deal more experimental confirmation of the theory is necessary. The main predictions of the modified logarithmic law are summarized below:

1. Diabatically stable, logarithmic profiles matching (5.35) should be observed on hilltops. This has been shown to be true in a small set of examples.
2. Diabatically unstable profiles matching (5.35) should be observed on the upwind slopes of hills. This has been shown to be true in an even smaller number of cases.
3. A substantial reduction in the surface drag coefficient $u_*/\overline{u}(z < l/3)$ relative to a flat surface of the same roughness should be observed on the hilltop and an increase in $u_*/\overline{u}$ in the upwind unstable region. This has been observed in several wind tunnel simulations.
4. A further prediction, not treated in detail above, is that no local equilibrium and logarithmic region will exist over a hill covered with tall roughness such as a forest canopy. This effect has been noted in wind tunnel models where it forms a major obstacle to reproducing inner layer behavior (Finnigan et al., 1990).

The small number of cases for which the theory has been tested so far reflects the difficulty of obtaining field data with the required spatial resolution to provide not only well-determined profiles in the z direction but also unequivocal values of z/L_a and R_c.

5.5 Scalar fields and fluxes

We have had little or nothing to say about the behavior of scalar quantities over hills; this is a fair reflection of the amount of experimental data available. Whereas some measurements have been made of mean temperature and heat flux over Bungedore Ridge and Cooper's Ridge (E.F. Bradley and P.A. Coppin, personal communication), they are incomplete and do little more than suggest yet one more set of interesting phenomena to study. Our most reliable guide at this stage is the theoretical work of Raupach et al. (1992) who have extended the linearized, small perturbation model of Hunt et al. (1988a) to scalar transport.

The conclusions of their work can be briefly summarized as follows:

1. Substantial changes in scalar fields and fluxes over low hills are confined to the inner layer.
2. Within the limits of linear theory, these effects can be split into three additive components associated, respectively, with streamline convergence and divergence, stress gradients above the surface, and changes in surface conditions, specifically surface flux and surface stress. Of these three components, the first is usually dominant close to the surface whereas the last is dominant above $z = l$.

This chapter has taken us on a rapid trip through the world of boundary layer flow over hills. Despite its relatively superficial treatment of so many topics and its restriction, primarily, to turbulent flow over low hills, it has proved to be one of the longer chapters in this volume. This emphasizes the vast range of this topic. Indeed, from a hilltop perspective, it seems fair to regard boundary layer processes over flat land as merely limiting cases of flow over topography. It is an exciting field to be in at the moment, with many more questions than answers and unlimited scope for innovative theory and wind tunnel modeling.

5.6 Special symbols

a, b	exponents in power laws for l_w and $\overline{u}_w$
a_z, a_y	exponents in the power laws for the parameters for 3-d wakes
c	constant (~ 1) in the definition of inner layer depth l
F_l	Froude number based on the inner layer depth
F_L	Froude number based on hill dimension L_h
F_h	Froude number based on hill height h

g_x, g_y, g_z	components of g in the x, y, z directions
h	height of hill
h_m	height above which wind shear is negligible
h_s	height of streamline dividing those that go over the hill and those that go around it in strongly stable air
l	depth of the inner layer
l_s^2	coefficient in linearized inviscid equation for $\overline{w}$ (in two dimensions)
l_w	wake depth
l_ϵ	distance eddy is advected in its lifetime
L_a	acceleration length scale that is identically the local radius of curvature of the z coordinate lines in two-dimensional flow
L_b, L_c	length scales characterizing departure of flow from two dimensionality
L_h	half-width of hill at half hill height
L_x, L_z	Monin-Obukhov lengths corresponding to g_x and g_z
r	local radius of curvature of the y lines
R	local radius of curvature of the x lines
$\hat{R}$	radius of curvature of z lines
R_c	curvature Richardson number
Δs	dimensionless increase in wind speed on the hilltop
Δs_{max}	value of Δs, where $\Delta\overline{u}$ is maximum
$\overline{u}_0(z)$	reference velocity profile in the undisturbed upwind flow
$\overline{u}_0(h_m)$	reference velocity at $z = h_m$
$\overline{u}_0(l)$	reference velocity at $z = l$
$\overline{u}_w$	wake velocity defect
$\Delta\overline{u}$	velocity speedup
U	characteristic velocity in the boundary layer
α, β	coefficients in Taylor series expansion of ϕ_m
δ	scale size in the theory of Hunt et al. (1988a)
$\epsilon_u, \epsilon_v, \epsilon_w$	dissipation rates for u, v, w
$\zeta(x)$	hill shape function in inner layer speedup formula
$\overline{\theta}_0$	upwind reference potential temperature, function of x
ρ_0	upwind air density
$\xi(x)$	hill shape function in the middle layer speedup formula

References

Arya, S.P.S., and P.S. Gadiyaram, 1986: An experimental study of flow and dispersion in the wakes of three-dimensional low hills. *Atmos. Environ., 20*, 729–740.

Arya, S.P.S., and M.S. Shipman, 1981: An experimental investigation of flow and diffusion in the disturbed boundary layer over a ridge. *Atmos. Environ., 15*, 1173–1184.

Bowen, A.J., and H.W. Teunissen, 1986: The Askervein Hill Project—Research Report No. 86-7. Wind tunnel simulation at length scale 1:2500. Part 1: Tests at the Atmospheric Environment Service, Environment Canada, Downsview, Ontario, 120 pp.

Bradshaw, P., 1967: "Inactive" motion and pressure fluctuations in turbulent boundary layers. *J. Fluid Mech., 30*, 241–258.

Bradshaw, P., 1969: The analogy between streamline curvature and buoyancy in turbulent shear flow. *J. Fluid Mech., 36*, 177–191.

Bradshaw, P., 1973: Effect of streamline curvature on turbulent flow. AGARDograph 169, AGARD (NATO), Neuilly-sur-Seine, France, 80pp.

Britter, R.E., J.C.R. Hunt, and K.J. Richards, 1981: Airflow over a two-dimensional hill: studies of velocity speedup, roughness effects and turbulence. *Quart. J. Roy. Meteor. Soc., 107*, 91–110.

Carruthers, D.J., and T.W. Choularton, 1982: Airflow over hills of moderate slope. *Quart. J. Roy. Meteor. Soc., 108*, 603–624.

Castro, I.P., and W.H. Snyder, 1982: A wind tunnel study of dispersion from sources downwind of three-dimensional hills. *Atmos. Environ., 16*, 1869–1887.

Chandrasekhar, S., 1961: *Hydrodynamic and Hydromagnetic Stability*. Dover Publications, Inc., New York, 654 pp.

Coppin, P.A., E.F. Bradley, and P.C. Katen, 1986: The evolution of atmospheric turbulence over a two-dimensional ridge. *Proc., 9th Australasian Fluid Mechanics Conf.*, 8-12 December 1986, Auckland, New Zealand, 569–572.

Finnigan, J.J., 1983: A streamline coordinate system for distorted turbulent shear flows. *J. Fluid Mech., 130*, 241–258.

Finnigan, J.J., 1988: Air flow over complex terrain. In *Flow and Transport in the Natural Environment: Advances and Applications*. (W.L. Steffen and O.T. Denmead, Eds.), Springer-Verlag, Berlin, Germany, 183–229

Finnigan, J.J., 1990: Streamline coordinates, moving frames, chaos and integrability in fluid flow. In *Topological Fluid Mechanics* (H.K. Moffatt and A. Tsinober, Eds.), *Proc., IUTAM Symposium*, 13-18 August 1989, Cambridge, U.K., 64–74.

Finnigan, J.J., 1992: The logarithmic wind profile in complex terrain. CSIRO Environmental Mechanics Technical Report No. T44, CSIRO, Canberra, Australia, 69 pp.

Finnigan, J.J., M.R. Raupach, E.F. Bradley, and G.K. Aldiss, 1990: A wind tunnel-study of turbulent flow over a two-dimensional ridge. *Bound.-Layer Meteor., 50*, 277–317.

Gong, W., and A. Ibbetson, 1989: A wind tunnel study of turbulent flow over model hills. *Bound.-Layer Meteor., 49*, 113–148.

Hanna, S.R., and D.G. Strimaitis, 1990: Rugged terrain effects on diffusion. In *Atmospheric Processes Over Complex Terrain. Meteorological Monographs, 35* (45), American Meteorological Society, Boston, MA, 109–142.

Hoffman, P.H., K.C. Muck, and P. Bradshaw, 1985: The effect of concave surface curvature on turbulent boundary layers. *J. Fluid Mech., 161*, 371–403.

Howarth, L., 1951: The boundary layer in three-dimensional flow. *Phil. Mag., 42*, 239–244.

Hunt, J.C.R., and D.J. Carruthers, 1990: Rapid distortion theory and the 'problems' of turbulence. *J. Fluid Mech., 212*, 497–532.

Hunt, J.C.R., S. Leibovich, and K.J. Richards, 1988a: Turbulent shear flow over low hills. *Quart. J. Roy. Meteor. Soc., 114*, 1435–1470.

Hunt, J.C.R., K.J. Richards, and P.W.M. Brighton, 1988b: Stably stratified shear flow over low hills. *Quart. J. Roy. Meteor. Soc., 114*, 859–886.

Hunt, J.C.R., C.J. Abell, J.A. Peterka, and H. Woo, 1978: Kinematical studies of the flow around free or surface-mounted obstacles: Applying topology to flow visualization. *J. Fluid Mech., 86*, 179–200.

Jackson, P.S., and J.C.R. Hunt, 1975. Turbulent wind flow over a low hill. *Quart. J. Roy. Meteor. Soc., 101*, 929–955.

Lamb, V.R., and R.E. Britter, 1984: Shallow flow over an isolated obstacle. *J. Fluid Mech., 147*, 291–313.

Lemberg, R., 1973. On the wake behind bluff bodies in a turbulent boundary layer. *Research Report BLWT-3-73*. Faculty of Engineering Sciences, University of Western Ontario, Canada, 163 pp.

Lighthill, J., 1989. *An Informal Introduction to Fluid Mechanics*. IMA Monograph Series 2, Clarendon Press, Oxford, U.K., 260 pp.

Mason, P.J., 1986: Flow over the summit of an isolated hill. *Bound.-Layer Meteor., 37*, 383–405.

Mason, P.J., and R.I. Sykes, 1979: Flow over an isolated hill of moderate slope. *Quart. J. Roy. Meteor. Soc., 105*, 383–395.

Morton, B.R., 1984: The generation and decay of vorticity. *Geophys. Astrophys. Fluid Dyn., 28*, 277–308.

Muck, K.C., P.H. Hoffmann, and P. Bradshaw, 1985: The effect of convex surface curvature on turbulent boundary layers. *J. Fluid Mech., 161*, 347–369.

Queney, P., 1948: The problem of air flow over mountains: a summary of theoretical studies. *Bull. Amer. Meteor. Soc., 29*, 16–26.

Raupach, M.R., J.J. Finnigan, and Y. Brunet, 1989: Coherent eddies in vegetation canopies. *Fourth Australasian Conference on Heat and Mass Transfer*, Christchurch, New Zealand, 9–12 May 1989, 75–90.

Raupach, M.R., W.S. Weng, D.J. Carruthers, and J.C.R. Hunt, 1992: Temperature and humidity fields and fluxes over low hills. *Quart. J. Roy. Meteor. Soc., 118*, 191–225.

Schlichting, H., 1968: *Boundary Layer Theory*. McGraw-Hill, New York, 747 pp.

Scorer, R.S., 1949: Theory of waves in the lee of mountains. *Quart. J. Roy. Meteor. Soc., 75*, 41–56.

Sheppard, P.A., 1956: Airflow over mountains. *Quart. J. Roy. Meteor. Soc., 82*, 528–529.

Smith, R.B., 1990: Why can't stably stratified air rise over high ground? In *Atmospheric Processes Over Complex Terrain. Meteorological Monographs, 35* (45), American Meteorological Society, Boston, MA, 105–107.

Smith, R.B., 1979: The influence of mountains on the atmosphere. *Adv. in Geophys., 21*, 87–230.

Smits, A.J., and D.H. Wood, 1985: The response of turbulent boundary layers to sudden perturbations. *Ann. Rev. Fluid Mech., 17*, 321–358.

Snyder, W.H., and R.E. Britter, 1987: A wind tunnel study of the flow structure and dispersion from sources upwind of three-dimensional hills. *Atmos. Environ., 21*, 735–751.

Snyder, W.H., R.S. Thompson, R.E. Eskridge, R.E. Lawson, L.P. Castro, J.T. Lee, J.C.R. Hunt, and Y. Ogawa, 1985: The structure of strongly stratified flow over hills: Dividing streamline concept. *J. Fluid Mech., 152*, 249–288.

Sreenivasan, K.R., 1985: The effect of contraction on a homogeneous turbulent shear flow. *J. Fluid Mech., 154*, 187–213.

Taylor, P.A., 1988: Turbulent wakes in the atmospheric boundary layer. In *Flow and Transport in the Natural Environment: Advances and Applications*. (W.L. Steffen and O.T. Denmead, Eds.), Springer-Verlag, Berlin, 270–292.

Taylor, P.A., and R.J. Lee, 1987: Simple guidelines for estimating wind speed variations due to small scale topographic features. *Climatol. Bull., 18* (2), 3–23.

Taylor, P.A., P.J. Mason, and E.F. Bradley, 1987: Boundary layer flow over low hills. *Bound.-Layer Meteor., 39*, 107–132.

Taylor, P.A., and H.W. Teunissen, 1987: The Askervein Hill project: overview and background data. *Bound.-Layer Meteor., 39*, 15–39.

Tennekes, H., and J.L. Lumley, 1972: *A First Course in Turbulence.* M.I.T. Press, Cambridge, MA, 300 pp.

Tobak, M., and D.J. Peake, 1982: Topology of three-dimensional separated flows. *Ann. Rev. Fluid Mech., 14*, 61–85.

Townsend, A.A., 1976: *The Structure of Turbulent Shear Flow* (2nd ed.). Cambridge University Press, Cambridge, U.K., 429 pp.

Townsend, A.A., 1980: The response of sheared turbulence to additional distortion. *J. Fluid Mech., 98*, 171–191.

Turner, J.S., 1973: *Buoyancy Effects in Fluids*, Cambridge Monographs in Mechanics and Applied Mathematics. Cambridge University Press, Cambridge, U.K., 367 pp.

Wygnanski, I., and H.E. Fiedler, 1970: The two-dimensional mixing region. *J. Fluid Mech., 41*, 327–361.

Wyngaard, J.C., 1983: Boundary-Layer Modelling. In *Atmospheric Turbulence and Air Pollution Modelling* (F.T.M. Nieuwstadt and H. Van Dop, Eds.). Reidel, Hingham, MA, 69–106.

Yaglom, A.M., 1979: Similarity laws for constant pressure and pressure gradient turbulent wall flows. *Ann. Rev. Fluid Mech., 11*, 505–540.

Zeman, O., and N.O. Jensen, 1987: Modification of turbulence characteristics in flow over hills. *Quart. J. Roy. Meteor. Soc., 113*, 55–80.

6

SENSORS AND TECHNIQUES FOR OBSERVING THE BOUNDARY LAYER

Sensors used for boundary layer measurements fall into two broad categories: in situ sensors that can be mounted on the ground, on masts, towers, tethered balloons, free balloons, or aircraft; and remote sensors, ground-based or aircraft-mounted, that infer atmospheric properties through their effects on acoustic, microwave, and optical signals propagating through the air. In situ sensors are the traditional instruments of choice for surface and lower boundary layer studies, being the only ones capable of the accuracy and resolution needed for quantitative work. A major portion of this chapter will therefore be devoted to discussions of their characteristics. Remote sensors have the advantage of increased range and spatial scanning capability, but the constraints on minimum range and spatial resolution limit their usefulness for surface layer measurements. Used in combination, however, the two types of sensors provide a more complete description of the flow field being studied than either of the two can provide separately. New remote sensors with shorter minimum ranges and finer range resolutions are now becoming available for boundary layer applications. A brief discussion of such devices is also included in this chapter.

The variables of greatest interest to boundary layer meteorologists are wind speed, temperature, humidity, and the fluxes of momentum, heat, mass, and radiant energy. Given suitable fast-response measurements of wind velocity and scalar fluctuations, we can calculate the eddy fluxes directly from the products of their fluctuating components (e.g., $\overline{u'w'}, \overline{w'\theta'}$) as explained in Chapter 1. If only the gradients of their means are available, however, then over a flat homogeneous surface the fluxes may be inferred from the Monin-Obukhov relationships of Chapters 1 and 3. Practical methods for doing that are described in many texts; see, for example, Monteith (1975, 1976). (Those simple relationships do not hold, as we know, under advective conditions, in plant canopies, and over hills.) There are also sensors in use that measure surface and near-surface fluxes directly, such as the drag plate (surface stress), the lysimeter (latent heat flux), flux plates (soil heat flux), and radiometers (radiant heat flux). We will discuss these and a few other types as well because of their application to studies of plant canopies.

We begin our survey with in situ sensors that measure the mean and fluctuating properties of air above the ground. The requirements for measurements of the two are different. Sensors for measuring the time-averaged properties of the flow require a high degree of absolute accuracy and long-term stability and should be able to provide good averages over periods 10 min and longer. Sensors for measuring the turbulent properties of the flow need a frequency response broad enough to cover all the eddy scales contributing to the process, but the accuracy need only be relative, that is, with respect to any arbitrary working mean. For measuring both means and fluctuations, the accuracy requirements are the same: ± 0.05 m s^{-1} for wind velocity, ± 0.05°C for temperature, and ± 0.05 mb for humidity. The response time for mean sensors can be as large as 1 min; but for fluctuation sensors it should be no longer than 0.1 s. These requirements are based on our present understanding of boundary layer structure over uniform terrain. In the case of technologies that are evolving (e.g., trace gas measurements), however, the accuracies and response times are what the state of the art can offer.

6.1 In situ sensors for profile measurements

6.1.1 *Mean wind sensors*

Cup anemometers, propeller anemometers, and vanes are the devices most commonly used to measure wind profiles. They are rugged, dependable, and relatively inexpensive. They depend on moving parts coming into dynamic equilibrium with the flow, so their response times are typically too long for turbulence work. But they are entirely adequate for mean profile measurements. Periodic calibrations and intercomparison checks are essential to ensure accuracy in the vertical gradient estimations.

Cup and vane systems are marketed in a variety of shapes and forms. The most common configuration is a three-cup anemometer and direction vane mounted side by side at the ends of a T-shaped horizontal boom. Both rotate on vertical axes, so they are separated horizontally to avoid mutual interference. The advantage of the cup anemometer is that it can accept winds from any direction. (Only the directions blocked by the wind vane and the supporting mast would be considered unfavorable.) For wind vanes with potentiometer systems that read vane position, the gap in the resistance element, typically a 10° sector, is often pointed in the direction of the mast to keep the number of unfavorable directions to a minimum. Starting speeds for cups are typically 0.5 m s^{-1}, and distance constant (63% recovery time converted to distance) is between 1.5 and 5 m. Properly designed cup anemometers can be calibrated in wind tunnels to an accuracy of $\pm 1\%$. There is a tendency in cup anemometers to overspeed, resulting partly from their nonlinear response to wind speeds (Fig. 6.1) and partly from sensitivity to the vertical component of the wind (see Fig. 6.2). Reports of overspeeding error range from 5% to 10%.

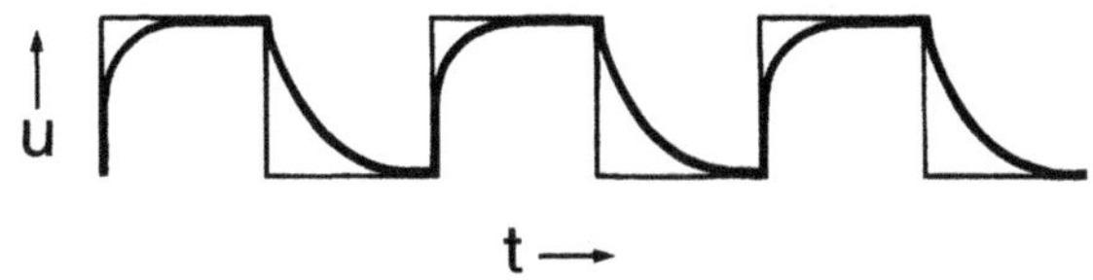

Fig. 6.1 Cup anemometer response (heavy curve) to square wave velocity function.

Propeller anemometers do not overspeed, but they operate dependably only when pointing directly into the wind. Propellers exhibit significant deviations from perfect cosine response (Fig. 6.3). When they are used in a fixed orthogonal configuration, corrections for deviation from cosine response have to be made based on an iterative approximation to the true wind direction (Horst, 1973). The measured wind components will otherwise be underestimated. A configuration well suited for profile measurements is the propeller-vane anemometer with its propeller mounted at the end of a vane with closely matched response characteristics. The vane keeps the propeller oriented into the wind. Propellers can be flat or helicoid. Two- and four-blade propellers are available, the former for high wind (>30 m s^{-1}). Accuracies and distance constants for propeller-vane anemometers are comparable to those of lightweight cup anemometers. Their calibrations should be checked in the wind tunnel; for most noncritical applications, periodic

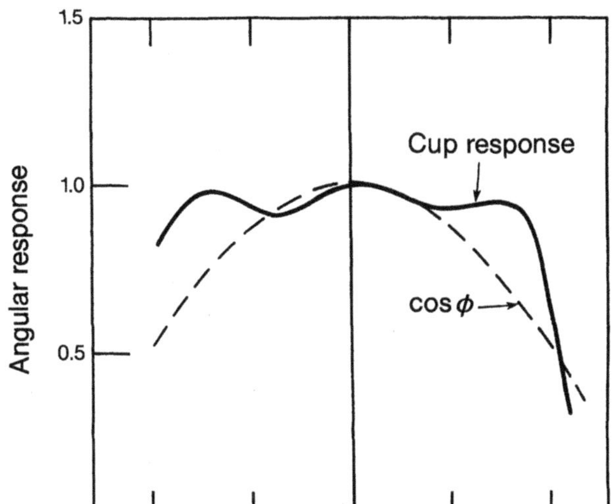

Fig. 6.2 Vertical velocity sensitivity in cup anemometers compared with ideal cosine response (after Busch et al., 1980). The inclination angle ϕ is defined positive for winds pointing downward.

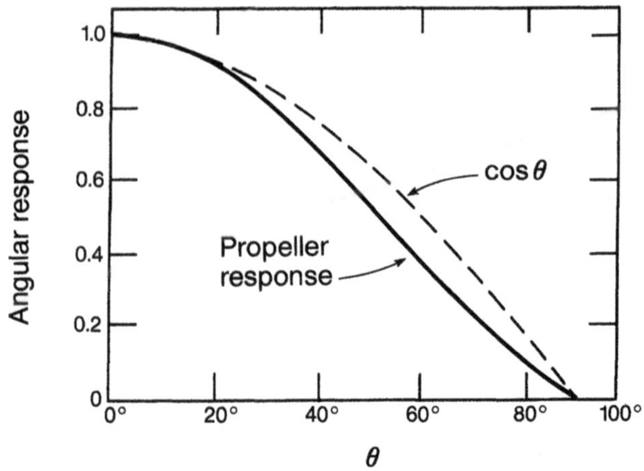

Fig. 6.3 Off-axis propeller response compared with ideal cosine response (after Busch et al., 1980). θ denotes the angle between the wind direction and the propeller axis.

calibration checks with a motor of known rotation rate (usually supplied by the manufacturer) are sufficient.

The optimum vertical spacing for wind profile measurement depends on the specific application. In the surface layer ($z \lesssim 50\,\mathrm{m}$), where the wind profile is roughly logarithmic, a vertical spacing that doubles with height (e.g., 1, 2, 4, 8, 16, 32 m) is often used. Above 50 m a fixed spacing is preferred, since the wind profile is no longer as sensitive to the presence of the ground below. For applications in which only one wind observation is required, as in surface wind networks, the World Meteorological Organization-recommended height is 10 m.

6.1.2 *Mean temperature sensors*

Temperature sensors most commonly used for measuring vertical gradients are platinum resistance thermometers, thermocouples, thermistors, and quartz thermometers. Accuracies of the order of 0.05°C and resolutions of 0.01°C can be achieved with careful calibration and attention to the design of leads, grounding, radiation shielding, and aspiration. The sensing elements are usually encapsulated in glass or metal for protection from atmospheric contamination and wear. The resulting increase in sensor mass and the attendant increase in time constant are not serious problems for mean profile measurements.

Platinum resistance thermometers are frequently used for both temperature and temperature difference measurements. The bridge circuit (see Fig. 6.4) is ideally suited for difference measurement. The bridge output V_o may be viewed as the difference between two voltage divider outputs:

$$V_o = \frac{E\alpha(T_2 - T_1)}{4(1 + \alpha T_a)}, \tag{6.1}$$

where E is the voltage across the bridge, α is the temperature coefficient of resistance (0.0039 at 0°C for platinum), T_1 and T_2 are the temperatures in degrees Celsius at the two elements, and T_a is the ambient temperature equal to $(T_1+T_2)/2$. The bridge sensitivity decreases with increasing T_a. For platinum, this amounts to a decrease of 12% over the range 0–30°C. Where long leads are involved, lead compensation should be provided to counterbalance the temperature and length differences (Fig. 6.4).

Thermocouples have relatively small outputs ($\sim$ 40 μV/°C for copper constantan), but this is not a serious disadvantage now that stable dc amplifiers are available. Absolute temperature measurement is less convenient with thermocouples because of the need for a stable temperature reference. Copper leads can be used for measurements at a distance from the amplifier, but care must be taken to keep the lead junctions at the same temperature (see Fig. 6.5).

Thermistors have high (negative) temperature coefficients of resistance, about 10 times those of platinum-resistance thermometers, but their thermal dissipation rates make self-heating a problem, tending to reduce the available sensitivity drastically. Nonlinearity in output and poor long-term stability are problems common to thermistor probes.

The ideal choice for a temperature profile system is the quartz thermometer. Its output is a beat frequency that varies linearly with temperature, and its basic

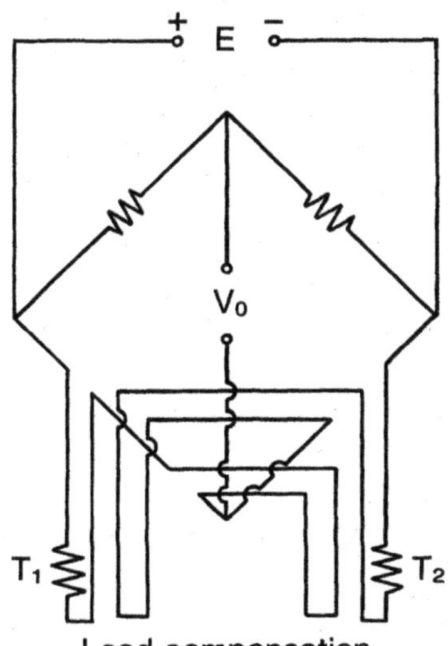

FIG. 6.4 Typical bridge circuit with lead compensation for a platinum-wire thermometer.

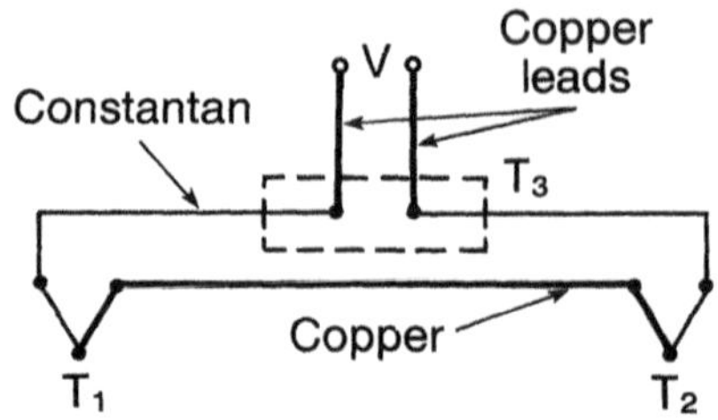

FIG. 6.5 Thermocouple configuration with external junctions kept at the same temperature.

accuracy (0.001°C) is degraded only by the errors (±0.05°C) introduced by the radiation shield and the aspirator. The main drawback is its cost, but the benefits of trouble-free performance may make it worth the cost for many applications.

Several types of aspirated radiation shields are available. They all perform reasonably well except at very low sun angles, where some designs do not provide adequate shielding. Wind-aspirated shields mounted on vanes have, in general, not proved satisfactory.

6.1.3 *Mean humidity sensors*

Humidity is among the more difficult variables to measure in the boundary layer. Many different definitions are used to specify the water vapor content of air: vapor pressure (in mb), absolute humidity (mass per unit volume), specific humidity (mass of water per mass of moist air), and relative humidity (mass of water in air per mass of water in saturated air at the same temperature and pressure).

Dewpoint (or frostpoint) hygrometers provide absolute measurement of humidity, whereas psychrometers that measure wet-bulb depression yield specific humidity through the psychrometric equation. Hygroscopic devices that respond to changes in relative humidity are simpler to use but usually depend on curves provided by the manufacturer for calibration.

For applications in which accuracy and long-term stability are of critical importance, the dewpoint hygrometer would be the most logical choice. In this device, a mirrored surface is maintained thermoelectrically at the temperature at which the moisture in the air begins to condense or freeze on it (Fig. 6.6). The temperature of the mirror is sensed by a platinum resistance thermometer, thermocouple, or thermistor. The device is generally housed in an aspirated shield. The sensor requires periodic cleaning of the mirror surface and checks of the calibration adjustments; with reasonable maintenance, absolute accuracies of ±0.5°C are attainable.

The psychrometric technique, used widely in field experiments, is inexpensive (compared with the cost of dewpoint devices), simple in concept, and relatively easy to maintain. It consists of two identical ventilated temperature sensors,

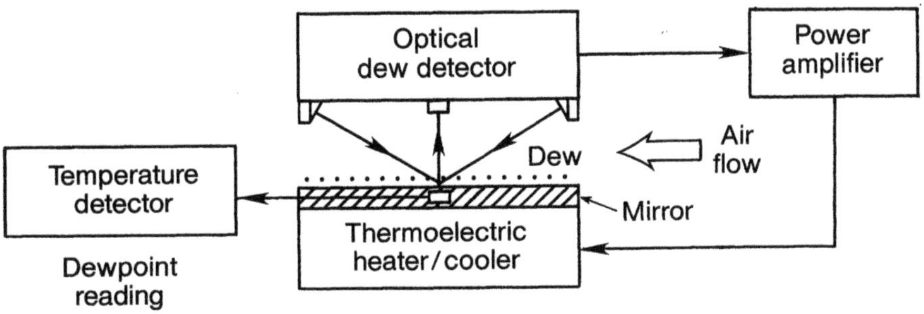

FIG. 6.6 Optical dewpoint hygrometer.

one of which is covered with a wick saturated with distilled water. Great care is needed in its design to ensure proper shielding from solar radiation, adequate ventilation, and wetting of the wick. The accuracy of this system is low at low humidities and temperatures, but is usually assessed in terms of the reading at 100% relative humidity. Accuracies of 0.5–1.0°C equivalent dewpoint temperatures can be maintained over a 20% to 80% relative humidity and a 0–25°C temperature range. The errors increase sharply below freezing because it is difficult to form and maintain an ice bulb on the wet thermometer. An example of an aspirated, radiation-shielded psychrometer of a type successfully used at Commonwealth Scientific and Industrial Research Organization (CSIRO) is shown in Fig. 6.7. This radiation shield is not affected by rays from low sun angles.

The hygroscopic sensors are also relatively inexpensive and easy to use. The Väisälä humicap is one of the more widely used sensors of this type. It responds to humidity with a capacitance change. The sensor provides a voltage directly proportional to relative humidity. Among its common failings are hysteresis effects, susceptibility to contamination, and loss of accuracy at high relative humidities. Because the humicap senses relative humidity, the air temperature must also be measured before absolute humidity can be obtained. Zero drifts caused by any of the effects mentioned above mean that sensors must be interchanged regularly if they are used for profile measurements. In practice, humicaps are often used not as in situ sensors but in the bench-mounted mode, held at a known reference temperature, taking measurements of air drawn along tubes from the sampling points. In this arrangement, it may be necessary to heat the tubes to avoid condensation when relative humidities are high.

6.1.4 *Trace gas sensors*

The most important trace gases in the atmosphere, from the biological point of view, are water vapor and CO_2. Mean concentrations of both these gases can be measured with great accuracy by infrared gas analyzers (IRGA). Standard IRGAs are bench-mounted instruments that must be shielded from vibrations and placed

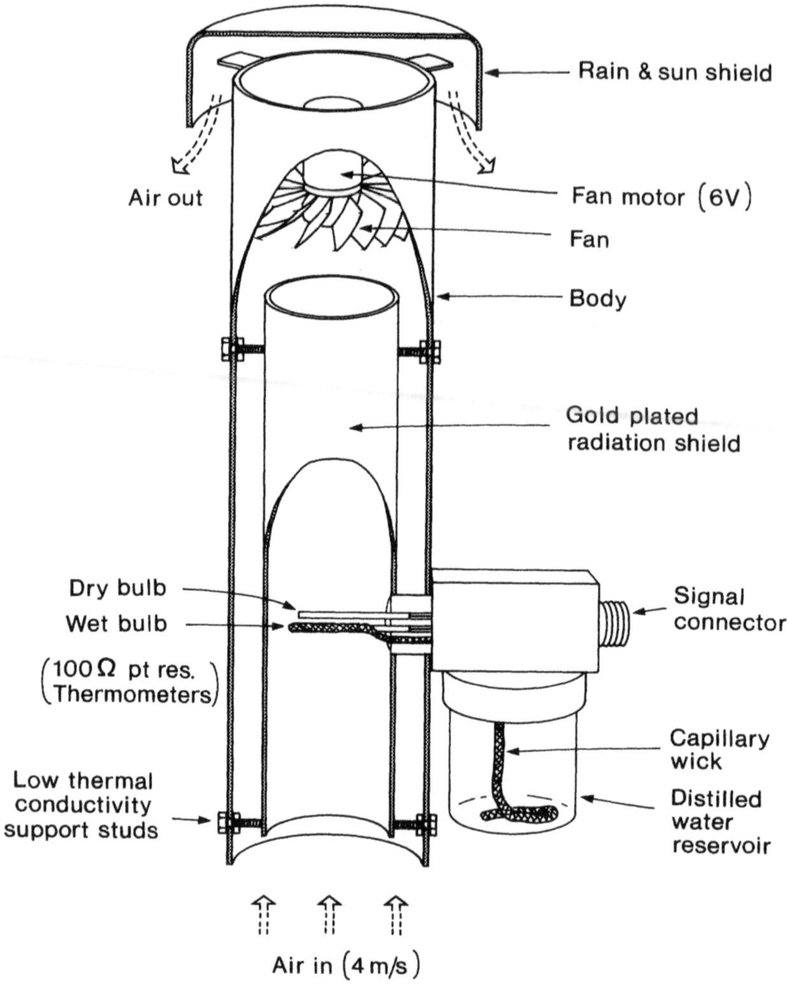

Fig. 6.7 CSIRO aspirated radiation-shielded psychrometer.

in a relatively stable temperature environment. They operate by comparing the absorption of infrared radiation in a sample of air drawn from the measurement point with that in a reference sample of accurately known concentration (Szeicz, 1975). For measuring CO_2, a reference sample containing a known amount of CO_2 would be used, but to measure water vapor a reference sample containing NH_3 or ethane, gases that absorb infrared in the same bands that absorb water vapor, would be used.

Other gases that can be conveniently measured with IRGAs are N_2O, CO, SO_2 (although not in trace amounts), and CH_4. The IRGA is a slow-response device requiring a minute or so to come to equilibrium with the sample. It can, however, be used in a continuous flow mode so that concentration variations are obtainable, albeit with this rather long time constant. Further detailed discussions of techniques used for trace gas measurements will be found in Andreae and Schimel (1989).

6.2 In situ sensors for turbulence measurements

6.2.1 *Velocity fluctuation sensors*

The hot-wire anemometer is the sensor traditionally used in turbulence work. It is best suited for laboratory and wind tunnel studies in which the dimensions of the sensor have to be small (on the order of millimeters) and the frequency response high (on the order of 10 kHz). For boundary layer flux measurements, the excellent frequency response of the hot-wire anemometer does not make up for its three major drawbacks: susceptibility to calibration shifts from atmospheric contamination, fragility, and narrow wind acceptance angle. The spatial scales of interest in the boundary layer range from 1 m to 10 km, and continuous measurements are sometimes needed for periods ranging from days to weeks. In such applications, the sonic anemometer is the preferred instrument; it has none of the response problems associated with rotating-type anemometers. The high cost of sonic anemometers, however, has led many researchers to look for simpler and less expensive options such as dynamic anemometers. Dynamic anemometers measure wind velocity by sensing either the pressure or the drag force on an object placed in the flow. They include thrust anemometers, anemoclinometers, and vortex anemometers. They respond to fluctuations in the wind more slowly than sonic anemometers but more quickly than the rotating types. Most of these devices are custom-made at various laboratories and are not available commercially. The types of sensors currently used for measuring wind fluctuations in most low-budget applications are configurations of lightweight propeller anemometers and bivanes offered by manufacturers of meteorological instruments.

Sonic anemometers measure wind velocity by sensing the effect of the wind on transit times of acoustic pulses traveling in opposite directions across a known path. The main limitation to their frequency response is imposed by line averaging along the path, which we will discuss later in the context of array geometries. Sonic anemometers can be of either the pulse type or the continuous-wave type. The former measures transit time differences directly to compute the velocity component along the path, whereas the latter measures phase differences that can be converted to time differences. Both measurements relate directly to wind velocity (see Appendix 6.1). The velocity component, V_d, along path length d, can be expressed as

$$V_d = \frac{c^2}{2d}(t_2 - t_1), \tag{6.2}$$

where c is the velocity of sound in air and t_1 and t_2 are the transit times for sound pulses traveling downwind and upwind along parallel paths. If c and d are known, V_d reduces to the measurement of $(t_2 - t_1)$, a relatively simple time interval measurement. For most applications, the velocity of sound can be approximated by

$$c^2 = 403\,T(1 + 0.32e/p), \tag{6.3}$$

where T is the absolute temperature, e is the vapor pressure of water, and p is the atmospheric pressure, all expressed in SI units. The contribution of e/p is small and usually neglected. Thus c^2 in (6.2) can be replaced by $403\overline{T}$, where $\overline{T}$ is the average temperature for the observing period. In the earlier sonic anemometers, $(t_2 - t_1)$ was measured electronically and adjustments in the calibration made during data analysis for variations in the mean temperature. Two separate paths were then used for simultaneous transmission of pulses in opposite directions.

The advent of integrated circuits and microprocessors opened the way for a more direct approach to measuring velocity through the relationship (Appendix 6.1):

$$V_d = \frac{d}{2}\left(\frac{1}{t_1} - \frac{1}{t_2}\right). \tag{6.4}$$

Newer sonic anemometers of the type described by Coppin and Taylor (1983), Hanafusa et al. (1980), and Kaimal et al. (1990), among others, exploit this fact; two of the above systems even extract temperature from the reciprocals of t_1 and t_2 (Appendix 6.1), as we shall see in the section to follow. All three use a single path to transmit sound pulses back and forth and measure t_1 and t_2 separately, thus reducing the number of transducers needed in the probes. (The transducers at the end of each path serve alternately as transmitters and receivers.) The first two, produced commercially by Kaijo Denki, Inc. (Japan) and Dobbie Instruments (Australia), have nonorthogonal path configurations (Figs. 6.8 and 6.9). The third, produced by Applied Technologies Inc. (US), offers an orthogonal path configuration (Fig. 6.10). They all have their paths arranged to measure one vertical and two horizontal wind components,[1] but the different arrangements of the paths reflect different needs and operating philosophies. (Kaijo Denki and Applied Technologies offer other probe configurations as well.)

The probes in Figs. 6.8 and 6.9, with their 120° separation between the horizontal axes, are designed to face the wind for best results. For winds within

[1]A different nonorthogonal configuration, using three intersecting paths tilted 30° from the vertical in an equilateral arrangement, has been described by Zhang et al. (1986). Commercial versions using that concept are available from Kaijo Denki, Inc., and Solent Instruments, Ltd. (UK).

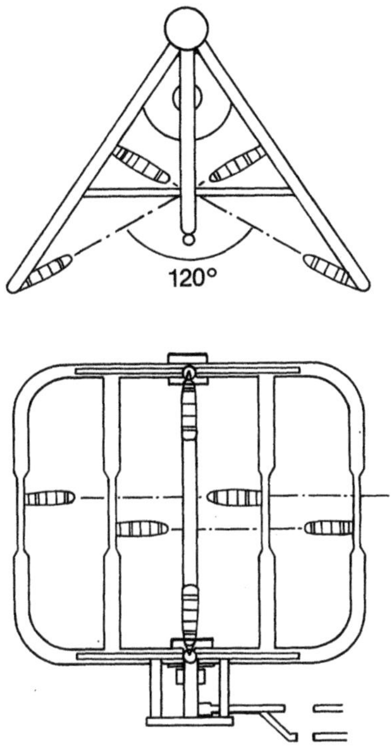

Fig. 6.8 Kaijo Denki single-path ($d = 0.2$ m) nonorthogonal sonic anemometer-thermometer probe.

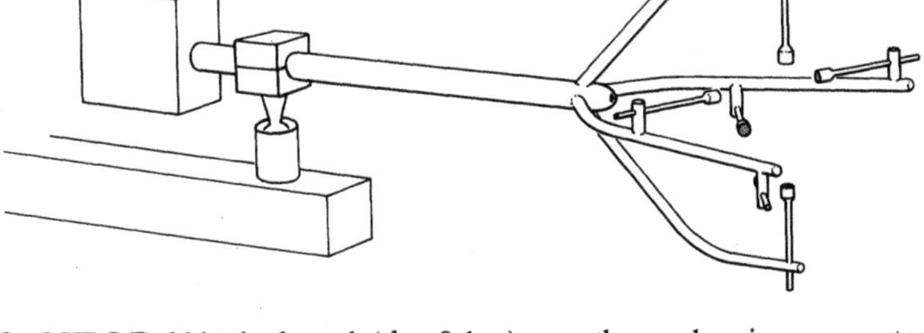

Fig. 6.9 CSIRO/Dobbie single-path ($d = 0.1$ m) nonorthogonal sonic anemometer probe.

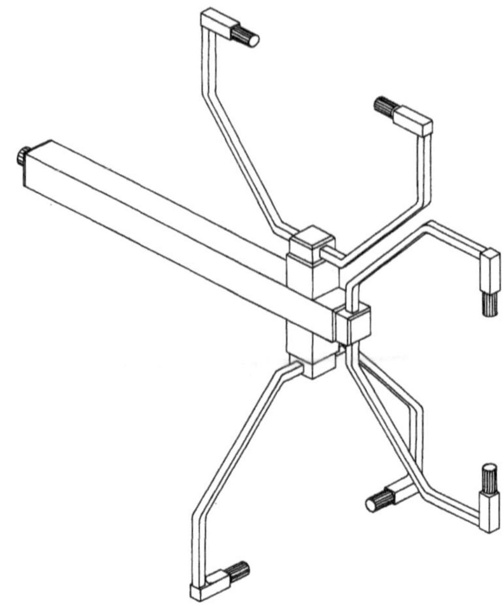

Fig. 6.10 Applied Technologies single-path ($d = 0.15$ m) orthogonal sonic anemometer-thermometer probe.

an azimuth range of ±30° from the probe's central axis (Grant and Watkins, 1989), the horizontal wind readings require no correction for shadowing by the structural supports or transducers at the upwind end of their paths. In order to benefit from this design, it is essential that the probe be reoriented periodically into the mean wind direction. The transformation equations for this array are given in Appendix 6.2. The CSIRO/Dobbie probe,[2] with its short path and vertically offset horizontal axes, operates on the same principle but offers a more compact measuring volume (~12-cm-diameter sphere) for use in plant canopies, close to branches and leaves and to within 1 m of the ground. Here, we accept a higher degree of flow distortion and transducer shadowing (Wyngaard and Zhang, 1985) as the price for compactness in the sampling volume and consequent improvement in frequency response.

The probe in Fig. 6.10, on the other hand, is designed for fixed operation at the end of a boom and aims to correct, in real time under microprocessor control, any deviations from cosine response due to upwind obstructions to the flow. By separating vertically the two horizontal paths, the error reduces to only the deficit due to shadowing by the transducers at the ends of horizontal paths. The form of the velocity deficit varies with the probe and the transducer shape (Wyngaard and Zhang, 1985); it is maximum for winds blowing along the axis

[2]Parallel versions of the original CSIRO and Dobbie versions are in use.

and drops off gradually as the wind deviates from that direction. Response curves for Applied Technologies' 1-cm-diameter transducer are presented in Appendix 6.3, plotted as a function of path length-to-transducer diameter ratio. The 37-cm vertical separation in the horizontal paths has been shown to have negligible effect on the observed mean and turbulence statistics at 10 m and 22 m over flat ground (Kaimal et al., 1990). The probe is not recommended for use below 4 m where the vertical gradients of wind speed are typically the largest and the eddy scales of interest too small to be resolved accurately by this probe.

As pointed out earlier, the main limitation to the frequency response of the sonic anemometer is the one imposed by line averaging along the acoustic paths. Spatial separation between the paths also introduces distortions at the high-frequency end; its effect is mostly felt in the transformed horizontal wind components u and v, which are derived primarily from measurements along spatially separated horizontal paths. A rough rule of thumb for calculating the onset of spectral distortion attributable to path length d, or separation distance s, is $\kappa_1 d$ or $(\kappa_1 s) = 1$, where κ_1 is the streamwise wavenumber ($\simeq 2\pi f/\overline{u}$), as defined in Chapter 1. The larger of the two effects determines the onset frequency, which we will define here as the 98% power cutoff.

The exact form for the spectral response in the wind components therefore depends on the length of the paths and orientation with respect to the mean wind direction. Kaimal et al. (1968) had shown that the line-averaging effect for the w component (from a single vertical path) approximates that of a single-pole, low-pass filter with half-power point at λ (or $\overline{u}/f$) = d, as shown in Fig. 6.11. In the nonorthogonal array of Fig. 6.8, the power spectral responses for u and v differ from that of w because of the spatial separation ($s \approx 0.6d$) between the horizontal paths. For all three, however, the roll-off starts at $\lambda \approx 2\pi d$ (or 1.25 m for d = 20 cm).

In the case of the orthogonal probe in Fig. 6.10, the 37-cm vertical separation between the horizontal axes determines the useful wavelength range for u and v measurements; it will be $\lambda > 2\pi \times 37$ cm, compared with $\lambda > 2\pi \times 15$ cm for w. For measurements above 4 m height over flat terrain this will not seriously degrade the estimations of $\overline{u'w'}$ and $\overline{w'\theta'}$ since the bandwidth requirements for the two fluxes also differ by a factor of 2.5, as seen in Fig. 2.18. Over complicated topography and in canopies, however, this benefit may not be realized because the w path may not be normal to the mean wind vector. In that case, it is wise to assume a sampling diameter that includes not only the sonic paths but also any scalar sensor contributing measurements for flux calculations.

It is instructive to compare the dynamic responses of the bivanes (vanes that can rotate vertically as well as horizontally) and propellers used in atmospheric work with those of the sonic anemometer. Manufacturers' specifications list delay distances for bivanes (50% recovery) and distance constants for propellers (63% of step change) as being close to 1 m. These distances, however, translate to much

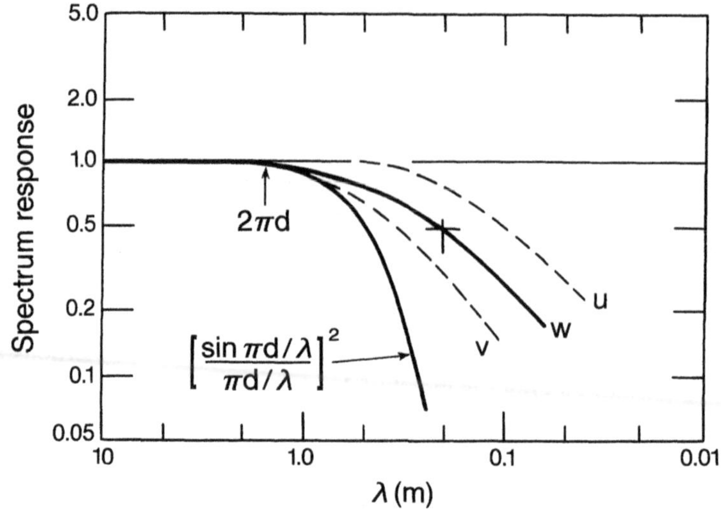

FIG. 6.11 Power spectral transfer functions for u, v, and w in a three-axis sonic anemometer with $d = 0.2$ m and 120° separation between the horizontal axes for wind blowing along the probe's central axis. The $(\sin^2 x)/x^2$ function representing simple line averaging along the direction of the flow is shown for comparison (after Kaimal et al., 1968).

longer cutoff wavelengths if we define the response in terms of departures from the ideal spectral response (Finkelstein et al., 1986). Figure 6.12 shows response functions for the vertical wind component w measured by R. M. Young's bivane system and the Gill propeller anemometer. These functions were derived from comparison tests with a sonic anemometer at 10 m height. Defining λ_c, the cutoff

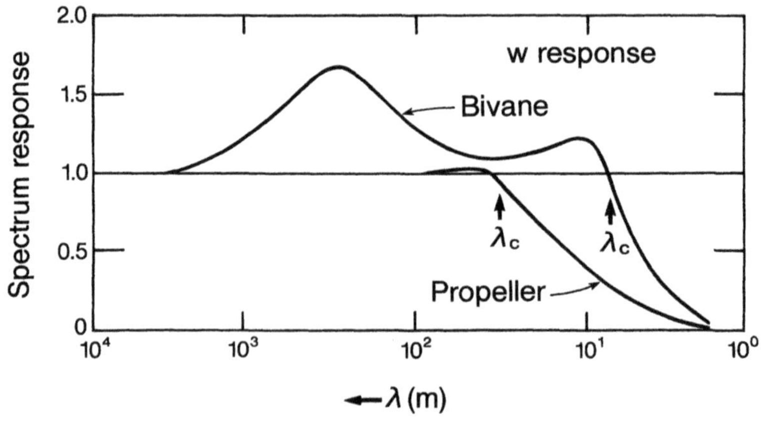

FIG. 6.12 Power spectral transfer functions for the w component measured by a propeller bivane and a single vertical propeller anemometer (after Finkelstein et al., 1986).

wavelength, as the point where the power spectrum drops to 98% of its true value, we find λ_c to be 32 m for the vertical propeller and 7 m for the propeller-bivane combination.

The bivane's response to w appears to be superior to that of the propeller for $\lambda < 32$ m, but the amplification at the middle frequencies raises concern as to its effect on flux calculations. To a moderate degree, the enhancement in the variance is beneficial because it compensates for the high-frequency cutoff; vertical velocity variances from the bivane compare very closely with sonic anemometer values. The propeller w variances, on the other hand, are severely underestimated: 25% lower under daytime conditions and up to 50% lower under nighttime conditions, at 10 m height.

The large λ_c value for the w propeller is not surprising when we consider the off-axis degradation in response and the relationships that exist between distance constant l, half-power wavelength λ_0, and λ_c. The distance constant for the Gill propeller increases from 1 to 2 m as the flow deviates from axial to 80° off axis. In principle, the distance constant approaches infinity at 90° off axis.

For a linear first-order system, the power transfer function can be written as

$$[H(\lambda)]^2 = \frac{1}{1 + (2\pi l/\lambda)^2}. \tag{6.5}$$

At the half-power point we have

$$\lambda_0 = 2\pi l, \tag{6.6}$$

and λ_c, defined here as the 98% power point, becomes

$$\lambda_c = 2\pi\lambda_0 \simeq 40l. \tag{6.7}$$

With $l = 2$ m (for the w measurement), λ_c would be 80 m for a linear first-order system. Fortunately, the propeller's response is better by a factor of 2.5 (Fig. 6.13), which brings the actual value of λ_c down to 32 m as observed.

6.2.2 *Temperature fluctuation sensors*

Fine-wire platinum-resistance thermometers, thermocouples, thermistors, and sonic thermometers have all been used for temperature fluctuation measurements. Thermistor and thermocouple sensing elements cannot be made as small as platinum wire probes, so they tend to be less responsive and not as well suited for flux measurements near the ground. All three types are normally exposed to free air without the benefit of protection from radiation shields. They are, therefore, vul-

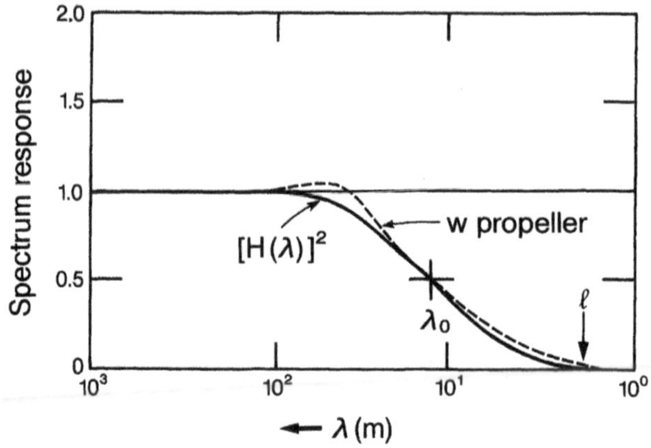

FIG. 6.13 Response of a single vertical propeller compared with first-order linear response showing the disparity between half-power wavelength λ_0 and distance constant l (after Finkelstein et al., 1986).

nerable to damage from winds, rain, snow, and flying debris. The chances of damage can be reduced significantly if the element is properly supported. Such supports increase the time constant of the sensor. The 12.5-μm platinum wire in the fast-response probe (also marketed by Atmospheric Instrumentation Research) used at the Boulder Atmospheric Observatory (BAO) (Kaimal and Gaynor, 1983) is wound around a small helical bobbin on threads strung across the windings of the helix. The spectral response of the wire is degraded by conduction through the support, but only at frequencies above 5 Hz. With this support, the sensing element can survive for weeks and even months without breaking. Freezing rain, wet snow, and very high winds are the events that usually break the wire. It should be pointed out that platinum wires should not be left exposed to the elements for too long. Contamination of the wire has a direct effect on its time constant. A convenient scheme for calibrating the platinum wire probe in the field is illustrated in Fig. 6.14. Switching in a fixed resistance changes the current flowing through the wire by an amount corresponding to a known temperature change (5°C). The amplifier gain is adjusted to set the corresponding step change in the output voltage of the temperature circuit to 5 V to match the input calibration requirement in the data acquisition system. A small flask is slipped over the platinum wire probe to suppress the fluctuations in the signal while this calibration is in progress. This procedure compensates for probe-to-probe variations in resistance and even changes due to aging.

The sonic thermometers incorporated in the Kaijo Denki and Applied Technologies sonic anemometers have the requisite frequency response for boundary layer flux measurements. Temperature is sensed along the same path as w, so its spatial averaging characteristic is compatible with that of w. The sonic temperature, which is proportional to $[(1/t_1) + (1/t_2)]^2$, is, however, contaminated by

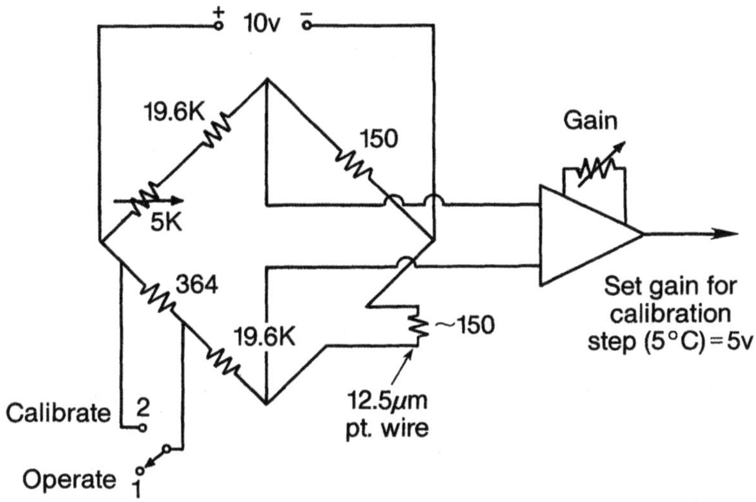

Fig. 6.14 Switching arrangement for field calibration of a fast-response platinum wire thermometer.

residual sensitivities to humidity and to the wind component normal to the path. These errors are negligible under daytime unstable conditions when temperature fluctuations are large. The wind contamination, in particular, however, can be significant under near-neutral and stable conditions (Kaimal, 1969). The nature of the cross-wind contamination on the evaluated heat flux is such that it introduces a bias in the estimation of the time the atmosphere passes through neutral stability, shifting it slightly into the stable regime. We show in Appendix 6.1 how the crosswind velocity (V_n) modifies the sonic thermometer equation:

$$\frac{1}{t_1} + \frac{1}{t_2} = \frac{2}{d}(c^2 - V_n^2)^{1/2}. \tag{6.8}$$

Substituting for c^2 from (6.3) we have, for the temperature measured along the vertical path,

$$T(1 + 0.32e/p) = \frac{d^2}{1612}\left(\frac{1}{t_1} + \frac{1}{t_2}\right)^2 + \frac{1}{403}\left(V_X^2 + V_Y^2\right), \tag{6.9}$$

where V_X and V_Y are the wind components along the orthogonal horizontal axes of the anemometer and all variables are expressed in SI units. In the Applied Technologies sonic thermometer (Fig. 6.9), V_X and V_Y, compensated for transducer shadowing, are used to correct, in real time, each temperature reading as it is processed for transfer to its output registers.

The humidity term and its effect on calculated variances and fluxes can be ignored, as is often done, or compensated for in the data analysis if concurrent measurements of humidity fluctuations are available. Kaimal and Gaynor (1991) argue that the sonic-derived temperature on the left in (6.9) should be treated as the true virtual temperature T_v, which it approximates very closely since

$$T_v = T(1 + 0.38e/p), \tag{6.10}$$

and could be used directly in its place in applications where the contribution to buoyancy from moisture should be included. A case in point is the Obukhov length L, where the temperature flux term should, in the presence of moisture, be $\overline{w'T_v'}$ (for most applications we assume $\theta_v' = T_v'$). The V_n^2 corrected temperature from the sonic thermometer approximates T_v to within 0.01°C, well within the bounds of experimental uncertainty.

6.3.2 *Humidity fluctuation sensors*

Three different techniques are currently used for measuring humidity fluctuations. They involve the absorption of ultraviolet radiation by water vapor (Lyman-alpha and krypton hygrometer), the absorption of infrared radiation by water vapor (infrared hygrometer), and the dependence of microwave refractivity on humidity (microwave refractometer).

The simplest of the three devices is the ultraviolet hygrometer. The Lyman-alpha version requires a source (excited hydrogen), a nitric oxide detector, and a space between the two where the absorption takes place (see Fig. 6.15). Magnesium fluoride windows are needed on the source and detector tubes since most other materials are opaque to ultraviolet radiation. The cutoff frequencies of the nitric oxide detector and the magnesium fluoride windows neatly bracket the Lyman-alpha emission line of atomic hydrogen (121.56 nm). The other emission lines in the hydrogen glow discharge produced by the source are thus filtered out.

The very strong absorption of this emission line by water vapor makes measurement possible over short path lengths (~1 cm). (By comparison, absorption by ozone and oxygen, the only other absorbers in the atmosphere, can be neglected in the boundary layer.) The dimensions of the sensor are small enough (15 cm long × 2 cm diameter) to permit installation close to a sonic anemometer. For uniform exposure in the azimuth, the sensor is mounted with its path oriented vertically. Two sources of drift in the calibration should be noted: aging of the Lyman-alpha source and window degradation from reaction of atmospheric constituents with the window material. The effect of aging is retarded to some extent by using higher hydrogen pressures and sealing uranium hydroxide in the source body. The window degradation is reversible, since it occurs on the outer surface; washing with alcohol and rubbing with a fine abrasive restores its transmission

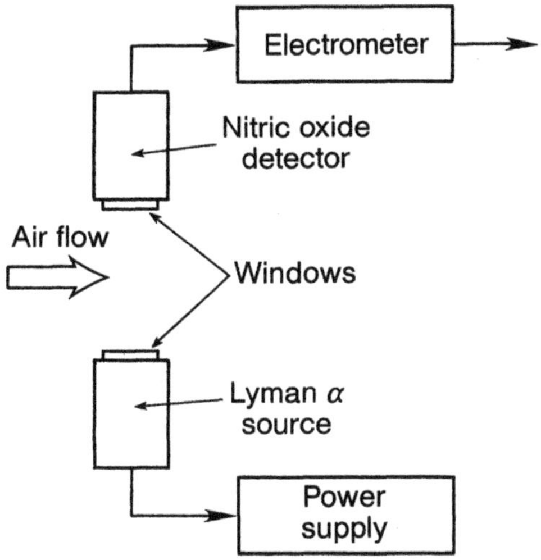

FIG. 6.15 Lyman-alpha hygrometer (after Buck, 1976).

properties. Nevertheless, for maximum accuracy, this device should be operated in conjunction with a dewpoint hygrometer, so its calibration can be continuously updated by comparing changes in its mean readings with those derived from the slower, more accurate, dewpoint measurements.

In the krypton hygrometer, the ultraviolet source is a krypton glow tube. This produces ultraviolet emission at 123.58 and 116.47 nm, rather than the 121.56 nm of the hydrogen line. In all other respects they operate the same way. The advantage of the krypton device is the long-term stability of the source in comparison with the degradation over time of the hydrogen source. The disadvantage is the greater absorption of the krypton lines by oxygen and ozone. In practice, this does not pose a severe problem, and reliable commercial krypton hygrometers are now available.

The infrared hygrometer detects humidity through differential measurement of infrared transmittance at two adjacent wavelengths, one located in a region of high water vapor absorption and the other where the absorption is negligible. The transmitting path is typically 0.2–1.0 m long, and the beams are usually modulated by a mechanical chopper to permit high-gain amplification of the detected signal. Optical components such as narrow-band filters and beamsplitters add to the complexity of this device.[3] At humidities approaching saturation, the calibration is undependable because of scattering from water adsorbed and absorbed by particles in the air.

[3]Recently, Y. Mitsuta and colleagues at the Kyoto University Disaster Prevention Research Institute have claimed success with a simpler single wavelength infrared system in which the filament temperature at the source is varied sinusoidally for stability in the humidity measurements (personal communication).

The microwave refractometer measures the refractive index of air in a cavity and depends on the relationship between refractivity, specific humidity, temperature, and pressure to derive humidity. Strictly speaking, simultaneous data on temperature and pressure fluctuations in the cavity are needed to remove their influence on the refractive index; in practice, pressure fluctuations are ignored since they are usually small (<1 mb) compared with the absolute pressure. The temperature measurement is made as close to the cavity as possible. The sensing element in the microwave refractometer is a resonating cavity with ventilating ports. The cavity dimensions are small, but the spatial resolution of the sensor is more a function of its flushing efficiency than of its actual size. For a 5-cm cavity, in moderately light winds, spectral attenuation starts at a wavelength of 1.5 m. In an experiment comparing various humidity sensors, Priestley and Hill (1985) found the refractometer response falling off rapidly at 2 Hz with wind speeds at 3–4 m s^{-1}. The Lyman-alpha hygrometer's response under the same conditions extends another decade, to 20 Hz.

6.2.4 *Trace gas fluctuation sensors*

Infrared absorption by CO_2 occurs at a different wavelength than water vapor. In recent years successful rapid-response open-path devices working on the same principle as infrared hygrometers have become available. The commercial versions measure both CO_2 and water vapor simultaneously. Recently the same technique has been extended to CH_4 at yet another infrared wavelength. Absorption of gases such as CO_2 and SO_2 at very low concentration is strongly modulated by density changes caused by temperature and water vapor fluctuations. The required corrections for measured eddy fluxes of the trace gases amount to 100% or more over freely transpiring vegetation (Webb et al., 1980), making such measurements over water problematic at best.

6.2.5 *Static pressure fluctuation sensors*

The most difficult dynamic variable to measure in a strongly turbulent flow is the static pressure fluctuation p' because any probe introduced into the airstream causes dynamic pressure fluctuations $p'_d = \rho(u'_p)^2/2$, where u'_p is the local velocity variation caused by the probe. p'_d is typically an order of magnitude larger than the desired static pressure fluctuation p'. If the probe can be aligned to within a few degrees of the wind vector, then sampling ports can be arranged so that contributions to p'_d from ports at different points on the probe cancel out, leaving the required signal p'. In turbulent flows where the instantaneous wind vector varies rapidly about the probe axis, an optimum configuration for the ports is much more difficult to achieve.

One successful probe design for surface layer studies is described by Elliot (1972). His spoon-like probe is relatively insensitive to wind angles of attack within ±135° in the azimuth and ±15° in elevation. This restriction limits its use to measurements in relatively low-intensity turbulence. A new omnidirectional probe claiming acceptance angles of ±30° in elevation has been described by Nishiyama and Bedard (1991). Their probe consists of a series of outer and inner ports arranged in rings around a vertical tube fitted with four discs as shown in Fig. 6.16. The width and placement of the discs provide cancellation of dynamic pressure changes at the inner ports over the ±30° range of elevation angles. The probe can operate in adverse environments. This probe has been deployed in field experiments and awaits confirmation of its usefulness for measuring p' in turbulent kinetic energy budget studies.

6.3 Sensors for surface measurements

Heading this list are sensing techniques that have been used for decades to measure the parameters needed to calculate energy balance at the earth's surface. The energy balance at the surface can be written, in its simplest form, as

$$R_n - G_s = H + \lambda_e E,$$

where R_n is the net radiation flux density at the surface, G_s is the flux of heat into the soil, H is the sensible and $\lambda_e E$ the latent heat flux from the surface as defined in Chapter 3. R_n and G_s are routinely measured with net radiometers and soil heat flux plates, respectively, whereas $\lambda_e E$ can be measured directly with a lysimeter. We also discuss the use of drag plates to measure the momentum flux at the surface directly. Its disuse in recent years notwithstanding, the drag plate remains the only sensor capable of measuring unequivocally the momentum flux at the surface, invaluable in situations in which the constant flux assumption cannot be invoked. Also included is radiometric sensing of surface temperature because of its increasing use in boundary layer studies.

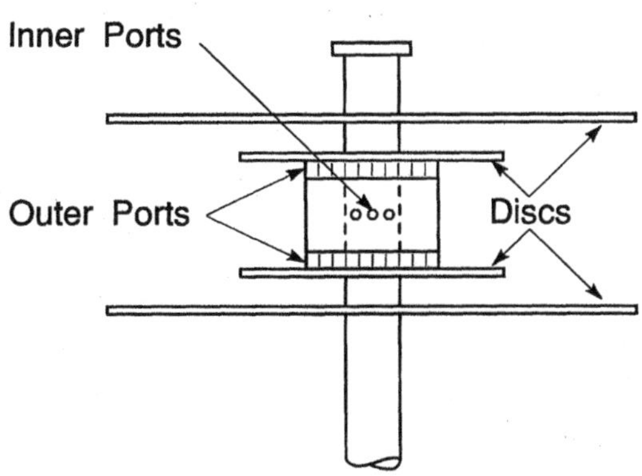

Fig. 6.16 Omnidirectional pressure fluctuation sensor (after Nishiyama and Bedard, 1991).

6.3.1 *Net radiometers*

Net radiometers measure the difference between total (all wavelength) incoming and outgoing radiation. The standard version of the instrument consists of two small ($\sim$ 20 mm square) black surfaces facing up and down, separated by an insulator and protected from the wind by polyethylene hemispherical domes, as shown in Fig. 6.17. The temperature difference between the upper and lower surfaces drives a thermopile so that the unamplified signal from the radiometer is typically a few millivolts. The whole assembly is mounted at the end of a tube 1 m long. In radiometers with thin polyethylene domes (Funk type), the space enclosed by the domes is ventilated with a supply of dry inert gas that also keeps them inflated; those with rigid polyethylene domes (Fritschen type) are usually not ventilated, but an external tube of desiccant is plumbed to the spherical space around the thermopile to prevent condensation.

Because R_n is usually the small difference between two large numbers, great care must be taken with calibration and installation of the instrument. The radiometer must be carefully leveled and the downward-looking hemisphere should "see" a representative section of the ground. Over short vegetation or bare soil, the radiometer is mounted about 2 m above the surface on its own mast, but over taller plants, as in forest canopies, it may be necessary to use a tall instrument tower. In such cases the radiometer should be mounted at the end of a long boom to remove as much of the tower as possible from the field of view of the radiometer. Ingenious arrangements are necessary to level an instrument that may be 4–5 m out from the tower. More discussion of net radiometers can be found in Szeicz (1975).

6.3.2 *Soil heat flux plates*

The sensible heat flux across the soil surface is commonly measured directly with soil heat flux plates. These are available commercially and consist of metal plates ($\sim$10 cm^2 in area) separated by a material of known conductivity. The temperature

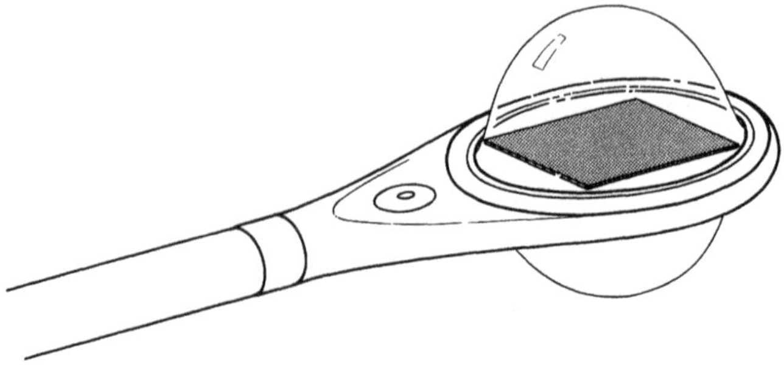

FIG. 6.17 Funk net radiometer.

differences between the plates drive a thermopile that provides the sensor output. Three or four such plates are usually connected electrically and distributed a few millimeters below the soil surface at representative locations. Errors arise through sampling problems if the surface is heterogeneous. More serious errors occur in a drying soil where the site of soil evaporation has moved below the surface and a significant fraction (~50%, at times) of energy flux to the ground is through diffusion of water vapor in the upper soil layers. These problems are discussed in detail by deVries and Philip (1986).

6.3.3 *Lysimeters*

In the weighing lysimeter, a block of soil large enough to be a representative sample of the surface and deep enough to contain the roots of any plants present is isolated from the surrounding soil in a watertight container mounted on a sensitive, continuously recording balance. The balance must be capable of resolving a few tens of grams, whereas the block itself may weigh several tons. Both the water running off at the surface during rain and the water percolating through the soil monolith past the roots are collected and recorded, while the time record of the change in the weight of the lysimeter itself gives the evaporation rate. A small correction may have to be made for plant growth over the long term. The accuracy obtainable from a lysimeter depends ultimately on the representativeness of the sample. This may be as small as 1 m^2 in an area over short grass or bare soil, whereas forest lysimeters covering areas more than 10 m^2 and containing mature trees have also been operated successfully. For the large lysimeters, excavating the soil monoliths in situ with minimum disturbance to the sample is a major undertaking. For more information see Dunin et al. (1991).

6.3.4 *Drag plates*

The same principle embodied in the lysimeter is used here to measure the flux of momentum to the surface, $\tau_0 = \rho u_*^2$. The drag plate consists of a representative sample of the ground surface mounted on a sensitive balance designed to measure the tangential force on the surface. The entire mechanism is buried beneath the ground so that no surface discontinuity or disturbance is apparent to the observer above. Typical drag plate configurations have circular platforms 1–2 m in diameter, deep enough (~10 cm) to hold sufficient soil to keep short vegetation alive for the duration of an experiment. The weight of the plate imposes severe demands on the mechanical structure supporting it. Also, the gap around the rim of the sample has to be very small (a few millimeters) to keep spurious aerodynamic pressure forces from swamping the small tangential forces due to τ_0, requiring essentially zero deflection under load. All these requirements lead to a heavy, high-precision device of the type described by Lynch and Bradley (1972). The drag plate approach is best suited for use over flat homogeneous surfaces with adequate fetch, bare or covered with short vegetation such as mowed grass or wheat stubble.

6.3.5 *Radiometric thermometers*

It is sometimes necessary to have a measure of surface temperature. Over bare soil, the surface temperature may be measured directly with an in situ thermometer; but over vegetated surfaces, infrared thermometers mounted on towers, aircraft, or satellites offer a means to measure that quantity remotely without the sampling limitations inherent in the in situ approach. Errors arise from many sources such as the dependence of soil surface emissivity on water content and surface texture, the "cool-skin" effect on open water bodies, and the difference between bulk aerodynamic temperature and radiometric temperature in plant canopies. A vast literature has recently bloomed on this subject.

6.4 Measuring from towers and masts

Towers, masts, booms, and mounts used for supporting a sensor can interfere with the flow, thus introducing errors in the measured gradients and fluxes. It is possible to reduce these errors to acceptable levels if care is exercised in the design of the supporting structure and placement of the sensors. There is considerable discussion on this subject in the literature. No simple solutions or relationships for flow distortion have emerged because the geometries of the obstacles discussed vary so greatly (Wucknitz, 1980). Upwind of the tower we need to consider two distinct regions of the disturbed flow: one close to the obstacle, at distances smaller than the diameter of the obstacle, where the flow is complicated and difficult to describe theoretically; the other at distances greater than the diameter, where potential flow can be assumed in order to calculate flow distortion.

The first region is one the experimenter would do well to avoid. For a tower with structural members that are small and widely separated (ratio of obstructed-to-unobstructed area < 0.1), we might consider installing sensors at upwind distances smaller than the outside dimensions of the tower. Most towers have structural member densities between 0.2 and 0.3. In that case, the sensors should be mounted no closer than 1.5 times the largest lateral dimension of the tower. (Up to 4 times the tower width is possible with certain portable and guyed TV-antenna-type towers that have lateral dimensions on the order of 0.3 m but are capable of reaching up to 60 m.) In applications where only one level of observation is needed (say at 10 m), the anemometer should be mounted on top of the tower (or mast) to avoid direct tower shadowing. To reach the region where potential flow can be assumed, the sensor has to be at least three lateral dimensions above the top of the tower and supported by a thinner mast.

Another region of the flow to be avoided is the downwind side of the tower. At the very least, readings from this region should be treated with caution. Here, the flow is strongly influenced by the wake structure. Even on towers with low structural member densities, the combined effect of wakes from the separate members can be greater than for a solid tower offering the same obstruction area to the flow. The wake intensity and width vary with Reynolds number and are sensitive to both the roughness elements on the tower and the turbulence intensity in the undisturbed flow. For the high Reynolds numbers appropriate to masts and towers in atmospheric flows, the wake is often nonstationary. Experimental results of different investigators therefore tend to be inconclusive.

In the absence of any firm guidance on the subject, we develop useful rules of thumb. For example, the data can be restricted to a sector that excludes wind directions through the tower and 30° on either side (as a safety factor). For sensors mounted at the end of a boom 1.5 times the tower width, as at the BAO (Fig. 6.18), a 270° sector is then available for useful measurements. Identical instrumentation on booms pointing in opposite directions is essential for full 360° coverage. In many research applications, the investigator can afford to wait for favorable wind directions or confine analyses only to periods when conditions are acceptable.

The amount of reduction in the measured average wind speed 1.5 tower diameters upwind of the tower is typically 5%. The wind direction deflection, on

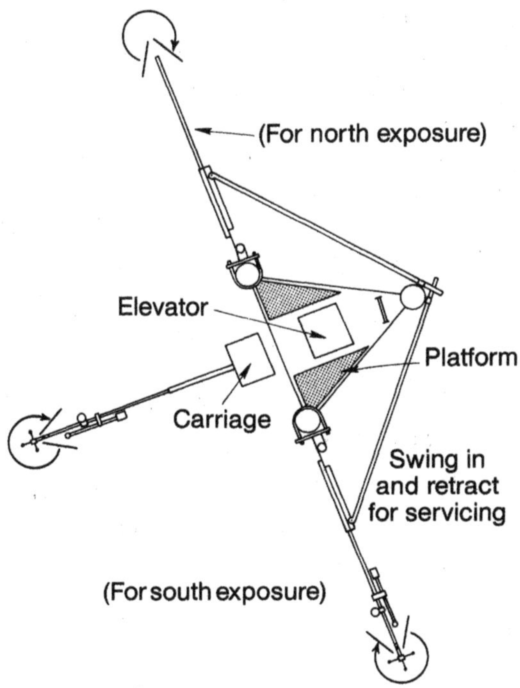

Fig. 6.18 Boom configuration on the Boulder Atmospheric Observatory tower (Kaimal and Gaynor, 1983).

the other hand, is small directly upwind but increases as the wind shifts to either side. The maximum deflection observed on the BAO tower is 5° (at a distance 1.5 times the tower width), for winds normal to the direction of the boom.

The influence of towers on flux measurements must be addressed. Investigators have examined errors in the measured stress from flow distortions caused by boxes stored on tower platforms, by the horizontal booms supporting the sonic anemometer and other sensors, and by the probe itself. Every effort should be made to avoid placement of obstacles above, below, and even downwind of the sensor and to incorporate as much vertical symmetry as possible in the probe design.

If we choose to work in standard Cartesian coordinates, the need for careful alignment of the measurement coordinates with respect to a known reference direction is critical. The momentum flux $(-\overline{u'w'})$ is especially sensitive to leveling errors (Kaimal and Haugen, 1969). Leveling accuracies of $\pm 0.1°$ can be achieved in a sonic anemometer by attaching to each probe an electrolytic level indicator with remote readout. An alternate approach is to analyze data in streamline coordinates as described in Chapter 5. Over complex terrain, this is the preferred choice, but even over nominally homogeneous terrain, working in streamline coordinates can remove the need for exact instrument alignment. (Practical procedures for using this approach will be discussed later in this chapter.) There are some applications (as in strongly distorted flows), however, where it is essential to know the orientation of sensors to an external reference frame.

When mounting two sensors side by side for flux measurements (e.g., vertical velocity and temperature), the spatial separation between them should be kept as small as possible. We find that the fluctuations measured become increasingly uncorrelated at wavelengths $\lambda < 2\pi s$ (where s is the separation distance). Thus the criterion for maximum separation distance is approximately the same as for spatial averaging in individual sensors: $s < \lambda_c/2\pi$, where λ_c is the 98% power point. One attempt to minimize this separation distance and to provide some degree of vertical symmetry in the w measurement is illustrated in Fig. 6.19.

6.5 Measuring in plant canopies

In plant canopies both the spatial heterogeneity and the very high turbulence intensities encountered impose extra constraints on the measurement techniques available.

6.5.1 *Mean velocity*

The very high turbulence intensities in plant canopies ($\sigma_u/\overline{u}$ often as high as 2 or 3) combine with the overspeeding properties of cup anemometers to register a significant and completely spurious mean flow in the lower canopy (Wilson et al.,

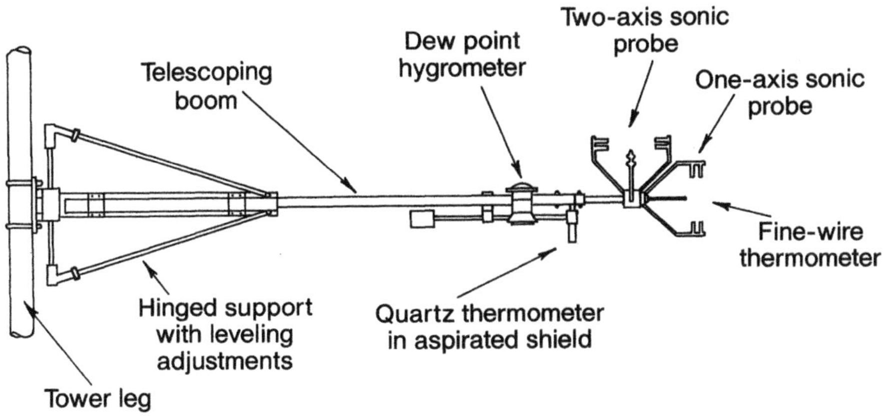

FIG. 6.19 Instrumentation at each observation level and adjustments for leveling the boom on the Boulder Atmospheric Observatory tower (Kaimal and Gaynor, 1983).

1982). Despite this, they remain in regular use in non-critical situations; however, accurate mean velocities are now usually obtained with sonic anemometers.

6.5.2 *Velocity fluctuations*

As in the surface layer over flat terrain, the sonic anemometer is the instrument of choice for measuring velocity fluctuations in a canopy. The effective sampling volumes of most commercial sonic anemometers are small enough for use in forest canopies but too large for smaller crop canopies. A significant portion of the variances and fluxes would be missed. Even in forest canopies, the large amount of small-scale, wake-generated turbulence present dictates the use of probes with the smallest effective path length, such as the CSIRO/Dobbie sonic anemometer. In reasonably dense canopies with plant height 1–2 m or less, even a 5-cm path sonic anemometer may miss significant amounts of the variance. The only small-scale instrument to have met this challenge to date is the servo-driven, split-film anemometer described by Shaw et al. (1973). It must be emphasized, however, that this solution requires considerable technical skill to implement. The research community still awaits a truly miniaturized sonic anemometer to use in plant canopies.

6.5.3 *Scalar fluctuations*

Temperature fluctuations seldom present a problem with thermocouple or platinum wire probes, which can be made very small. The short path length of Lyman-alpha and Krypton hygrometers makes them suitable for use in most canopies, but the longer path of the infrared devices limit their use to larger forest canopies where sonic anemometers may also be used.

6.5.4 *Net radiation*

Within the canopy, absorption of radiation on plant surfaces produces vertical divergence in the net radiation flux R_n. At the same time, the penetration of sunlight through gaps in the foliage makes mandatory some kind of spatial averaging in the R_n measurements. This can be accomplished by moving standard spherical net radiometers on tracks through the canopy or, more conveniently, by using cylindrical net radiometers. These may be several meters long, depending on the particular requirements of the canopy in question. Unlike spherical net radiometers, their response function depends on solar azimuth as well as inclination so that the orientation of the device relative to the sun's position at zenith becomes important. The problems involved are discussed by Lang (1978).

6.5.5 *Ground heat flux*

In forests containing large trees, the energy storage term G_s is no longer associated with the flux into the soil alone but also into the more massive trunks and branches. The term is often estimated from measurements of tree surface temperature and an assumption for the thermal conductivity of the green wood.

In Chapter 3 we emphasized the requirement for spatial averaging in canopies. This means, in practice, that multiple towers may be needed to sample properly the turbulent flow in general and to deduce the dispersive flux in particular. It also means that objective placement of sensors is required if biased samples of the flow are to be avoided. In simple terms, siting every anemometer in a clearing between plants is not acceptable. With objective placement of instruments, some end up in spaces, whereas others are thrust into tree crowns. This is most important in open forests and row crops and least important (also least practical) in dense crop canopies.

Finally, it is useful to remember the pitfalls that exist when standard methods for inferring fluxes from mean gradients are applied too close to the canopy top. Most such techniques rely on assuming equality between the eddy diffusivities for heat, water vapor, and CO_2 (the Bowen ratio method) or between scalars and momentum (the aerodynamic method). In the roughness sublayer above the top of the canopy, these equalities are lost even at neutral stability, as we saw in Chapter 3. Nevertheless, the rapid decrease in scalar gradients, as we move away from the surface, tempts many workers to use these methods in the roughness sublayer, where gradients are large and more easily measured. The literature is littered with the anomalous and inexplicable results of such adventures.

6.6 Measuring over sloping terrain

We have already mentioned the errors that arise in measured fluxes when instruments are not carefully leveled (Section 6.3). To be more precise, problems occur when vector quantities like velocities or fluxes are measured in a reference framework that does not coincide with that of the equations used to analyze them. In the most familiar case of homogeneous flow over flat ground described in Cartesian coordinates, such difficulties do not occur. In more complex flows, we must be aware of them and deal with them. Two different strategies can be adopted to avoid this problem:

The first is to choose a set of axes that will be used to interpret the measurements, then to align all instruments carefully with the local direction of these axes. A familiar example is the choice of rectangular Cartesian coordinates with the z axis aligned with the geopotential gradient, x in the E-W direction and y in the N-S direction. The difficulties are those of accurately knowing the x, y, z directions at every instrument and aligning the instrument with them. On a single rigid tower such as the BAO, this is the usual choice. In experiments with many instruments mounted on portable towers this strategy is much more difficult to follow.

The second choice is to allow the flow to set the coordinate directions. For example, x may be taken in the direction of the local mean wind vector $\overline{\mathbf{u}}$. The instrument is aligned at its optimum orientation to the prevailing wind and measurements transformed mathematically from the reference frame of the instrument into the chosen frame *after* they have been collected. The use of streamline coordinate equations, recommended in Chapter 5 as the preferred choice for complex terrain studies, relies on this strategy. The problems with this approach are threefold:

1. *Choosing an appropriate set of equations in which to analyze the data.* The streamline coordinate equations are one such set, but in using them we encounter the second problem.
2. *Choosing the direction of the* y *and* z *axes.* (We assume x is taken along a streamline.) In streamline coordinates, y and z are in the directions of the binormal and principal normal to the streamline. To define these directions we need to know how the streamline is varying in space, whereas usually we only have a measurement at a single point. This is a crucial question, and we return to it below.
3. *Knowing the direction of the accelerations due to gravity* **g** *relative to the axes of the instrument.* (This may be easier sometimes than aligning the instrument's vertical path with that direction.) If the buoyancy forces are important, we need this information.

The simplest way to illustrate these problems is to go through the steps of producing data in streamline coordinates. Three mathematical transformations are required to get from the instrument's reference frame to the streamline reference frame. We assume that digitally sampled velocity data are recorded in the instrument's rectangular Cartesian reference frame which we denote by x_1, y_1, z_1. The velocity components in this frame are u_1, v_1, w_1, whereas the components of the acceleration due to gravity vector **g** are $g_{x_1}, g_{y_1}, g_{z_1}$.

The first step is to average the time series of velocities to produce a mean wind vector $\overline{\mathbf{u}}$ with components $\overline{u}_1, \overline{v}_1, \overline{w}_1$. It is immediately clear that the mean vector depends upon both the averaging time chosen and the spatial resolution of the instrument. McMillen (1988), in partially adopting the streamline strategy we describe here, used a 200-s running-mean filter to detrend incoming variables and applied a coordinate rotation at the end of each 30-min period.

The first transformation swings the x_1 and y_1 axes around z_1 to produce a new set of axes (x_2, y_2, z_2) with x_2 in the plane spanned by $\overline{\mathbf{u}}$ and z_1 (Fig. 6.20). The new velocity components are given by

$$\begin{aligned} u_2 &= u_1 \cos\theta + v_1 \sin\theta, \\ v_2 &= -u_1 \sin\theta + v_1 \cos\theta, \\ w_2 &= w_1, \end{aligned} \tag{6.11}$$

where

$$\theta = \tan^{-1}\left(\frac{\overline{v}_1}{\overline{u}_1}\right).$$

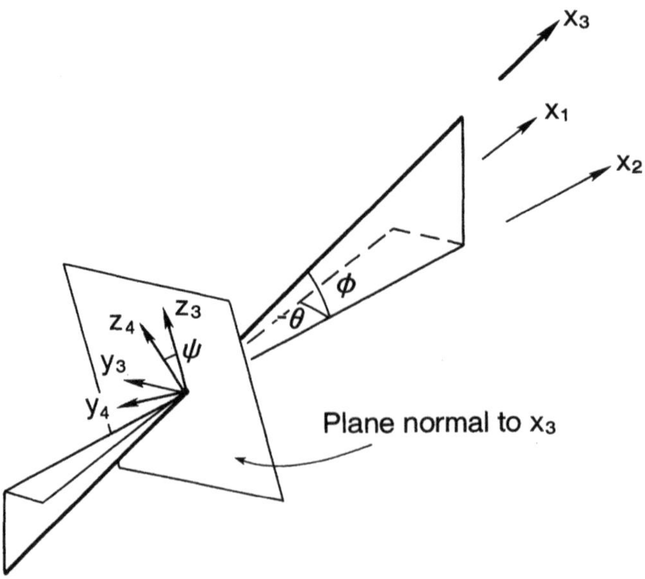

FIG. 6.20 Three-dimensional coordinate rotations for alignment of coordinate axes to the flow field over sloping terrain.

This rotation forces $\overline{v_2} = 0$, but the vertical component is unaffected. The components of the gravity vector become

$$
\begin{aligned}
g_{x_2} &= g_{x_1} \cos\theta + g_{y_1} \sin\theta, \\
g_{y_2} &= -g_{x_1} \sin\theta + g_{y_1} \cos\theta, \\
g_{z_2} &= g_{z_1}.
\end{aligned} \tag{6.12}
$$

The second rotation swings x_2 and z_2 about y_2 into new directions x_3, y_3, z_3 so x_3 now points along the mean wind direction (Fig. 6.20). The new velocity components are

$$
\begin{aligned}
u_3 &= u_2 \cos\phi + w_2 \sin\phi, \\
v_3 &= v_2, \\
w_3 &= -u_2 \sin\phi + w_2 \cos\phi,
\end{aligned} \tag{6.13}
$$

where

$$\phi = \tan^{-1}\left(\frac{\overline{w_2}}{\overline{u_2}}\right).$$

This rotation forces $\overline{w_3} = 0$. The gravity vector becomes

$$
\begin{aligned}
g_{x_3} &= g_{x_2} \cos\phi + g_{z_2} \sin\phi, \\
g_{y_3} &= g_{y_2}, \\
g_{z_3} &= -g_{x_2} \sin\phi + g_{z_2} \cos\phi.
\end{aligned} \tag{6.14}
$$

At this point the x_3 axis points in the direction of the mean wind vector, but the directions of the y_3 and z_3 axes depend on the initial choice of instrument alignment. Before we decide on a means to define the y and z directions, we must discuss the physical basis for the choice of axes in streamline coordinates.

In Chapter 5 we pointed out that the z axis in streamline coordinates points in the direction in which the streamline is curving most rapidly, the principal normal direction. In two-dimensional flows, as in winds blowing normal to a ridge, the principal normal direction is always perpendicular to the stream surfaces and so is the natural generalization of the z direction over flat ground. This is true for axially symmetric flows also, but in three-dimensional flow the behavior of stream surfaces is often much more complicated. In fact, the stream function, whose constant surfaces are the stream surfaces, does not exist in three dimensions except in some restricted types of flows.

In principle, we could identify a stream surface over flat ground somewhere upwind of our observing site and follow it as it is distorted over a hill, keeping our z_3 direction normal to its surface. In practice this is impossible. Instead we continue to identify the z_3 direction with the principal normal. We do this for two reasons: first, for compatibility with the simpler two-dimensional case; second, and more fundamentally, because this choice is what forces the canonical separation of the momentum equation in streamline coordinates into equations for linear and angular momentum.

The principal normal direction is defined by the spatial rate of change of **t**, the unit tangent to the streamline as we move a distance s along the streamline. This dependence is expressed in the Serret-Frenet formula:

$$\frac{d\mathbf{t}}{ds} = \frac{\mathbf{n}}{R}, \tag{6.15}$$

where $\mathbf{t} = \overline{\mathbf{u}}/|\overline{\mathbf{u}}|$, $\mathbf{n}$ is the unit vector in the principal normal direction, and R is the local radius of streamline curvature. Equation (6.15) can be rewritten in terms of derivatives taken in the Cartesian reference frame of the instrument (x_1, y_1, z_1) but to do this would not be helpful since a single instrument does not provide the information on the spatial gradients of the velocity field necessary to deduce the components of $\mathbf{n}$ from (6.15).

We are, however, familiar with one means of connecting spatial velocity gradients with single point measurement, that is, by an eddy diffusivity. In rectangular Cartesian coordinates the connection between the mean strain and turbulent stress tensors through an eddy diffusivity K can be written as

$$\overline{u_i'u_j'} - \frac{1}{3}\overline{u_i'u_i'} = -\frac{K}{2}\left(\frac{\partial \overline{u}_i}{\partial x_j} + \frac{\partial \overline{u}_j}{\partial x_i}\right). \tag{6.16}$$

If we transform the off-diagonal $(i \neq j)$ components of (6.16) into streamline coordinates, we obtain

$$\overline{u'w'} = -K\left(\frac{d\overline{\mathbf{u}}}{dz} + \frac{\overline{\mathbf{u}}}{R}\right), \tag{6.17a}$$

$$\overline{v'w'} = -KC_{123}, \tag{6.17b}$$

$$\overline{u'v'} = -K\frac{d\overline{\mathbf{u}}}{dy}. \tag{6.17c}$$

The quantity C_{123} that appears in (6.17b) is called the commutator of the field of streamlines. It is a measure of the degree that streamlines twist as well as curve. It is zero in two-dimensional and axially symmetric flow fields and on the centerline of axially symmetric hills because streamlines are plane curves in these situations. It is likely to be very small close to the surface over low topography, which we might define as hills that are gentle enough to avoid separation. In fact, it is a reasonable rule of thumb that C_{123} will be negligible in any region where we can apply the modified surface similarity theory proposed for use over hills in Chapter 5, Section 5.4. We should, however, be wary of assuming $C_{123} = 0$ over steep topography, on the flanks of axially symmetric hills, or far from the surface in any kind of three-dimensional complex terrain.

With these caveats, we see that, if $C_{123} = 0$ and an eddy diffusivity can be used (its actual form is immaterial), equation (6.17b) provides a means of choosing the z direction; we simply rotate the y_3, z_3 axes around x_3 until $\overline{v'w'} = 0$, as indicated in Fig. 6.19. The new velocity components are

$$\begin{aligned} u_4 &= u_3, \\ v_4 &= v_3 \cos\psi + w_3 \sin\psi, \\ w_4 &= -v_3 \sin\psi + w_3 \cos\psi, \end{aligned} \tag{6.18}$$

where

$$\psi = \frac{1}{2}\tan^{-1}\left[\frac{2\overline{v_3 w_3}}{(\overline{v_3^2} - \overline{w_3^2})}\right].$$

If $\overline{v_2 w_2} = 0$, no rotation is needed. If $\overline{v_2^2} = \overline{w_2^2}$, we have $\psi = 45°$. The components of the gravity vector become

$$\begin{aligned} g_{x_4} &= g_{x_3}, \\ g_{y_4} &= g_{y_3} \cos\psi + g_{z_3} \sin\psi, \\ g_{z_4} &= -g_{y_3} \sin\psi + g_{z_3} \cos\psi. \end{aligned} \tag{6.19}$$

Two notes of caution must be sounded at this point. Typical run-to-run variations in measured $\overline{v_3'w_3'}$ from any of the causes discussed in Chapter 7 may lead to apparently erratic behavior in the direction of the z axis. Careful choice of smoothing filters may be necessary to rectify this. Second, rotating into streamline coordinates does not mean that momentum transfer in the y direction vanishes. Equation (6.17c) demonstrates this even for the restrictive case where an eddy diffusivity may be used.

In situations where we cannot assume that K exists and $C_{123} = 0$, there is no alternative to direct measurement of the spatial variation in the velocity field. In such situations it is tempting to choose the z direction arbitrarily. For example, it could be taken as lying in the plane containing $\overline{\mathbf{u}}$ and $\mathbf{g}$. It must be emphasized, however, that data collected in such a reference frame would not form the dependent variables of the streamline coordinate equations and a new set of equations would have to be derived in order to perform the analysis. Such equations would be linear combinations of the streamwise and cross-stream momentum equations but the interpretive power of the streamline equations, for which $(\overline{u}_4 + u_4'), v_4'$, and w_4' are automatically the dependent variables, would be lost.

In order to apply streamline coordinate analysis, the important length scales L_a and R must be deduced. The acceleration length scale L_a can be inferred if velocity data from a series of masts are available. See, for example, Finnigan and Bradley (1983). Close to the surface, it is usually reasonable to take $R = R_0 + z$, where R_0 is the surface curvature. This could be measured directly by surveyor's methods or, on the large scale, deduced from a contour map. It is not advisable to attempt to deduce R from (6.17a).

To summarize, over gentle hills or close to the surface on steeper topography, as long as separation regions are avoided, working in streamline coordinates can avoid the necessity of accurately aligning anemometers relative to some external reference frame. In flows with complex distortions, although streamline coordinates may still be the preferred choice for analysis, information on the spatial evolution of the flow is required to fix the coordinate directions and to deduce the curvature and acceleration length scales. In such cases the relative orientation of the sensors, at least, must be accurately known. It is worth noting that most surface layer analyses undertaken to date fall into the first category; very few field experiments dealing with strongly distorted flows, as defined above, have yet been attempted.

6.7 In situ measurements above tower heights

Extending measurements made near the ground with in situ sensors into the upper boundary layer is by no means a simple task. The techniques available for reaching those heights to measure mean winds, temperatures, and humidities are many. They include free balloons, tethered balloons, kytoons (inflatable kites), aircraft, and drones. Measuring fluxes and variances above tower heights poses an even greater challenge. The difficulty arises because turbulence measurements require stable platforms and precise leveling for best results. Two techniques have so far provided dependable turbulence data above 100 m, but they are both costly and complex and involve specialized equipment and trained personnel to operate them. In one, the tethering cable of a very large balloon is used to suspend sensors at different heights (Caughey and Palmer, 1979; Readings et al., 1974). In the other, an aircraft is instrumented for level flights at different heights over the site (Lenschow et al., 1980; Nicholls and Readings, 1981).

The tethering cable of a large balloon offers the closest approximation to a tall tower. Barrage balloons of the type used in the Minnesota experiment (Readings et al., 1974) are ideal for this purpose, being capable of ascents up to 3 km with large payloads. This type of balloon is relatively insensitive to turbulent fluctuations in the wind but does oscillate laterally as it hovers downwind of the tethering point. The oscillation, although very slow (5–10-min period), introduces overestimations in the measured wind speed (~10%) and its variance (~30%) without seriously affecting the vertical velocity and temperature variances or even the fluxes of momentum and heat (Haugen et al., 1975). The turbulence probe has to be specially designed to maintain its verticality and orientation to the wind while attached to the tethering cable.

The instrumented aircraft measures the spatial-temporal distribution of variables which differs from the temporal measurements obtained from sensors on the balloon cable. (If the aircraft speed U is large compared with the turbulent velocity scales, we can use Taylor's hypothesis, writing $\kappa_1 = 2\pi f/U$.) The aircraft can probe the entire depth of the CBL above 50 m in less than 2 h, flying in combined horizontal- and vertical-profiling modes. The aircraft itself, however, responds to turbulent motions, and this contamination has to be measured accurately and removed from the data. Sensors for measuring wind, temperature, and humidity fluctuations are usually mounted on specially designed booms or supports on the front of the aircraft. These sensors have to respond more rapidly to the fluctuations than their counterparts on balloon cables because aircraft speeds are higher than eddy translation speeds in the boundary layer. Corrections also need to be applied for compressibility and aircraft flow distortion effects on the measurements. These corrections complicate subsequent analysis of the aircraft data. Details of the instrumentation and the various corrections applied to the measurements are described by Lenschow (1986). Accuracies of 10% in the second moments can be achieved with flight legs of 20 min (80–120 km) or more.

6.8 Remote sensors for the boundary layer

Remote sensors are now used increasingly in boundary layer experiments. They have the range to get above tower heights, but their accuracies and spatial resolutions are still too coarse for detailed turbulence work. Turbulence parameters such as C_T^2, C_V^2 and the fluxes of heat and momentum can be inferred indirectly from backscattered and forward-propagated signals with varying degrees of accuracy. In many applications they are valued for the broad spatial coverage they offer in the forms of vertical profiles of winds and temperatures; for time-height plots showing evolution of such features as the daytime capping inversion, gravity waves, elevated inversion layers, and SBL depths; and for flow fields over complicated terrain. Such remote sensors fall into three major categories depending on the type of signal used and are commonly referred to as sodars, radars, and lidars, acronyms for *so*und, *ra*dio, and *li*ght *d*etection *a*nd *r*anging, respectively.

6.8.1 *Sodars*

The standard Doppler sodar, operating at frequencies around 2 kHz, is available in different versions from several manufacturers. In its monostatic version, the sodar measures radial wind components along three narrow beams, one pointing vertically and two tilted 30° (typically) from zenith pointing E (or W) and N (or S). The wind components are derived from Doppler shifts measured in the backscattered sound pulses transmitted sequentially in the three directions. (Each antenna switches from transmitting to receiving mode to sense the back-scattered signal.) The three beams may be produced either from three separate collocated antennas or from a phased array of transducers that steer the beam electronically into those directions. The range resolution for such a sodar (i.e., the length of the acoustic pulse transmitted) is typically 30 m over a height range 50–500 m. (Radial velocities measured along the two tilted beams are converted into horizontal components along the cardinal directions.) The minimum cycling time for one set of transmissions is typically 6 s; the minimum averaging time for a reliable estimate of the radial wind component is 6 min. Expected accuracy in these sodars is ± 1 m s^{-1}.

Doppler minisodars operating in the 5–8-kHz range, with similar antenna options, claim better accuracies ($\pm$ 0.5 m s^{-1}), better range resolutions ($\sim$10 m), shorter averaging times ($\sim$1 min), and greatly reduced operating range (10–200 m). Only a few research versions exist (Coulter, 1990). Because of their small size and portability, they can be placed in difficult-to-reach locations; their low cost and short range permit deployment of several units in the same area for studies of complicated flows within valleys and on sloping terrain.

Time-height facsimile records of the monostatic echo intensities, familiar to sodar enthusiasts for over two decades, provide useful records of temperature structures in the ABL and their evolution with time. In the monostatic mode these backscattered signals are produced entirely by temperature inhomogeneities (C_T^2). Many ABL structures [e.g., convective plumes, waves, gust fronts, capping inversions, and ground-based stable turbulent layers (used often as an indicator of SBL depth)] contain temperature inhomogeneities that scatter sound, and each can be identified by its own special signature (Neff and Coulter, 1985). Because of the sodar's limited maximum range, the capping inversion can be tracked through only part of its rise, until about 1000 local time on most days.

In the bistatic mode (in which separate tilted transmitter and receiver beams intersect at designated heights above ground), scattering is produced primarily by velocity inhomogeneities (C_V^2); the contribution from C_T^2, usually two orders of magnitude smaller, is ignored (Thomson et al., 1978). Bistatic Doppler systems (e.g., Kaimal and Haugen, 1977), common a decade ago, need more area on the ground than the monostatic systems to accommodate their widely separated antennas and, moreover, perform less well. The trend today is toward increasingly compact transducer groupings, such as multiple monostatic antennas and phased arrays.

6.8.2 *Radar*

Doppler wind profiling radars operating at UHF (915 and 404 MHz) and VHF (50 MHz) frequencies currently used for monitoring winds in the troposphere (Frisch et al., 1986; Strauch et al., 1984), have minimum ranges that are too high ($\sim$1 km) for boundary layer measurements. The new 915-MHz boundary layer profiler described by Ecklund et al. (1988), with its highly portable (0.9×0.9 m) microstrip antenna panels and compact electronics, appears, however, ideally suited for that application. It has a height resolution of $\sim$50 m, a minimum range of $\sim$100 m, and a maximum range conservatively rated at 1.5 km. It adequately covers the CBL depth except under very dry conditions when moisture fluctuations also happen to be very low. In a moist environment, the maximum range can be as high as 4.5 km.

Like monostatic Doppler sodars, these radar wind profilers are ground-based, three-beam systems: one beam vertical and two tilted 15° from zenith in the N (or S) and E (or W) directions. The antennas are pulsed rapidly ($\sim$100-μs repetition rate), the propagation speed being much higher for the electromagnetic wave than for sound. This rapid sampling is necessary because the signal-to-noise ratio in radar returns is very low compared to sodars, and a large number of samples is needed to achieve comparable accuracy. Typically, a single reading from any range gate (time gate corresponding to calculated range interval along the beam) is an average of $\sim$ 150,000 successive returns (transmission every 100 μs for, say, 15 s). Counting the time needed for switching between antennas, the total cycling time for one sequence of data points is about 1 min. A minimum of 20 such data points is needed to construct a wind profile accurate to within ± 0.5 m s^{-1}.

The strongest radar returns are predominantly from refractive index variations of scale size $\lambda/2$, where λ is the radar wavelength (33 cm at 915 MHz). If the $\lambda/2$ scale sizes are in the inertial subrange of the turbulence spectrum, the radar reflectivity is given by 0.38 $C_n^2\lambda^{-1/3}$ (Ottersten, 1969). (C_n^2 is the refractive index structure parameter mentioned in Chapter 2.) The abrupt drop observed in backscatter intensity above the CBL is attributed to a sharp attenuation of turbulent energy in the $\lambda/2$ scale sizes in the relatively smooth flow in that region. Lack of moisture also plays a role. For radar signals, C_n^2 is most strongly influenced by C_Q^2, with lesser contributions from C_T^2 and C_{TQ}. This new term, C_{TQ}, is the structure parameter derived from the temperature-humidity cospectrum (Friehe et al., 1975) through the relationship:

$$C_{\theta q}(\kappa_1) = 0.25\, C_{TQ}\, \kappa_1^{-5/3}. \tag{6.20}$$

The C_{TQ} contribution may be positive or negative, depending on the sign of the correlation between temperature and humidity, so it could either enhance or diminish C_n^2. In any case, radar reflectivity drops off abruptly at the top of the CBL, and the level where this drop begins can, under ideal conditions, be used as a measure of z_i.

6.8.3 *Lidars*

Pulsed Doppler lidars are optical analogs of the more familiar Doppler radars. The lidar's shorter wavelength makes it sensitive to aerosols and cloud particles that follow air motions well and therefore serve as good targets for wind sensing. The aerosol loading in the boundary layer is usually sufficient for many lidars to operate with good signal-to-noise ratio. This ratio can drop to very low values above the CBL if the air is very "clean." Obscurants like smoke will also limit its range, whereas thick fog and clouds could block it entirely.

The lidar can be deployed in a three-beam mode as in radar wind profilers. It is, however, most effectively used in the velocity azimuth display (VAD) mode making conical scans through 360° azimuth at constant elevation angle (say 30°). The lidar beam has no side lobes and therefore suffers no complication from ground clutter, except for direct strikes. The CO_2 Doppler lidar, operating in the infrared (10.6 μm), is ideal for wind profiling. It is eyesafe and can therefore be used in field experiments to provide low-angle conical and elevation scans of the wind field over a 10-km horizontal range. In the VAD mode, the radial winds in each range gate execute a sinusoidal pattern as the beam moves through its conical scan. The azimuth of the velocity peak yields the wind direction and the amplitude of the peak, the speed. This assumes that the flow is horizontally homogeneous. The range resolution is poor ($\sim$300 m) because of the minimum record length needed for Doppler processing; however, at shallow scan angles this range resolution would translate to much better height resolutions.

The lidar may also be used as an indicator of z_i. The top of the CBL is often clearly delineated in lidar backscatter displays because of strong aerosol gradients in that region. C_n^2 for optical waves is produced primarily by C_T^2, with much smaller contributions from C_Q^2 and C_{TQ} (Friehe et al., 1975). Nevertheless, the C_T^2 maximum at z_i (Chapter 2), although strong enough to be identified in the sodar returns, is not apparent in lidar plots.

Present CO_2 lidars operating in the Doppler mode are large, complicated systems (Hardesty, 1984). They have proved valuable in complex terrain studies, especially for mapping flows through canyons. Because of their complexity and high operating cost, they are seldom available for low-budget boundary layer studies. Doppler lidar technology is, however, rapidly moving toward less expensive, more compact, and more robust systems with hopes also for better accuracy and resolution; when those systems are developed, we can expect to see lidars used more widely in boundary layer experiments.

6.8.4 *Radio acoustic sounding system*

Radio Acoustic Sounding System (RASS) combines both radar and acoustic techniques to sense temperature profiles remotely. The concept is not new, but it was not widely used in the United States until the arrival of sensitive Doppler radar systems (such as those developed for wind profiling). These new radars are capable of tracking Doppler shifts in the weak signals scattered back from radar refractive index undulations induced by sound waves propagating upward from a source near the radar. The Doppler shift in the signal is a measure of the speed of sound, which is directly related to the virtual temperature, T_v. [From our definitions in (6.3) and (6.10), we have $c^2 \simeq 403T_v$.] The heights of observation would be those corresponding to the radar range gates where the receding refractive index undulations (accompanying the sound) provide strong enough returns for Doppler processing.

Strong backscattered signals are obtained when the wavelength of the transmitted acoustic signal is matched to half the radar wavelength (the Bragg condition) and the radar and acoustic phase fronts are matched. For the 915-MHz radar, the matching acoustic frequency is $\sim$2 kHz at temperatures typically encountered in the ABL. But exact matching is difficult to achieve with a fixed frequency acoustic source, given the spatial and temporal temperature variabilities in the ABL. So a frequency-modulated, continuous-wave (FM-CW) acoustic signal is transmitted (upward) to ensure Bragg match with the radar at all heights that give a detectable radar signal. The range gates where both Bragg match and phase match are achieved, and where the T_v readings are the most reliable, would vary from one radar scan to the next. However, the temperature readings thus obtained, when consolidated, yield profiles with fine details that include sharp inversions (Currier et al., 1988; May et al., 1990) that compare very well (within ± 1°C) with radiosonde data.

In all RASS systems (404- and 50-MHz systems included), the maximum range achieved is limited by strong winds that advect the acoustic wavefronts so they no longer match the radar wavefronts, by turbulence and temperature gradients that distort the acoustic wavefronts, and by acoustic attenuation. In the 915-MHz RASS, acoustic absorption is particularly serious. The maximum range of this RASS system is typically 750 m, although observations have been reported up to 1.2 km in moist air (Currier et al., 1988). The presence of moisture enhances RASS performance because acoustic absorption in the 2-kHz range (at temperatures encountered in the ABL) decreases with increasing humidity. The height resolution and minimum range for RASS are the same as for the 915-MHz wind profiler, $\sim$50 m and $\sim$100 m, respectively.

6.9 Special symbols

a	transducer diameter
c	velocity of sound in air
C_{123}	commutator of the field of streamlines
E	voltage across bridge
g_x, g_y, g_z	components of gravity vector along x, y, z directions
$\mathbf{g}$	gravity vector
$H(\lambda)$	power transfer function, function of λ
l	distance constant of anemometer
L_a	acceleration length scale
M_a, M_w	molecular weights for air and water vapor
$\mathbf{n}$	unit vector in the principal normal direction
p_d	dynamic pressure
R	local radius of curvature of streamline
R_0	surface curvature
s	separation distance
s	distance along streamline
$\mathbf{t}$	unit tangent vector to a streamline
t_1, t_2	sonic anemometer transit times
u_p	local pressure variation caused by the pressure probe
$\overline{\mathbf{u}}$	local mean wind vector
U	aircraft speed
V_d	velocity component along acoustic path d
V_n	velocity component normal to V_d
V_o	bridge output in volts
V_A, V_B	velocity components along nonorthogonal sonic anemometer axes, A and B
V_H	horizontal wind vector
V_X, V_Y	velocity components along orthogonal sonic anemometer horizontal axes, X and Y
γ_a, γ_w	ratio of specific heats for air and water vapor
θ	azimuth angle
λ_c	cutoff wavelength
λ_0	half-power wavelength
ϕ	elevation angle
ψ	angle of coordinate rotation around the $\overline{\mathbf{u}}$ direction

Appendix 6.1 Principle of the sonic anemometer and thermometer

The sonic anemometer measures wind components from transit times of acoustic signals traveling in opposite directions along a fixed path. Figure 6.21 illustrates the effect of wind V on the sound ray vectors for a single-axis anemometer. If t_1 and t_2 are the transit times for pulses traveling along wind directions defined positive and negative, respectively, we have

$$t_1 = \frac{d}{(c\cos\gamma - V_d)}, \tag{6.21}$$

$$t_2 = \frac{d}{(c\cos\gamma + V_d)}, \tag{6.22}$$

where d is the path length, c is the velocity of sound, V_d is the wind component along the path, and $\gamma = \sin^{-1}(V_n/c)$, V_n being the component normal to the path.

Two methods are used to extract V_d. One, which involves measuring $(t_1 - t_2)$, is simpler to implement if the pulses travel simultaneously along two close, but separate, parallel paths. The measurement has a residual temperature sensitivity, which, although small, cannot be ignored. Assuming $V_n^2 \ll c^2$, we have

$$\begin{aligned} t_2 - t_1 &= \frac{2d}{c^2} V_d \\ &\simeq \left[\frac{2d}{403T(1 + 0.32e/p)}\right] V_d \\ &\approx \left(\frac{d}{201.5T_v}\right) V_d, \end{aligned} \tag{6.23}$$

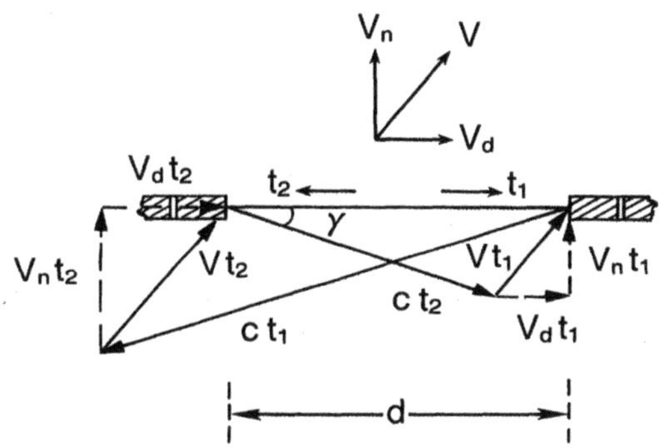

FIG. 6.21 Sound ray vectors for a single-axis, single-path sonic anemometer.

on substituting for c^2 from (6.3) and assuming $T(1 + 0.32e/p)$ is an adequate approximation of the virtual temperature T_v [defined as $T(1 + 0.38e/p)$]. The error involved in assuming the measured temperature equals T_v is on the order of 0.01°C (Kaimal and Gaynor, 1991), well within the bounds of experimental uncertainty. [The difference arises from the fact that the coefficient of e/p in the expression for T_v is $(1 - M_w/M_a)$, where M_w and M_a are the molecular weights for water vapor and air, whereas that for the sonic-derived temperature is $(\gamma_w/\gamma_a - M_w/M_a)$, where γ_w and γ_a are the ratio of specific heats for water vapor and air.]

The other method computes the reciprocals of t_1 and t_2 to give

$$\frac{1}{t_1} - \frac{1}{t_2} = \frac{2}{d} V_d, \tag{6.24}$$

or

$$V_d = \frac{d}{2}\left(\frac{1}{t_1} - \frac{1}{t_2}\right), \tag{6.25}$$

which is free of any temperature sensitivity. This approach is implemented in the newer single-path sonic anemometers that transmit signals back and forth sequentially along the same path and measure t_1 and t_2 separately, instead of the difference between them.

Sonic thermometry benefits as well from the separate determinations of t_1 and t_2. We can write

$$\begin{aligned}\frac{1}{t_1} + \frac{1}{t_2} &= \frac{2}{d} c \cos\gamma \\ &= \frac{2}{d} (c^2 - V_n^2)^{1/2}\end{aligned} \tag{6.26}$$

or

$$c^2 = \frac{d^2}{4}\left(\frac{1}{t_1} + \frac{1}{t_2}\right)^2 + V_n^2. \tag{6.27}$$

Substituting for c^2 from (6.3) and assuming again that the temperature derived from the speed of sound approximates T_v, (6.21) becomes

$$T_v = \frac{d^2}{1612}\left(\frac{1}{t_1} + \frac{1}{t_2}\right)^2 + \frac{1}{403} V_n^2. \tag{6.28}$$

If $1/t_1$ and $1/t_2$ are determined from the w axis of the sonic anemometer, V_n would be the magnitude of the horizontal wind vector V_H.

In the orthogonal array of Fig. 6.9,

$$V_n^2 = V_X^2 + V_Y^2, \tag{6.29}$$

where V_X and V_Y are wind components measured along the horizontal X and Y axes, corrected for transducer shadowing as described in Appendix 6.3.

Appendix 6.2 Nonorthogonal array

In an orthogonal array (e.g., Fig. 6.9), coordinate transformation to resolve the horizontal wind components is fairly straightforward. For the nonorthogonal 120° array (Fig. 6.8), however, the transformation involves more terms. If V_A and V_B are the velocities measured along the horizontal axes A and B (Fig. 6.22) and u and v are the velocity components along the vector mean coordinates,

$$u = V_A\left(\cos\theta - \frac{1}{\sqrt{3}}\sin\theta\right) + V_B\left(\cos\theta + \frac{1}{\sqrt{3}}\sin\theta\right), \tag{6.30}$$

$$v = -V_A\left(\frac{1}{\sqrt{3}}\cos\theta + \sin\theta\right) + V_B\left(\frac{1}{\sqrt{3}}\cos\theta - \sin\theta\right), \tag{6.31}$$

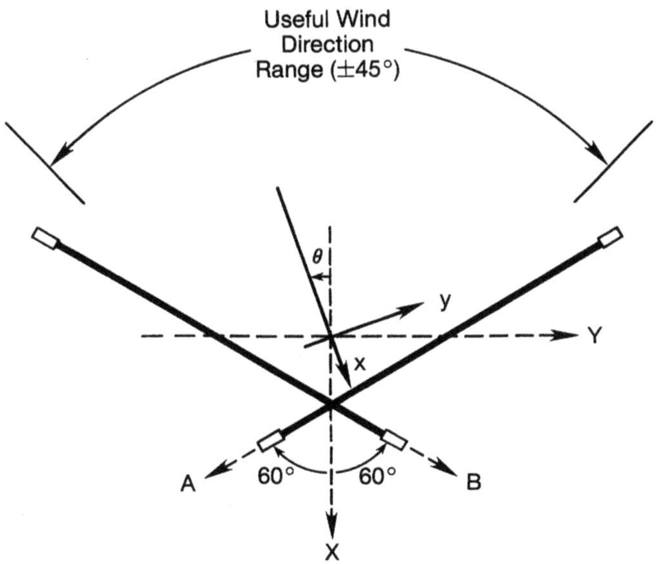

FIG. 6.22 Horizontal vector-mean wind coordinates (x, y) and nonorthogonal sonic anemometer axes (A, B).

where

$$\theta = \tan^{-1}\left[\frac{\overline{V}_B - \overline{V}_A}{\sqrt{3}(\overline{V}_B + \overline{V}_A)}\right],$$

where θ is the angle between the horizontal wind vector and the axis of symmetry of the array.

Appendix 6.3 Correction for transducer shadowing

In the orthogonal array of Fig. 6.9, partial shadowing of the acoustic path by the transducers causes the velocity readings to be underestimated. Results of wind tunnel and atmospheric tests indicate a linear drop in response as the wind direction approaches the direction of the acoustic path. The degree of attenuation is a strong function of the aspect ratio d/a for $d/a < 50$, where d is the path length and a the diameter of the transducer (Fig. 6.23). The measured velocity component $(V_d)_m$ for $d/a = 15$ becomes

$$(V_d)_m = \begin{cases} V_d(0.82 + 0.18\theta/75), & 0° \leq \theta \leq 75° \\ V_d, & 75° \leq \theta \leq 90°, \end{cases} \tag{6.32}$$

where

$$\theta = \cos^{-1}\left(\frac{|V_d|}{|V_H|}\right).$$

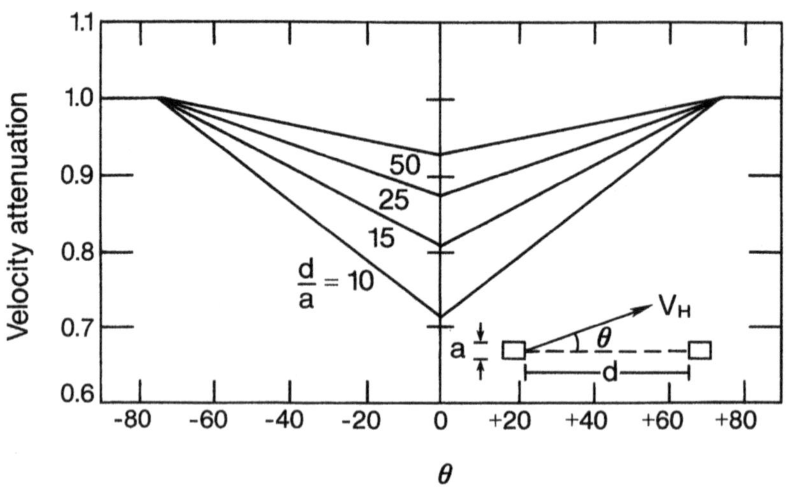

Fig. 6.23 Velocity attenuation from transducer shadowing in a sonic anemometer shown as a function of wind direction for various values of d/a.

θ is the angle between V_d and the horizontal wind vector V_H. The attenuation for d/a other than 15 can be estimated from the curves in Fig. 6.19. Note that Fig. 6.19 represents the shadow effect produced by the Applied Technologies' 1-cm diameter × 1-cm long transducer mounted at the end of a short 1-cm diameter transducer holder (Fig. 6.9). The shapes will be different for other types of transducers and mounts (Kaijo Denki and CSIRO/Dobbie arrays), as pointed out by Wyngaard and Zhang (1985).

References

Andreae, M. O., and D. S. Schimel, 1989: Exchange of trace gases between terrestrial ecosystems and the atmosphere, Report of the Dahlem Workshop, 19-24 February 1989, Berlin, Germany, Wiley-Interscience, New York, 346 pp.

Buck, A. L., 1976: A variable path Lyman-alpha hygrometer and its operating characteristics. *Bull. Amer. Meteor. Soc., 57*, 1113–1118.

Busch, N. E., O. Christensen, L. Kristensen, L. Lading, and S. E. Larsen, 1980: Cups, vanes, propellers and laser anemometers. In *Air-Sea Interaction—Instruments and Methods*. Plenum Press, New York, 11–46.

Caughey, S. J., and S. G. Palmer, 1979: Some aspects of turbulence structure through the depth of the convective boundary layer. *Quart. J. Roy. Meteor. Soc., 105*, 811–827.

Coppin, P. A., and K. J. Taylor, 1983: A three-component sonic anemometer/thermometer system for general micrometeorological research. *Bound.-Layer Meteor., 27*, 27–42.

Coulter, R. L., 1990: Minisodars—Applications and potential. In *Acoustic Remote Sensing* (S. P. Singal, Ed.), Tata McGraw-Hill, New Delhi, 88–96.

Currier, P. E., W. L. Ecklund, D. A. Carter, J. M. Warnock, and B. B. Balsley, 1988: Temperature profiling using a UHF wind profiler and an acoustic source. *Extended Abstracts, Lower Tropospheric Profiling—Needs and Technologies*, May 31–June 3, 1988, Boulder, CO; American Meteorological Society, Boston, MA, 121–122.

deVries, D. A., and J. R. Philip, 1986: Soil heat flux, thermal conductivity, and the null alignment method. *Soil Sci. Soc. Am. J., 50*, 12–18.

Dunin, F. X., W. Reyenga, and I. C. McIlroy, 1991: Australian lysimeter studies of field evaporation, *Proc. International Symp. on Lysimetry* (R. G. Allen, T. A. Howell, W. O. Pruitt, I. A. Walter, and M. E. Jensen, Eds.), 23-25 July 1991, IR Div./ASCE, Honolulu, HI, 237–245.

Ecklund, W. L., D. A. Carter, and B. B. Balsley, 1988: A UHF wind profiler for the boundary layer: Brief description and initial results. *J. Atmos. Oceanic Tech., 5*, 432–441.

Elliott, W. P., 1972: Instrumentation for measuring static pressure fluctuations within the atmospheric boundary layer. *Bound.-Layer Meteor., 2* , 476–495.

Finkelstein, P. L., J. C. Kaimal, J. E. Gaynor, M. E. Graves, and T. J. Lockhart, 1986: Comparison of wind monitoring systems. Part I. In situ sensors. *J. Atmos. Oceanic Tech., 3* , 583–593.

Finnigan, J. J., and E. F. Bradley, 1983: The turbulent kinetic energy budget behind a porous barrier: An analysis in streamline coordinates. *J. Wind Eng. Ind. Aerodyn., 15*, 157–168.

Friehe, C. A., J. C. LaRue, F. H. Champagne, C. H. Gibson, and C. F. Dreyer, 1975: Effects of temperature and humidity fluctuations on the optical refractive index in the marine boundary layer. *J. Opt. Soc., 65*, 1502–1511.

Frisch, A. S., B. L. Weber, R. G. Strauch, D. A. Merritt, and K. P. Moran, 1986: The altitude coverage of the Colorado wind profilers at 50, 405, and 915 MHz. *J. Atmos. Oceanic Tech., 3*, 680–692.

Grant, A. L. M., and R. D. Watkins, 1989: Errors in turbulence measurements with a sonic anemometer. *Bound.-Layer Meteor., 46* , 181–194.

Hanafusa, T., Y. Kobori, and Y. Mitsuta, 1980: Single head sonic anemometer-thermometer. In *Instruments and Observing Methods, Report No. 3*, World Meteorological Organization, Geneva, Switzerland, 7–13.

Hardesty, R.M., 1984: Coherent DIAL measurements of range-resolved water vapor concentration. *Appl. Optics, 23*, 2545–2553.

Haugen, D. A., J. C. Kaimal, C. J. Readings, and R. Rayment, 1975: A comparison of balloon-borne and tower-mounted instrumentation for probing the atmospheric boundary layer. *J. Appl. Meteor., 14*, 540–545.

Horst, T. W., 1973: Corrections for response errors in a three-component propeller anemometer. *J. Appl. Meteor., 12*, 716–725.

Kaimal, J. C., 1969: Measurement of momentum and heat flux variations in the surface boundary layer. *Radio Sci., 4*, 1147–1153.

Kaimal, J. C., and J. E. Gaynor, 1983: The Boulder Atmospheric Observatory. *J. Clim. Appl. Meteor., 22*, 863–880.

Kaimal, J. C., and J. E. Gaynor, 1991: Another look at sonic thermometry. *Bound.-Layer Meteor., 56*, 401–410.

Kaimal, J. C., and D. A. Haugen, 1969: Some errors in the measurement of Reynolds stress. *J. Appl. Meteor., 8*, 460–462.

Kaimal, J. C., and D. A. Haugen, 1977: An acoustic Doppler sounder for measuring wind profiles in the lower boundary layer. *J. Appl. Meteor., 16*, 1298–1305.

Kaimal, J. C., J. C. Wyngaard, and D. A. Haugen, 1968: Deriving power spectra from a three-component sonic anemometer. *J. Appl. Meteor., 7*, 827–837.

Kaimal, J. C., J. E. Gaynor, H. A. Zimmerman, and G. A. Zimmerman, 1990: Minimizing flow distortion errors in a sonic anemometer. *Bound.-Layer Meteor., 53*, 103–115.

Lang, A. R. G., 1978: Note on the cosine response of cylindrical net radiometers. *Agric. Meteor., 19*, 391–397.

Lenschow, D. H., 1986: Aircraft measurements in the boundary layer. In *Probing the Atmospheric Boundary Layer*, American Meteorological Society, Boston, MA, 39–55.

Lenschow, D. H., J. C. Wyngaard, and W. T. Pennell, 1980: Mean-field and second-moment budgets in a baroclinic, convective boundary layer. *J. Atmos. Sci., 37*, 1313–1326.

Lynch, R. A., and E. F. Bradley, 1972: Shearing stress meter. *J. Appl. Meteor., 13*, 588–591.

May, P. T., R. G. Strauch, K. P. Moran, and W. L. Ecklund, 1990: Temperature sounding by RASS with wind profiler radars: A preliminary study. *IEEE Trans. Geosci. Remote Sens., 28*, 19–28.

McMillen, R. T., 1988: An eddy correlation technique with extended applicability to non-simple terrain. *Bound.-Layer Meteor. , 43*, 231–245.

Monteith, J. L., 1975: *Vegetation and the Atmosphere, Vol. 1. Principles.* Academic Press, New York, 278 pp.

Monteith, J. L., 1976: *Vegetation and the Atmosphere, Vol. 2. Case studies.* Academic Press, New York, 439 pp.

Neff, W. D., and R. L. Coulter, 1985: Acoustic remote sensing. In *Probing the Atmospheric Boundary Layer*, American Meteorological Society, Boston, MA, 201–239.

Nicholls, S., and C. J. Readings, 1981: Spectral characteristics of surface layer turbulence over the sea. *Quart. J. Roy. Meteor. Soc., 107*, 591-614.

Nishiyama, R. T., and A. J. Bedard, 1991: A "Quad-Disc" static pressure probe for measurement of adverse atmospheres: With a comparative review of static pressure probe designs. *Rev. Sci. Instr., 62*, 2193–2204.

Ottersten, H., 1969: Atmospheric structure and radar backscattering in clear air. *Radio Sci., 4*, 1179–1193.

Priestley, J. T., and R. J. Hill, 1985: Measuring high-frequency humidity, temperature and radio refractive index in the surface layer. *J. Atmos. Oceanic Tech., 2*, 235–251.

Readings, C. J., D. A. Haugen, and J. C. Kaimal, 1974: The 1973 Minnesota atmospheric boundary layer experiment. *Weather, 29*, 309–312.

Shaw, R. H., G. E. Kidd, and G. W. Thurtell, 1973: A miniature three-dimensional anemometer for use within and above plant canopies. *Bound.-Layer Meteor., 3*, 359–380.

Strauch, R. G., D. A. Merritt, K. P. Morgan, K. B. Earnshaw, and D. van de Kamp, 1984: The Colorado wind profiling network. *J. Atmos. Oceanic Tech., 1*, 37–49.

Szeicz, G., 1975: Instruments and their exposure. In *Vegetation and the Atmosphere* , Vol. 1 (J. L. Monteith, Ed.), Academic Press, New York, 229-274.

Thomson, D. W., R. L. Coulter, and Z. Warhaft, 1978: Simultaneous measurements of turbulence in the lower atmosphere using sodar and aircraft. *J. Appl. Meteor., 17*, 723–734.

Webb, E. K., G. I. Pearman, and R. Leuning, 1980: Correction of flux measurements for density effects due to heat and water vapor transfer. *Quart. J. Roy. Meteor. Soc., 106*, 85–100.

Wilson, J. D., D. P. Ward, G. W. Thurtell, and G. E. Kidd, 1982: Statistics of atmospheric turbulence within and above a corn canopy. *Bound.-Layer Meteor., 24*, 495–519.

Wucknitz, J., 1980: Flow distortion by supporting structures. In *Air-Sea Interaction—Instruments and Methods.* Plenum Press, New York, 605–626.

Wyngaard, J. C., and S-F. Zhang, 1985: Transducer-shadow effects on turbulence spectra measured by sonic anemometers. *J. Atmos. Oceanic Tech., 2*, 548–558.

Zhang, S. F., J. C. Wyngaard, J. A. Businger, and S. P. Oncley, 1986: Response characteristics of the U.W. sonic anemometer. *J. Atmos. Oceanic Tech., 3*, 315–323.

7

ACQUISITION AND PROCESSING OF ATMOSPHERIC BOUNDARY LAYER DATA

Much of what we know about the structure of the boundary layer is empirical, the result of painstaking analysis of observational data. As our understanding of the boundary layer evolved, so did our ability to define more clearly the requirements for sensing atmospheric variables and for processing that information. Decisions regarding choice of sampling rates, averaging time, detrending, ways to minimize aliasing, and so on, became easier to make. We find we can even standardize most procedures for real-time processing. The smaller, faster computers, now within the reach of most boundary layer scientists, offer virtually unlimited possibilities for processing and displaying results even as an experiment is progressing.

The information we seek, for the most part, falls into two groups: (1) time-averaged statistics such as the mean, variance, covariance, skewness, and kurtosis and (2) spectra and cospectra of velocity components and scalars such as temperature and humidity. We discuss them separately because of different sampling and processing requirements for the two. A proper understanding of these requirements is essential for the successful planning of any experiment.

In this chapter we discuss these considerations in some detail with examples of methods used in earlier applications. We will assume that sensors collecting the data have adequate frequency response, precision, and long-term stability and that the sampling is performed digitally at equally spaced intervals. We also assume that the observation heights are chosen with due regard to sensor response and terrain roughness.

7.1 Time-averaged statistics

For calculations of means and higher order moments we need time series that are long enough to include all the relevant low-frequency contributions to the process, sampled at rates fast enough to capture all the high-frequency contributions the sensors are able to measure. Improper choices of averaging times and sampling rates can indeed compromise our statistics. We need to understand how those two factors affect our measurements in order to make sensible decisions on how long and how fast to sample.

7.1.1 *Choice of averaging time*

Ideally, the statistical quantities we would use to describe the properties of turbulent flow would be those obtained by "ensemble" averaging (i.e., averaging over many realizations under identical conditions). In reality we are forced to describe them in terms of averages over time, making the "ergodic" hypothesis that time averages are, for all practical purposes, equivalent to ensemble averages. For this assumption to be true, the fluctuations must be statistically stationary over the period chosen for analysis. Only then can we justify the application of the Reynolds averaging rules (Appendix 7.1), which specify conditions the data must satisfy for proper separation of the fluctuating components from the mean.

Statistical stationarity of a time series $\alpha(t)$ demands that variances and covariances approach stable values as the averaging time T is increased, which implies that an integral time scale $\mathcal{T}_\alpha$ exists for $\alpha(t)$. The requirements for averaging time T with $T \gg \mathcal{T}_\alpha$ can then be expressed in terms of $\sigma^2_{\overline{\alpha}}$, the variance of the measured time mean $\overline{\alpha}$ about the expected ensemble mean, and σ^2_α, the ensemble variance of α. From Lumley and Panofsky (1964), we have

$$\sigma^2_{\overline{\alpha}} \simeq \frac{2\sigma^2_\alpha \mathcal{T}_\alpha}{T}. \tag{7.1}$$

For averaging times $T \gg \mathcal{T}_\alpha, \sigma^2_{\overline{\alpha}}$ becomes negligible. We can now specify an acceptable level of error $\sigma_{\overline{\alpha}}/\overline{\alpha}$ (say ϵ) and express the averaging time required to keep $\sigma_{\overline{\alpha}}/\overline{\alpha}$ within that level as

$$T \simeq \frac{2\sigma^2_\alpha \mathcal{T}_\alpha}{\overline{\alpha}^2\epsilon^2}. \tag{7.2}$$

In order to use (7.2) to estimate suitable averaging times, we have to replace the ensemble variance σ^2_α by the more accessible time average variance—a further application of the ergodic hypothesis. For typical daytime values of $\sigma_u = 1 \text{ m s}^{-1}, \mathcal{T}_u = 10 \text{ s}$, and $\overline{u} = 5 \text{ m s}^{-1}$ and specifying $\sigma_{\overline{u}} = 0.1 \text{ m s}^{-1}$ (i.e., $\epsilon = 0.02$), we get $T = 2000 \text{ s} \simeq 30$ min, a reasonable averaging time for mean horizontal winds, corresponding to the passage of two or three of the large convective cells that extend through the depth of the CBL. Lumley and Panofsky (1964) offer a clear discussion of ergodicity and averaging requirements. Using the expression (7.2), they show how T increases with the order of the moment, estimating that it takes roughly five times longer to measure the fourth moment to the same accuracy as the second moment, assuming $\mathcal{T}_\alpha$ is the same for both (here α represents the higher powers or products of measured variables). The reason for this is the tendency for $\sigma_\alpha/\overline{\alpha}$ to increase with the order of the moment. The averaging time requirement can also be different for different second moments, depending on the magnitude of $\sigma_\alpha/\overline{\alpha}$.

If the averaging period can be increased indefinitely, ϵ can be kept to within negligible levels. $T \simeq 1$ h is, however, about as long as we can extend the averaging period without encountering nonstationarity in the form of diurnal variations in surface heat flux and boundary layer depth. With an averaging period this long we can expect errors on the order of 3.5%, 5%, and 10%–50% for near-surface σ_w^2, $\overline{w'\theta'}$, and $\overline{u'w'}$, respectively (Haugen, 1978).

7.1.2 *Choice of sampling rate*

Having established a suitable averaging time based on the accuracies we desire in the results, we turn to the next important consideration: how frequently to sample the signal. We assume the sampling process itself is virtually instantaneous, that is, the sensors are at least fast enough to respond to all the frequencies contributing to the process. (In the boundary layer, this means a high-frequency cutoff somewhere in the inertial subrange where spectral energy is falling off rapidly.) The conservative approach would be to sample at a rate commensurate with the frequency response of the sensors. But this is really not necessary because the accuracy in our statistics should, under stationary conditions, be a function only of the number of samples, N.

For equally spaced instantaneous samples with sampling interval $\Delta t > \mathcal{T}_\alpha$, the error variance can be expressed in a form similar to (7.1):

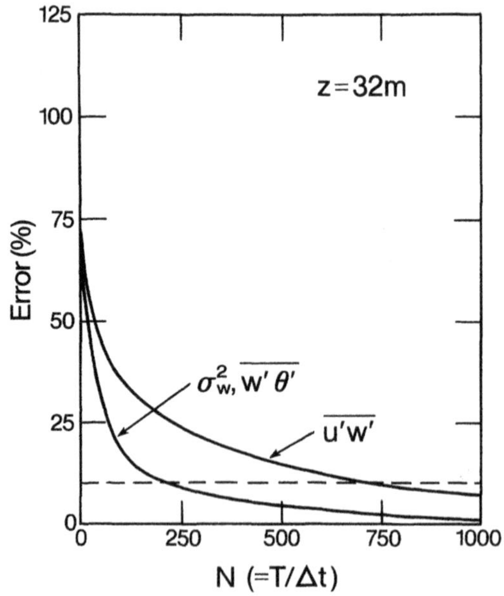

FIG. 7.1 Percentage error in calculated variances and fluxes as functions of the number of discrete samples used (after Haugen, 1978).

$$\sigma_{\overline{\alpha}}^2 = \frac{\sigma_\alpha^2}{N}, \tag{7.3}$$

where $N = T/\Delta t$. As with T in (7.2), we can estimate the number of samples needed to maintain a given level of confidence in $\overline{\alpha}$. For $\sigma_u = 1$ m s^{-1}, we need $N = 100$ to ensure that $\sigma_{\overline{u}} \leq 0.1$ m s^{-1}. For $T = 2000$ s, calculated from (7.2) for the same set of conditions, this value of N suggests a sampling interval $\Delta t =$ 20 s. Here too, it can be shown that the number of samples needed to keep ϵ within desired limits increases with the order of the moment.

The results of Haugen's (1978) calculations, summarized in Fig. 7.1, confirm the dependence of ϵ on N. Using various combinations of T ($\gg \mathcal{T}_\alpha$) and Δt, he derived empirical curves from field data that show the number of samples for 10% accuracy in σ_w^2 and $\overline{w'\theta'}$ to be 200 and in $\overline{u'w'}$ to be 750. The implication is that Δt can be increased without loss of accuracy as long as T is increased proportionally to maintain N at the same level. Note that we still require the sampling to be near instantaneous. The accuracy in the moment calculations depends on our ability to capture, at least occasionally, the peaks and valleys in the signal, as illustrated in Fig. 7.2. In the frequency domain, the process is explained in terms of spectral folding (see Section 7.2).

7.2 Spectra and cospectra

Conversion of data collected in the form of equally spaced samples in the time domain to spectral and cospectral estimates in the frequency domain involves a series of calculations we refer to as spectrum analysis or Fourier analysis. Several methods for accomplishing this conversion exist, but the one most widely used today is the fast Fourier transform (FFT), which accepts data points in groups of powers of 2 (e.g., $N = 2^{10} = 1024$) and calculates the Fourier components in an elegant, time-efficient manner ideally suited to low-level computer assembly language. [For details on FFT processing see Bendat and Piersol (1971) and

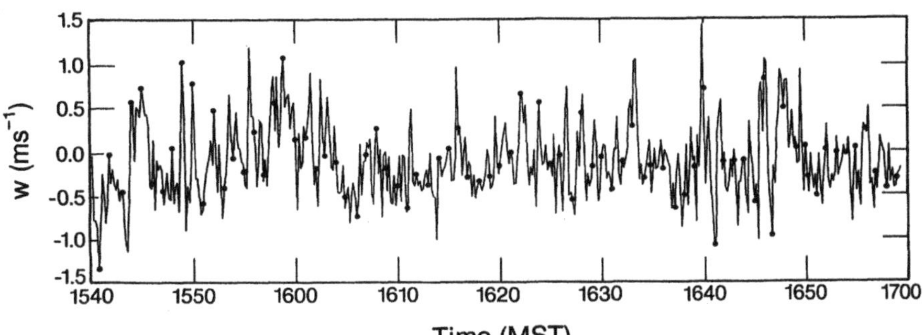

Fig. 7.2 Time series of w showing how even infrequent (but instantaneous) sampling (represented by the dots) can capture enough values at the peaks and valleys to approximate the true variance. The time indicated is Mountain Standard Time (MST).

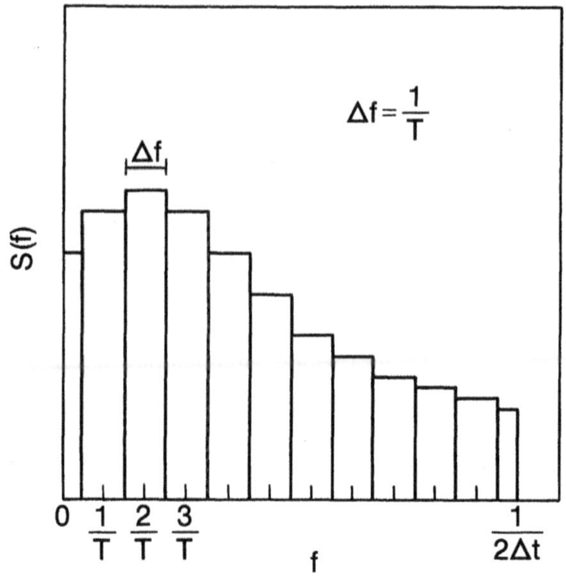

Fig. 7.3 Averaging period T and its relationship to spectral resolution and to the center frequency of spectral estimates.

Ramirez (1985).] The choices we make for T and Δt have particular relevance for FFT analysis because they directly influence the range and resolution of the spectral information the analysis provides. We discuss below the implications of those choices.

7.2.1 *Choice of record length*

For spectrum analysis we usually require record lengths somewhat longer than for variance calculations because of the clearer definition they promise in the energy peak and the rolloff on the low-frequency side of the spectrum. The selection of T automatically establishes the lowest frequency ($1/T$) we can resolve as well as the width of each elementary frequency band Δf in the spectrum output:

$$\Delta f = \frac{1}{T} = \frac{1}{N\Delta t}. \tag{7.4}$$

The elementary bands (excluding the first and last, which are only $\Delta f/2$ wide) are centered on frequencies $1/T, 2/T, 3/T \ldots, (1/2\Delta t - 1/T)$, as shown in Fig. 7.3. The number of spectral (or cospectral) estimates we obtain from an FFT analysis[1] is $N/2$, which gives us a total bandwidth of $1/2\Delta t$ in our spectral data.

[1] The FFT yields $N/2$ estimates that are complex numbers, so there is no loss of information. The squared magnitudes of the spectral estimates comprise the power spectrum, whereas the phase information is used in forming the cospectra and quadrature spectra (Appendix 2.1, Chapter 2).

With spectra (and cospectra) it is easy to determine whether the chosen T is long enough. A well-defined spectral peak in the $fS(f)$ [or $fC(f)$] plot and a rolloff over a decade or so on the low-frequency side are indications of an adequate T and a reasonably good approach to stationarity. To achieve this for wind and temperature spectra over most of the ABL requires averaging times of 60–90 min, which often conflict with nonstationarity caused by diurnal variations. If there is no evidence of a peak or even a leveling of the spectrum (or cospectrum) at the low end, the time series should be examined, even plotted out, to determine if T should be increased (to include a significant peak that was not originally anticipated) or if the data need to be filtered (to remove a long-term trend that is masking a real peak within the existing spectral range). These questions and more will be discussed in the next section dealing with the preparation of time series for spectrum analysis.

7.2.2 *Choice of sampling rate*

Here, the bandwidth desired in the calculated spectra and cospectra determines the rate at which the variables should be sampled. The decision is simple if the spectrum of the variable has a clearly defined cutoff frequency f_c. Shannon's sampling theorem tells us that at least two samples per cycle are needed to define a frequency component in the original signal completely; for this band-limited signal the sampling interval Δt should be

$$\Delta t = \frac{1}{2f_c}. \tag{7.5}$$

Sampling at this rate is referred to as "critical sampling." Sampling more frequently ($\Delta t < 1/2f_c$) yields no additional benefits[2], since no useful spectral information exists between f_c and $1/2\Delta t$. Sampling less frequently ($\Delta t > 1/2f_c$) yields a spectral bandwidth inadequate to cover the full range of frequencies contained in the signal. The spectral information at frequencies above $1/2\Delta t$ is not lost but folded back into the spectrum in an accordion fold, as shown in Fig. 7.4. This process is called "aliasing," since it results in high-frequency components emerging as low-frequency components within the resolved bandwidth. The frequency $f_0(= 1/2\Delta t)$ is referred to as the folding frequency or Nyquist frequency.

We rarely find sharp cutoffs in boundary layer turbulence. The inertial subrange, however, presents a logical place to establish the cutoff for data acquisition purposes. There are three benefits:

1. Being in a region of rapidly falling spectral energy, the folding is confined to no more than the first fold and its effects seldom extend below $0.5f_0$ (Fig. 7.5).

[2]Note that, in theory, (7.5) applies to an infinite record. With a finite record we must sample more frequently to properly resolve f_c. In practice, this would only be a problem for very short records.

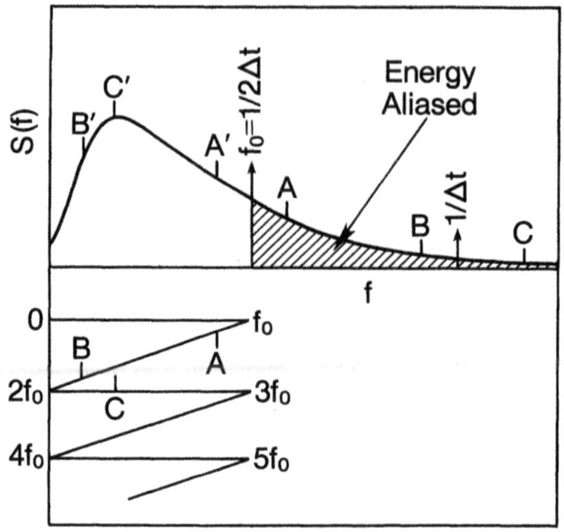

FIG. 7.4 Folding of spectral energy in an aliased spectrum. Frequencies A, B, and C appear at A', B', and C' in the resolved spectrum.

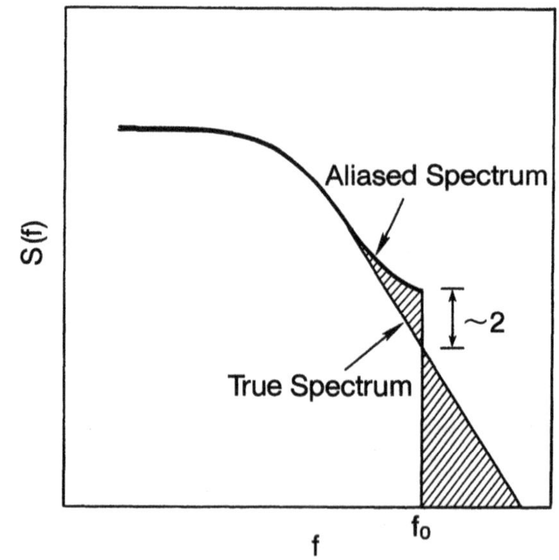

FIG. 7.5 Aliased energy shown raising spectral levels at the Nyquist frequency by a factor of 2. (Log scales on both axes.)

2. With the slope of the spectrum known, the true spectrum can be approximated by simple extrapolation. In the absence of any high-frequency noise, the spectrum at f_0 is raised by a known factor (~ 2).
3. No significant transport of momentum and heat occurs in this region of the spectrum. Therefore, a cutoff here is ideal from the point of view of flux measurements.

The ideal choice for f_0 is one that would reveal at least two octaves of the inertial subrange (allowing for aliasing errors) to provide a basis for extrapolation. This implies $f_0 > 4f_i$, where $f_i(\sim 2\overline{u}/z$, when M-O scaling applies) is the low-frequency limit of the inertial subrange and $\overline{u}$ is the mean wind speed at height z. This criterion is often stretched for measurements close to the ground. (In that case, particular care should be taken to minimize aliasing by using methods discussed in the following section.) The sampling frequency for turbulence data over reasonably flat terrain is usually set at 10 or 20 Hz depending on whether the observation height is above or below 5 m.

In selecting the sampling frequency for turbulence sensors, it is wise to consider where the power line frequency (50 or 60 Hz) would fall within the spectrum. Since line frequency contamination is present to some extent in most analog information transmitted over signal cables, it is important to place the aliased line frequency where it will do the least harm, that is, at $f = 0$. For $\Delta t = 1/10$ s, both 50 and 60 Hz frequencies should fold back to $f = 0$. The choice of $\Delta t = 1/20$ s, although ideal for the 60-Hz line frequency, would, however, place the 50-Hz line frequency at $f = f_0$ as seen in Fig. 7.6. Hence, the logical choice for Δt in countries with 50-Hz line frequency would be 1/25 s in applications where sampling at 1/10 s is not fast enough.

7.3 Preparing data for spectrum analysis

Measurements in the real atmosphere may not always satisfy our definitions of stationarity, or the sampling rate may not be fast enough to avoid serious aliasing

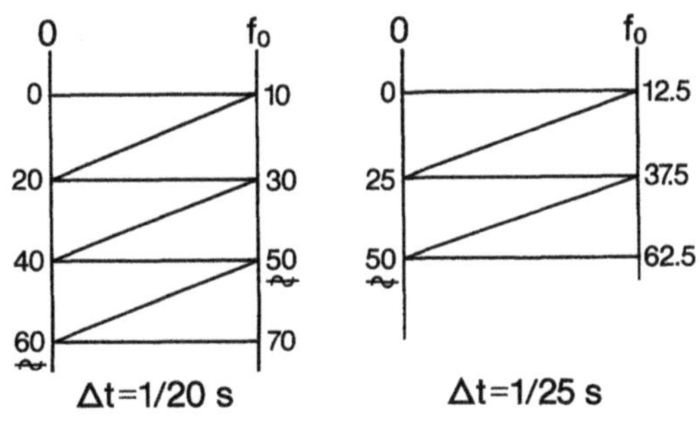

Fig. 7.6 Aliasing of power line frequency for 20- and 25-Hz sampling rates. Line frequencies are given in Hz.

in the part of the inertial subrange we are able to cover. These deficiencies can be met through proper intervention before the spectrum analysis. Some of the techniques most commonly used to minimize their effects are discussed below.

7.3.1 *Reduction of aliasing*

Since the very act of sampling at discrete intervals causes energy above frequency f_0 to be aliased back to lower frequencies, any scheme to minimize aliasing has to be applied very early in the data handling stage. There are two simple approaches; both depend on low-pass filtering to reduce the energy above f_0 that is available for aliasing. Both can be incorporated in the sensor electronics.

1. *Analog prefiltering.* Before digital conversion, the analog signal is low-pass filtered with an analog filter whose half-power point is set at $f_0 = 1/2\Delta t$. This will reduce the energy at the Nyquist frequency f_0 by 50%. When this filtered signal is sampled at interval Δt, aliasing restores the energy to nearly its true value at f_0 (assuming energy only in the first fold counts) and to slightly lower values in the range $0.5f_0 < f < f_0$ as shown in Fig. 7.7. The steeper the filter rolloff, the narrower the frequency band affected. The advantage of this method is that multiplexing and analog-to-digital (A/D) conversion speeds can be maintained at a lower rate. The disadvantage is the possibility of additional noise and drift from low-pass filters inserted ahead of the multiplexer contaminating the data. No filters are needed, however, if the sensor response drops off naturally to half-power at f_0.

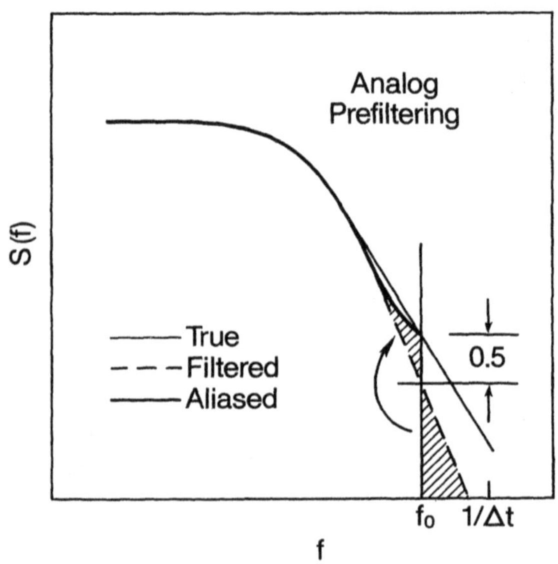

FIG. 7.7 Analog prefiltering aimed at reducing aliasing and restoring full energy at the Nyquist frequency. Filtering reduces energy at Nyquist frequency by a factor of 0.5. (Log scales on both axes.)

2. *Digital prefiltering.* When sampling speed is not a limiting factor (but the digital recording capacity is), it is more convenient to sample the variable at rates 10 to 20 times higher than f_0 and have the low-pass filtering performed digitally at the sensor output by constructing nonoverlapping block averages of the time series. Taking nonoverlapping block averages is equivalent to subjecting the time series to a moving average filter of width Δt but using only points Δt apart. With the first step, we impose the power transfer function $(\sin^2 \pi f \Delta t)/(\pi f \Delta t)^2$ on the original spectrum; with the second we alias back into the spectrum whatever energy remains above f_0 after filtering, as shown in Fig. 7.8.

The block averaging is accomplished by accumulating successive readings in the output buffer of the sensor (if it has a digital output) or of the local A/D converter sampling the signal (in the case of analog signals). Note that in this simple digital prefiltering scheme, aliasing restores only part of the energy lost through filtering (Fig. 7.8) because of the 60% drop in power at f_0. (Aliasing introduced by the original sampling at the higher rate will not normally affect the spectrum, unless large noise spikes exist at frequencies that could fold back into the frequency band between 0 and f_0.) The advantage of this approach is its simplicity. More sophisticated digital filters can be used if recording capacity is not a limiting factor; the low-pass filtering would then be performed at a later time, after the experiment.

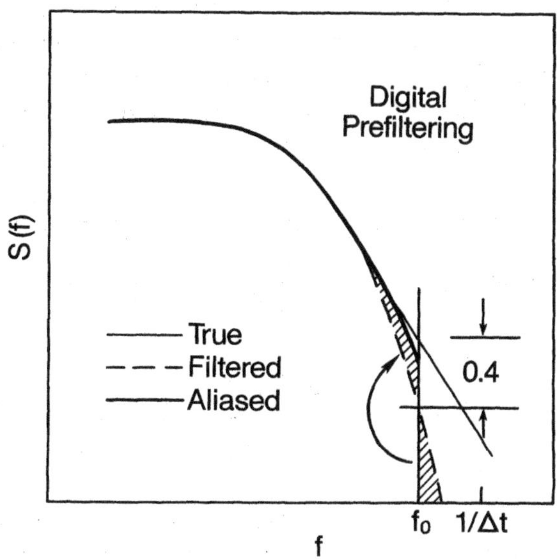

FIG. 7.8 Digital prefiltering achieved by computing nonoverlapping block averages with oversampled data. Energy at the Nyquist frequency is only partially restored because filtering reduces it by a factor of 0.4. (Log scales on both axes.)

7.3.2 *Trend removal*

The presence of a trend in the time series makes the data nonstationary and therefore less suitable for analysis. We define a trend as any frequency component with a period longer than the record length T. In the limit of very long periods (e.g., diurnal variation in temperature), the trend may appear to be linear within that period. Least-squares methods are often used to remove both linear and polynomial trends. For many applications, digital high-pass filtering is preferred because it is simpler and better understood. The same digital filter should be applied to all the variables processed, ensuring uniform low-frequency treatment of the signals.

Trends in the time series produce distortions at the low-frequency end of the spectrum (Fig. 7.9). This distortion, if large enough, could totally mask the true maximum in the $fS(f)$ spectrum and replace it with a spectrum that continues to rise with decreasing f (Fig. 7.10a). Detrending offers no guarantee that the true shape of the spectrum can be retrieved. The particular detrending method used often determines the shape of the detrended spectrum and the location of its maximum (Fig. 7.10b). Trend removal should be performed only if trends are physically expected or clearly apparent in the time series. Automatic detrending is not recommended, except for certain variance and flux calculations where the presence of trends can be highly detrimental (e.g., $\overline{u'\theta'}$ with trends likely in u and θ).

One indication of a linear trend in the time series is an f^{-1} slope at the low end of the $fS(f)$ versus f plot of the spectrum (Fig. 7.10a) best seen in temperature spectra. A linear trend would appear as a replicated ramp function in the Fourier analysis, producing harmonics that decrease in power as f^{-2}, hence the f^{-1} slope in the $fS(f)$ plot. Autocorrelation plots also announce the presence of trends (Fig. 7.9). The correlation function fails to drop to zero even at large lags; trend removal brings it down to zero very quickly.

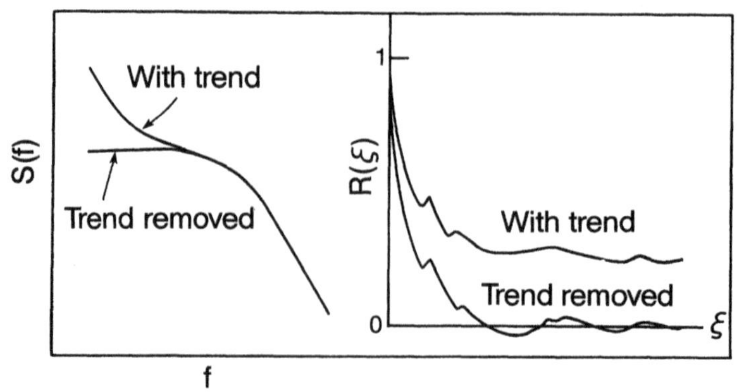

FIG. 7.9 Effects of trend on the spectrum (left) and on the autocorrelation function (right). (Log scales for spectrum axes.)

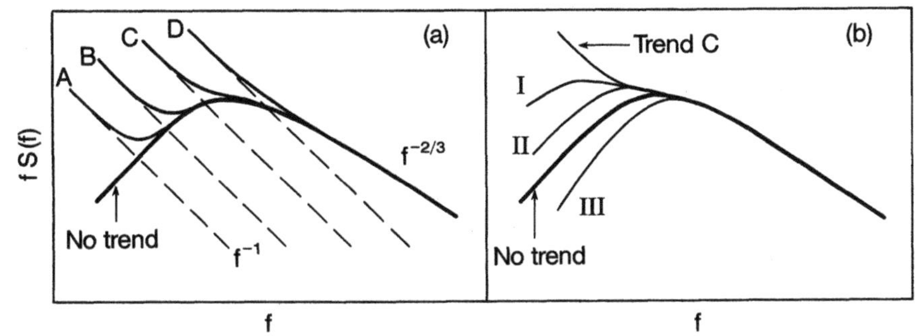

FIG. 7.10 (a) Spectrum distortion from trends of varying magnitudes (A-D). (b) Sensitivity of low-frequency spectrum shape to three hypothetical detrending procedures (I-III). (Log scales on all axes.)

There are arguments both for and against detrending. Scientists who work with data from aircraft find it essential to detrend. Others argue that trends, especially those in w associated with wave motions in stable air (curves A and B, Fig. 7.10a) are physically significant and should be included in the variance and flux calculations because they contribute to vertical transport. The undetrended spectrum should always be available for comparison with the detrended one to ensure that the detrending process removes only the suspected trend and no more. Autocorrelation functions and integral time scales calculated from such data can be particularly misleading because of their strong sensitivity to the detrending procedures used.

The simplest high-pass filter for detrending is one in which the original time series x_i is differenced from an equally weighted running mean of width τ_L. The process can be viewed as a sequence of low-pass, moving-average filtering and subtraction. The filtered time series x_i' becomes

$$x_i' = x_i - y_i,$$

where y_i represents the low-pass filtered time series. The power spectral transfer function $[K(f)]^2$ for this high-pass filter is well known:

$$[K(f)]^2 = [1 - H(f)]^2, \tag{7.6}$$

where

$$H(f) = \frac{\sin \pi f \tau_L}{\pi f \tau_L}. \tag{7.7}$$

$H(f)$ is the low-pass filter function. All frequencies with periods longer than τ_L are attenuated when the power spectrum of x_i is multiplied by $[K(f)]^2$.

Much sharper detrending filters, with less oscillation near $f = 1/\tau_L$, can be constructed using recursive techniques in which the results of successive filtering operations are fed back to the input terms. The operation involves fewer terms than the above "boxcar" approach, and the filter characteristics can be tailored to approximate those of electronic filters (e.g., R-C, Butterworth). McMillen (1988) describes one that simulates an ideal R-C filter with an easily specified time constant. With x_i and y_i representing, as before, the original and low-pass-filtered time series, respectively, a simple recursive filter can be constructed for which

$$y_i = ay_{i-1} + (1-a)x_i, \tag{7.8}$$

where

$$a = e^{\Delta t/\tau_c}. \tag{7.9}$$

Δt is the time interval between data points and τ_c is the time constant of the desired low-pass filter. When $\Delta t/\tau_c \ll 1$, we have $a \simeq 1 - (\Delta t/\tau_c)$. The new time series is $x' = (x_i - y_i)$, as in the simple moving-average detrending filter.

It is important to point out in this context that departures from the running mean cannot be used directly for computing the fluxes as we would use departures from the conventional time average. Running means do not satisfy Reynolds averaging rules (see Appendix 7.1) so we cannot assume, as we do with simple deviations from the time average, that

$$\overline{\underline{w}'\underline{\theta}'} = \overline{\underline{w}\,\underline{\theta}} - \overline{\underline{w}\,\underline{\theta}},$$

where the underbarred terms represent the running means and the deviations from them.

Integral time scales $\mathcal{T}_\alpha$ calculated from high-pass filtered time series also need to be treated with caution. This is because the integral to infinity of the autocorrelation $R(\xi)$ is zero for any high-pass filtered signal. The form of $R(\xi)$, over the range of time lags we normally use to compute the function, is very sensitive to τ_H, the time constant of the high-pass filter, for $\tau_H < 10\mathcal{T}_\alpha$ (see Appendix 7.2). For $\tau_H > 10\mathcal{T}_\alpha$, the time lag at the 1/e point on the correlation curve is usually a good estimate of $\mathcal{T}_\alpha$. At $\tau_H \simeq \mathcal{T}_\alpha$, the time lag corresponding to the first zero-crossing can be taken as an estimate of $\mathcal{T}_\alpha$. At $\tau_H < \mathcal{T}_\alpha$, the correlation curve will be too compromised for any calculation of $\mathcal{T}_\alpha$. All these points are treated in detail in Appendix 7.2.

7.3.3 *Tapering the time series*

When the sampling duration is too short to satisfy the condition $T \gg \mathcal{T}_\alpha$, a tapering window is applied to the time series to bring the values down to zero, or close to it, at both ends of the sampling period. It minimizes the adverse effect of finite sampling[3] on the magnitude of the computed spectrum and improves our ability to resolve discrete contributions to the spectrum from waves in the signal.

The process of limiting the data to a finite period is equivalent to multiplying the time series with a rectangular time window of unit height and width T. In the frequency domain, this translates to a convolution of the true transform of the process with the function (sin πfT) / πfT. Convolution, being a smoothing operation, smears out details present in the original transform and extends its range along the frequency axis. The effect on the power spectrum is loss of resolution and overestimation in regions where the power spectrum is dropping off rapidly, as in the inertial subrange. The smaller the T, the wider the lobes of the (sin πfT) / πfT function and the greater the smearing and overestimation. For a discussion of the effects of finite sampling on power spectra, see Kaimal et al. (1989).

By multiplying the time series (from which the mean has been removed) with a tapered window we are, in effect, replacing (sin πfT) / πfT with a different function, one with a slightly wider main lobe but with greatly suppressed side lobes (the negative and positive oscillations on either side of the main lobe). In the time domain, the tapering reduces the discontinuity at the boundaries of the data when viewed as a replicated sequence of the same time series. Kaimal and Kristensen (1991) tested a number of tapered windows and found the Hamming window brings the measured spectrum to within 1% of the spectral levels we might expect from a long enough record. The window has the form

$$w(n) = 0.54 + 0.46 \cos\left(\frac{2\pi n}{N}\right), n = -\frac{N}{2} \cdots 0 \cdots + \frac{N}{2}, \tag{7.10}$$

where $w(n)$ is the window function and N is the number of equally spaced data points in the sample. The taper in the window reduces the variance in the time series, so the spectrum has to be compensated for that loss by multiplying it with the ratio of the squares of the areas of the two windows. For the Hamming window, this ratio is 2.52. (Note that, after windowing, the finite time series may have acquired a non-zero mean which must be removed before applying the FFT.)

A striking example of its application is the shipboard measurement of fluxes in the marine boundary layer through the "inertial-dissipation" method (Fairall et al., 1990), where it is necessary to compute the dissipation rates of turbulent kinetic energy, temperature variance, and humidity variance from very short samples of data ($T \approx T_\alpha$). Use of the Hamming filter on samples of the order of seconds yields dependable inertial-subrange spectral intensities for the flux extimates.

[3] In earlier discussions of the effects of finite sampling (e.g., Pasquill and Smith, 1983) the process is treated as being analogous to high-pass filtering with a spectral transfer function $[1 - H(f)^2]$, where $H(f) =$ (sin $\pi fT)/\pi fT$; the actual transfer function is more complicated (Kaimal et al., 1989) and approaches the above function only as $T \rightarrow \infty$.

Other windows tested by Kaimal and Kristensen (1991) fared less well. The Hanning (cosine squared), Gaussian, and Bartlett (triangle) windows performed almost as well as the Hamming window, but the Poisson (exponential) and the 25% cosine taper windows were clearly unacceptable. [See Harris (1978) for definitions of above windows.] The last two caused overestimations approaching 50%, about the same as reported for the untapered (rectangular) window. Clearly, the commonly used cosine taper is not recommended for very short samples.

The other important benefit to be derived from tapering is the reduction in "side-band leakage" (the leakage of energy to neighboring frequencies through the side lobes in the convolving function), which is especially bad with the $(\sin \pi fT) / \pi fT$ function. Spurious spikes could appear in the spectrum that may be difficult or impossible to distinguish from spikes representing discrete waves. Finnigan et al. (1984) successfully used a Hamming window to identify gravity waves in their studies of wave-turbulence interactions.

7.3.4 *Addition of zeros*

When the time series available falls short of the length required for the spectrum analysis program on hand (power of 2, if an FFT is used), it is common practice to add zeros to the data sequence to make up the required number of points. This procedure, referred to as "padding", should be approached with caution because the consequences of improper application can be severe. The addition of zeros reduces the spectral estimates by a factor $(N - N_z)/N$, where N_z is the number of zeros added and N is the total number of points including the zeros. The spectral estimates should be corrected for this reduction. We can assume that the spectral estimates are diminished uniformly across the bandwidth as long as $N_z < N/3$. It is essential that means and trends be removed from the time series before adding zeros.

7.3.5 *Block averaging*

When the number of data points in the period selected for analysis is too long for a designated spectrum analysis program, it is customary to block average the time series in unweighted, nonoverlapping blocks to reduce the original time series to the number of points desired. Such compression of data enables us to examine the low-frequency behavior in spectra and cospectra quickly. (FFT programs that can handle an entire hour of data, 36,000 data points for instance, in one pass certainly exist, but the computing time needed to produce spectra and cospectra with all the points would be longer and the data handling more cumbersome.) The block averaging introduces some attenuation at the high-frequency end (see Section 7.1), but the loss is easily restored in the spectral output. A good reason for block averaging is to keep aliasing under control.

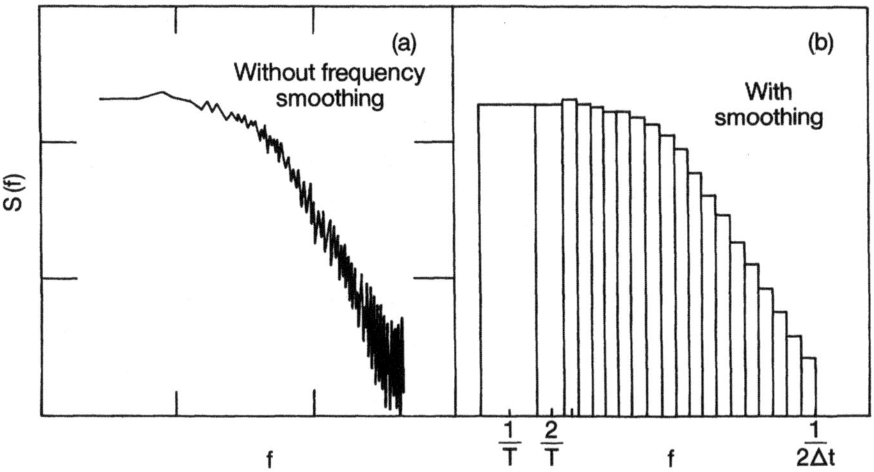

FIG. 7.11 (a) Scatter in the unsmoothed high-frequency spectral estimates. (b) Scatter reduced by smoothing with a frequency window that expands in width with frequency. (Log scales on all axes.)

7.4 Processing spectral data

The raw spectral density estimates derived from spectrum analysis programs, such as the FFT, are too numerous and too ragged to be of direct use to meteorologists.[4] Atmospheric turbulence covers a spectral range of five decades; this is best represented on a logarithmic scale not a linear one. The spectral density estimates generated by the digital programs, however, appear at equally spaced intervals ($\Delta f = 1/T$) on the frequency scale (Fig. 7.3). On a logarithmic scale this spacing results in excessive crowding and large scatter of spectral estimates at the high-frequency end, as illustrated in Fig. 7.11a. Some form of frequency smoothing is needed to extract a representative spectral curve from the estimates. Spectra derived through separate computations covering different segments of the desired spectral bandwidth are easier to splice if they are properly smoothed.

7.4.1 *Frequency smoothing*

An effective smoothing procedure for boundary layer work is one in which the averaging interval keeps expanding with frequency f. If m is the number of estimates in each nonoverlapping block, m is systematically increased as a function

[4]The power spectral estimates obtained directly from the FFT program (which when summed equal the variance of the original time series) have to be divided by $\Delta f (= 1/T = 1/N\Delta t)$ to get them in the form of spectral densities $S(f)$ used in the spectral plots. $S(f)$ has units of variance per Δf, and the area under that curve equals the variance. The frequency-weighted spectrum $fS(f)$, on the other hand, has units of variance, since multiplication by $f (= i\Delta f$, where $i = 1, 2, 3, \ldots N/2)$ removes the Δf dependence.

of frequency (roughly exponentially) to yield about seven to eight estimates per decade, as in Fig. 7.11b. Each smoothed estimate (of width $m\Delta f$) is assigned to the center frequency of the band. In practice the first few estimates are accepted as they are; then m is increased in steps 3, 5, 7, and so on, until the density of smoothed estimates per decade reaches seven (or eight). The expected power law in the inertial subrange can only be tested dependably after such smoothing.

Other types of frequency smoothing are sometimes applied to spectral data. A three-point Hanning window (weighted 1/4, 1/2, 1/4) is one of many recommended in earlier treatises on the subject (Blackman and Tukey, 1958). Filtering the spectral estimates with such fixed bandwidth windows may provide the smoothing needed in many engineering applications where the spectral range of interest is often narrow, but is not particularly useful for boundary layer work because it still causes crowding of estimates at the high-frequency end.

7.4.2 *Spectral splicing*

When dealing with a very long time series ($N \gg 2^{10}$), it is advisable to split the record into two sets:

1. A set of r short records each with N/r data points
2. A series made up of N/r nonoverlapping block averages (r point averages) of the original time series

The former yields a bandwidth $r/N\Delta t$ to $1/2\Delta t$ and the latter $1/N\Delta t$ to $1/2r\Delta t$ (Fig. 7.12). They overlap over the range $r/N\Delta t$ to $1/2r\Delta t$ which, on a log scale, corresponds to a frequency ratio of $N/2r^2$. [For the Kansas spectra, Kaimal et al. (1972) had $\Delta t = 1/20\,\text{s}$, $T = 3600\,\text{s}$, $N = 72{,}000$, and $r = 16$ with an $N/2r^2 = 136$, which corresponds to an overlap of more than two decades. With

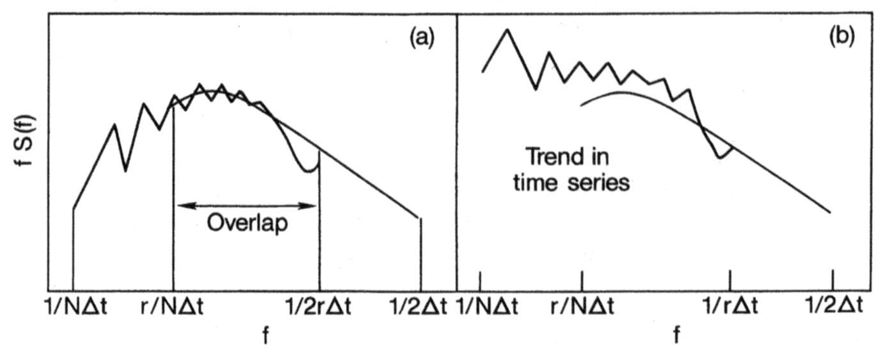

Fig. 7.12 (a) Splicing spectra from the shortened and the block-averaged time series in the absence of significant trends in the data. (b) Effect of trend in the time series that precludes matching in the regions of overlap. (Log scales on all axes.)

this choice of parameters, the same 4096-point FFT routine was used for both the high-end and the low-end spectral computations.]

The r successive spectra from the short records are first averaged to produce a single high-frequency spectrum. This "incoherent averaging" within each frequency band produces a spectrum that is very smooth. By comparison, the low-frequency spectrum will typically have more scatter. Figure 7.12a illustrates how the two usually combine. The match in the region of overlap is often very good so the analyst can decide where to drop the estimates from one spectrum and start with the other. If there is a mismatch, as in Fig. 7.12b, it is invariably the result of a long-term trend in the data. When that occurs, either the data should be detrended or the run discarded as unsatisfactory.

Unacceptably high noise levels at the high-frequency end may also be grounds for rejecting the spectrum. They usually appear as a large rise in the spectrum, approaching an f^{+1} slope, stretching over a decade or more at the high end, where the spectrum should be falling off (Fig. 7.13). This type of noise can be traced to one of the following:

- Sensor noise rising above signal levels as the signal drops below the sensor's noise threshold
- Sporadic spikes in the signal from radio frequency interference, a faulty cable, or mistriggering in the sensor (e.g., sonic anemometer or thermometer)

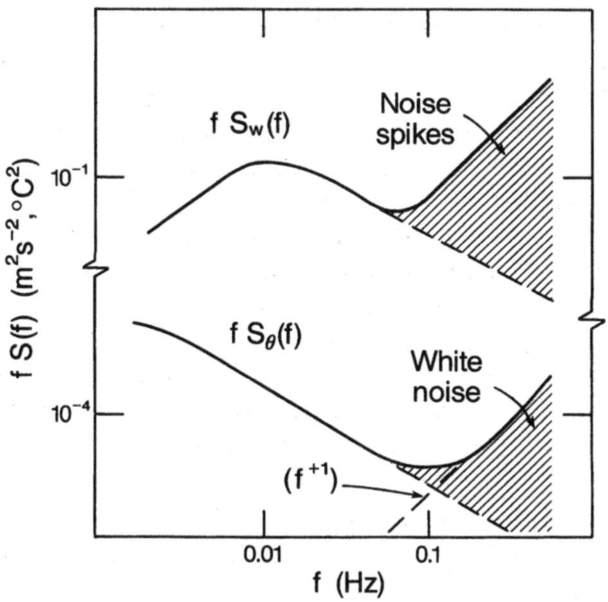

FIG. 7.13 High-frequency distortions arising typically from spikes in a sonic anemometer w signal and white noise in a platinum wire thermometer. (Log scales on both axes.)

Their effects on spectra are illustrated in Fig. 7.13. The noise threshold effect is usually observed in platinum-wire fast-response thermometer spectra when the temperature $fS(f)$ levels drop to 10^{-4} °C^2 or lower. The noise spectrum has the characteristics of white noise: flat in the $S(f)$ versus f plot, f^{+1} slope in the $fS(f)$ versus f plot (Fig. 7.13). This noise is often accepted as inevitable. Spikes, on the other hand, cannot be ignored, unless they are isolated and infrequent, as their effect can extend to all frequencies. Their contribution to the high end of the spectrum can cause it to exceed in magnitude the energy in the true turbulence peak. Corrective action (adjustments and repairs) should be taken to prevent the spikes from contaminating the signal.

In the context of spectral splicing we have to consider the possible implications of deriving the high-frequency portion of the spectrum by averaging spectra from relatively short segments of the time series. The overestimation predicted by Kaimal et al. (1989), discussed earlier, has negligible effect on the spectral estimates in most applications unless we choose the duration T of those segments to be smaller than the period of the spectral peak (τ_m). In the Kansas data analysis T was 3.75 min with τ_m typically 3–4 s for daytime w spectra. In Kaimal et al.'s (1989) study, the effects of finite sampling became apparent only when T dropped below about $10\tau_m$.

7.5 Archiving strategies

In field experiments of limited duration, raw data can be easily stored on digital magnetic tapes and played back as needed for analysis. Different strategies are needed for data collection over extended periods. In an operation similar to the Boulder Atmospheric Observatory where data collection is continuous, it is imperative that a system of archival and retrieval be worked out so the information is standardized and accessible to future users (Kaimal and Gaynor, 1983).

At the BAO, slow-response channels are sampled once a second and fast-response channels 10 times a second. A schematic of the operations at the BAO is shown in Fig. 7.14. In the fast response sonic anemometer channels, aliasing is minimized through digital prefiltering (20-point block average); in the fast-response platinum-wire temperature channels the natural rolloff from thermal conduction losses provides the needed high-frequency attenuation. Three types of data are prepared by the data acquisition computer at the BAO site for transmission to the central computer in Boulder, 25 km away:

- 10-s averaged values of readings on all channels, fast and slow
- 10-s grab samples (last data point in each 10-s block) of all fast-response channels
- Frequency-smoothed spectral estimates (35 estimates) over the frequency range 0.01–5 Hz, updated every 20 min

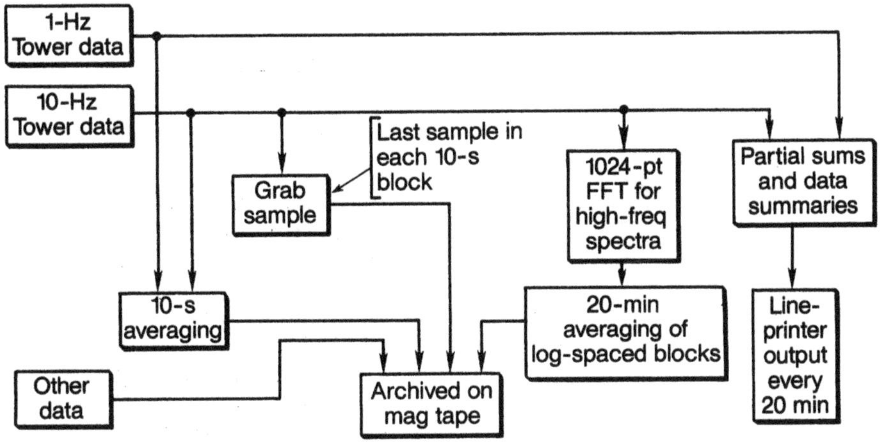

Fig. 7.14 Schematic of data acquisition and processing at the Boulder Atmospheric Observatory (BAO).

The on-site computer also generates partial sums of fast-response data and summaries of slow-response data for presentation of profiles, variances, fluxes, Obukhov lengths, and so on, at the end of each 20-min archiving period.

The 10-s averaged data are saved for future reconstruction of time plots and for filling in the low-frequency end of the spectra (0–0.05 Hz) as discussed in the section above. The 10-s grab samples are the "decimated" data points needed for recomputing fluxes, variances, and third moments over periods longer than 20 min. They are also useful for detecting the presence of noise in the original signal, which may not show up in plots of the 10-s averaged data.

Each 20-min spectrum is an average of ten 2-min (1024 point) FFT spectra. The choice of 20 min as the basic archiving block was arbitrary, selected as a compromise between the need for stability in the statistics and the need to track mesoscale variations in the boundary layer. Frequency smoothing is performed over blocks of increasing width, as discussed in Section 7.4. To fill in the low-frequency end of the spectrum, a 512-point FFT program is used on the 10-s averaged data with the option of adding zeros when the number of data points available falls short of 512. (Even combining four 20-min periods, the time series will be 32 points short.) How well this scheme fits in with spectral shapes in stable and unstable air can be seen in Fig. 7.15.

The partial sums generated in real time for the calculation of data summaries at the end of each 20-min period significantly reduces the processing time at the end. Individual variables and their products (e.g., $\overline{w}, \overline{\theta}, \overline{w^2}, \overline{\theta^2}, \overline{w\theta}$) are accumulated in separate registers as they are sampled. At the end of the 20-min period, variances and fluxes are generated by simple subtractions:

$$\overline{w'^2} = \overline{w^2} - (\overline{w})^2, \tag{7.11}$$

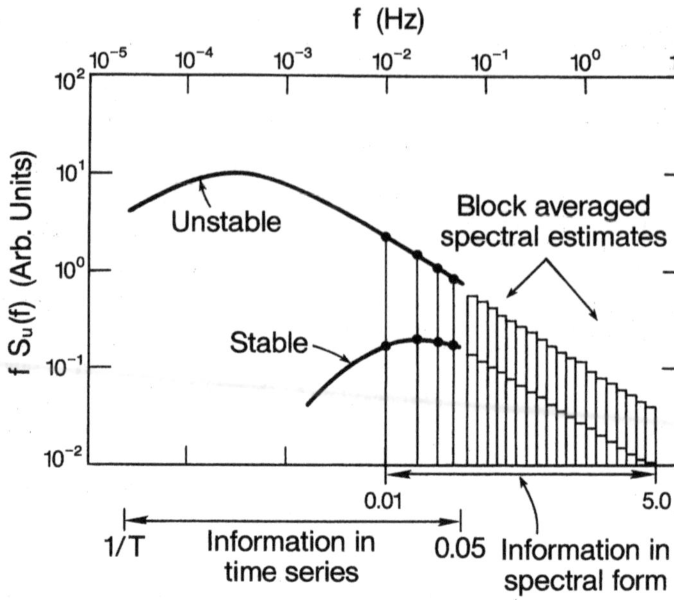

FIG. 7.15 Scheme for compressing turbulence data: high-frequency information stored as smoothed spectral estimates, low-frequency information as time series.

$$\overline{\theta'^2} = \overline{\theta^2} - (\overline{\theta})^2, \tag{7.12}$$

$$\overline{w'\theta'} = \overline{w\theta} - \overline{w}\,\overline{\theta}. \tag{7.13}$$

In deriving (7.11) and (7.12) we assume that $\overline{w'\overline{\theta}} = \overline{\overline{w}\theta'} = 0$, which is valid if $w = \overline{w} + w'$ and $\theta = \overline{\theta} + \theta'$. The same principle can be used for more complicated calculations involving coordinate transformation of the horizontal velocity measurements to the longitudinal (u) and lateral (v) components. If u_m and v_m are horizontal wind components (right-handed) measured along the probe axes and w_m is the measured vertical component, we should be able to compute the variances and fluxes in single operations immediately following ingest of the last data point in that averaging period in order to have them ready for display before the start of the next period. The algebraic operations involved are

$$\begin{aligned}\overline{u'^2} = {} & \left[\overline{u_m^2} - (\overline{u}_m)^2\right] \cos^2\theta_r + \left[\overline{v_m^2} - (\overline{v}_m)^2\right] \sin^2\theta_r \\ & + 2\left[\overline{u_m v_m} - \overline{u}_m\overline{v}_m\right] \sin\theta_r \cos\theta_r,\end{aligned} \tag{7.14}$$

$$\overline{v'^2} = \overline{v_m^2} \cos^2\theta_r - 2\overline{u_m v_m} \sin\theta_r \cos\theta_r + \overline{u_m^2} \sin^2\theta_r, \tag{7.15}$$

$$\overline{u'w'} = [\overline{u_m w_m} - \overline{u}_m \overline{w}_m] \cos\theta_r + [\overline{v_m w_m} - \overline{v}_m \overline{w}_m] \sin\theta_r, \tag{7.16}$$

$$\overline{v'w'} = \overline{v_m w_m} \cos\theta_r - \overline{u_m w_m} \sin\theta_r, \tag{7.17}$$

$$\overline{u'v'} = \left[\overline{v_m^2} - \overline{u_m^2}\right] \sin\theta_r \cos\theta_r + \overline{u_m v_m} \cos 2\theta_r, \tag{7.18}$$

where the angle of rotation (counterclockwise) is

$$\theta_r = \tan^{-1}\left(\frac{\overline{v}_m}{\overline{u}_m}\right). \tag{7.19}$$

Similar expressions can be derived for a full three-dimensional coordinate rotation if a tilt correction for sloping terrain or for an incorrectly mounted sensor makes it necessary. In the above transformation we assumed $\overline{w}_m = 0$.

7.6 Special symbols

f_c	cutoff frequency
f_i	low-frequency limit of the inertial subrange
f_0	folding (Nyquist) frequency
n	number of data points
N	number of data points in period T
N_z	number of zeros added for padding
r	factor by which number of data points is reduced
Δt	sampling interval
T	averaging time
u_m, v_m, w_m	measured wind components along instrument axes
$w(n)$	window function, function of n
x_i, y_i	time series of variables x and y
$\alpha(t)$	any time series
θ_r	azimuth rotation angle
$\rho_\alpha(\xi)$	autocorrelation function of α
τ_c	time constant for recursive low-pass filter
τ_H	time constant for high-pass filter
τ_L	cutoff period for boxcar low-pass filter

Appendix 7.1 Reynolds averaging and running mean filters

In standard operations of turbulence analysis, such as the derivation of flow equations or the computation of eddy fluxes, the averaging operator (denoted by an overbar) is assumed to separate the mean and fluctuating parts of a variable according to certain rules. These are known as the Reynolds averaging conditions and may be summarized as follows:

1. All primed (fluctuating) quantities must average to zero ($\overline{w'} = 0$).
2. The correlations between primed and averaged quantities must vanish ($\overline{w'\overline{\theta}} = 0$).
3. The average of an average must be equal to the same average ($\overline{\overline{w}} = \overline{w}$).
4. Differentiation in space and time commutes with the averaging operations

$$\overline{\frac{\partial w}{\partial x_i}} = \frac{\partial \overline{w}}{\partial x_i} \quad ; \quad \overline{\frac{\partial w}{\partial t}} = \frac{\partial \overline{w}}{\partial t}.$$

For example, when computing fluxes by the eddy correlation method, we usually write

$$\overline{w'\theta'} = \overline{w\theta} - \overline{w}\overline{\theta}, \tag{7.20}$$

neglecting the terms $\overline{w'\overline{\theta}}$ and $\overline{\overline{w}\theta'}$. These terms are, by definition, zero if the overbar represents a true Reynolds average. In fact, disappearance of these terms can be used as a test of its validity for the data being used.

Only ensemble averages can be expected to obey the Reynolds averaging rules precisely, but in practice we usually use time averages, making the ergodic hypothesis that they are equal to ensemble averages. (If the time series is not statistically stationary over relevant time scales, this assumption must be made cautiously.) In many situations, a running-mean low-pass filter is applied to the time series to approximate the "background" variations; the filtered time series is then subtracted from the original signal (i.e., the original is high-pass filtered) to derive the fluctuating components. We should point out that the terms $\overline{\underline{w'}\,\underline{\overline{\theta}}}$ and $\overline{\underline{\overline{w}}\,\underline{\theta'}}$ do not necessarily vanish (underbar denotes running-mean quantities) because $\underline{w'}$ and $\underline{\theta'}$ may not have zero means and $\underline{\overline{w}}$ and $\underline{\overline{\theta}}$ have spectral energy spread over finite bandwidths. Thus, the spectra of $\underline{w'}$ and $\underline{\overline{\theta}}$, for example, can be expected to overlap in the region where the high-pass and low-pass filters also overlap (Fig. 7.16). As a result, we can assume $\overline{\underline{w'}\,\underline{\overline{\theta}}} = 0$ only if a spectral gap exists in either w or θ in the region where the two filters overlap. The same comments apply to the computation of variances, since, by the same argument, $\overline{\underline{w'}\,\underline{\overline{w}}}$ and $\overline{\underline{\theta'}\,\underline{\overline{\theta}}}$ cannot be expected to vanish. These considerations are important if moments computed in this way are used in budget calculations that are based on equations derived by Reynolds averaging.

Appendix 7.2 High-pass filtering and integral time scales

In Section 7.4 we discussed the use of digital high-pass filtering to remove low-frequency trends in the data and its consequences for the estimation of integral time scales $\mathcal{T}_\alpha$. These consequences arise because the procedure has the effect of forcing the spectrum at zero frequency $\hat{S}_\alpha(0)$ to zero, causing $\mathcal{T}_\alpha$ to be zero, if the standard definition of the integral time scale is used (Appendix 2.1). Here we will examine how high-pass filtering affects the autocovariance function and through it the integral time scale.

For illustration, we use an example provided by Leif Kristensen (personal communication), which starts with a near-realistic spectrum for variable α

$$\hat{S}_\alpha(\omega) = \frac{\mathcal{T}_\alpha \sigma_\alpha^2}{\pi} \left(\frac{1}{1 + (\omega \mathcal{T}_\alpha)^2} \right). \tag{7.21}$$

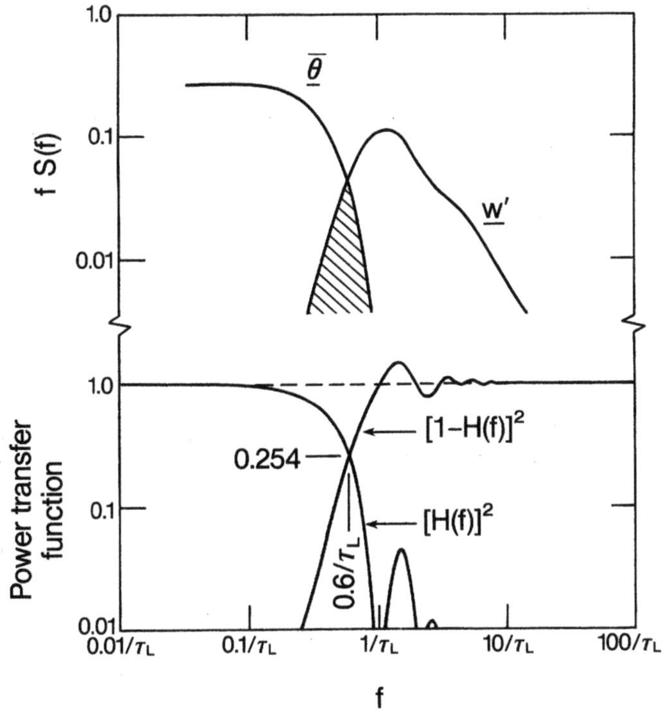

Fig. 7.16 Overlap of $\underline{w'}$ and $\underline{\overline{\theta}}$ spectra resulting from overlap in the transfer functions for the running mean high-pass and low-pass filters.

(For convenience, we retain the two-sided spectral convention used in Appendix 2.1.) Its autocovariance $R_\alpha(\xi)$ has the simple exponential form

$$\begin{aligned} R_\alpha(\xi) &= \int_{-\infty}^{\infty} \hat{S}_\alpha(\omega) e^{i\omega\xi} d\xi \\ &= \sigma_\alpha^2 e^{-|\xi|/\mathcal{T}_\alpha} \end{aligned} \tag{7.22}$$

and an integral time scale, by our standard definition,

$$\frac{1}{\sigma_\alpha^2} \int_0^\infty R_\alpha(\xi) d\xi = \int_0^\infty e^{-|\xi|/\mathcal{T}_\alpha} d\xi = \mathcal{T}_\alpha. \tag{7.23}$$

At $\omega = 0$, the spectrum reduces to the form derived in Appendix 2.1:

$$\hat{S}_\alpha(0) = \frac{\mathcal{T}_\alpha \sigma_\alpha^2}{\pi}. \tag{7.24}$$

We now apply a first-order high-pass filter with time constant τ_H and a power transfer function $(\omega\tau_H)^2/[1+(\omega\tau_H)^2]$ that ensures zero contribution at $\omega = 0$. The filtered spectrum will be

$$[\hat{S}_\alpha(\omega)]_{\text{filt.}} = \frac{\mathcal{T}_\alpha \sigma_\alpha^2}{\pi}\left[\frac{1}{1+(\omega\mathcal{T}_\alpha)^2}\right]\cdot\left[\frac{(\omega\tau_H)^2}{1+(\omega\tau_H)^2}\right], \tag{7.25}$$

and the corresponding autocovariance function becomes

$$\left[R_\alpha(\xi)\right]_{\text{filt.}} = \frac{\tau_H\sigma_\alpha^2}{\mathcal{T}_\alpha^2-\tau_H^2}\left(\mathcal{T}_\alpha e^{-|\xi|/\tau_H} - \tau_H e^{-|\xi|/\mathcal{T}_\alpha}\right). \tag{7.26}$$

$$= \frac{\sigma_\alpha^2}{2}(1-|\xi|/\mathcal{T}_\alpha)\,e^{-|\xi|/\mathcal{T}_\alpha} \quad \text{for} \quad \tau_H = \mathcal{T}_\alpha. \tag{7.27}$$

The effect of filtering on the autocorrelation function $\rho_\alpha(\xi)$ (the normalized autocovariance function) is shown in Fig. 7.17 for different ratios of $\tau_H/\mathcal{T}_\alpha$:0.1, 1.0, 10, and ∞. The effect of high-pass filtering is to introduce a negative lobe (visible, at least, in the first two curves), large enough to make the integral to infinity equal zero. For $\tau_H/\mathcal{T}_\alpha = 1$, the zero crossing occurs at $\xi = \mathcal{T}_\alpha$. For $\tau_H/\mathcal{T}_\alpha < 1$, the zero crossing is shifted closer to the origin and for $\tau_H/\mathcal{T}_\alpha > 1$, farther away from $\xi/\mathcal{T}_\alpha = 1$.

For $\tau_H/\mathcal{T}_\alpha > 10$, the zero crossing occurs outside the plot, at some distant ξ and $\rho_\alpha(\xi)$ is only slightly altered in the range we usually observe. Fortunately, this is the case in most applications, for otherwise the experimenter would be collecting highly compromised data.

The shapes of the correlation functions will, no doubt, vary with the type of filter chosen, but the above exercise does offer some useful insights on high-pass filtering.

1. It tells us roughly how far τ_H has to be separated from $\mathcal{T}_\alpha$ to ensure that important turbulence information is not lost along with the unwanted trend. Note that at $\tau_H/\mathcal{T}_\alpha = 1$,

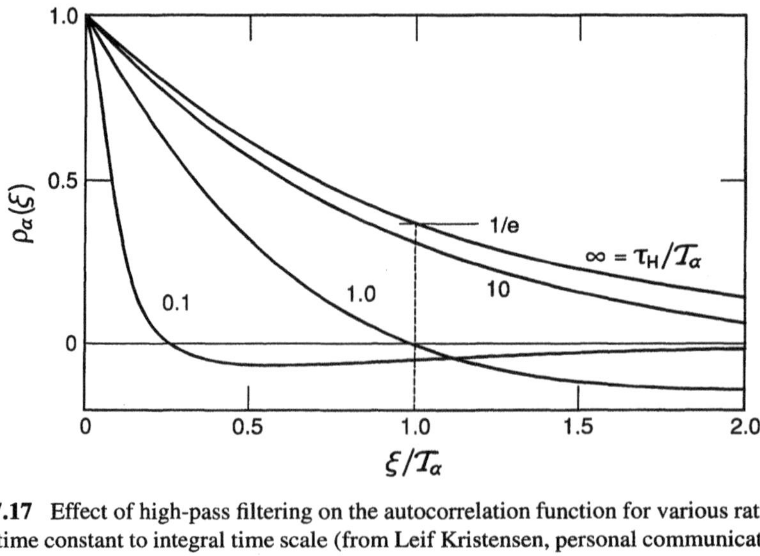

FIG. 7.17 Effect of high-pass filtering on the autocorrelation function for various ratios of filter time constant to integral time scale (from Leif Kristensen, personal communication).

the filter half-power point ($f = 1/2\pi\tau_H$) corresponds approximately to the frequency where the $fS(f)$ spectrum peaks [which roughly equals $1/2\pi\mathcal{T}_\alpha$ as shown in (2.82), Appendix 2.3]. So, a filter with $\tau_H/\mathcal{T}_\alpha > 10$ would preserve much of the energy at the spectral peak and the frequencies above.

2. Designing a high-pass filter for a given application requires prior knowledge of the magnitude of $\mathcal{T}_\alpha$, or where $fS(f)$ might peak. The relationships in Chapter 2 could serve as a guide because the collected data may be too contaminated by the trend to provide that information. The experimenter, by choosing τ_H in this manner, runs the risk of excluding from consideration conditions that deviate significantly from the norm. (The choice of record length T can be limiting in the same way.)

3. Practical alternatives to obtaining $\mathcal{T}_\alpha$ by integrating $\rho_\alpha(\xi)$ to infinity are needed because it is, first, impossible to implement with real data and, second, of questionable value, given the theoretical expectation of $\mathcal{T}_\alpha = 0$ for filtered data. One approach is to choose the value of ξ at which $\rho_\alpha(\xi) = 1/e \simeq 3.7$ ($e \simeq 2.72$ being the base for natural logarithms), which, for an exponential autocorrelation function, is precisely $\mathcal{T}_\alpha$, as pointed out in Chapter 2. Another alternative is to integrate to the first zero crossing or to the point where $\rho_\alpha(\xi)$ is very close to zero. Both these methods will systematically underestimate $\mathcal{T}_\alpha$, seriously so if $\tau_H/\mathcal{T}_\alpha$ is much less than 10.

The concept of the Eulerian integral time scale, or length scale, is intimately related to the question of ergodicity—whether time averages converge to constant values as the averaging time approaches infinity. As ξ increases, an ergodic variable not only becomes uncorrelated with itself, it also becomes statistically independent of itself. The integral time scale is a measure of the time for which $\alpha(t)$ "remembers" itself. At $\xi \gg \mathcal{T}_\alpha$, it approaches statistical independence. A well-designed high-pass filter would preserve the scales of motion that contribute to this decorrelation and remove only those that interfere with it. As a final caution, note that analyses such as those involved in the estimate of errors in moments from finite time averages (Section 7.3) are based on the standard definition of the integral time scale.

References

Bendat, J. S., and A. G. Piersol, 1971: *Random Data: Analysis and Measurement Procedures*. Wiley-Interscience, New York, 407 pp.

Blackman, R.B., and J.W. Tukey, 1958: *The Measurement of Power Spectra*. Dover Publications, Inc., New York, 190 pp.

Fairall, C. W., J. B. Edson, S. E. Larsen, and P. G. Mestayer, 1990: Inertial-dissipation air-sea flux measurements: A prototype system using realtime spectral computations. *J. Atmos. Oceanic Tech., 7*, 425–453.

Finnigan, J. J., F. Einaudi, and D. Fua, 1984: The interaction between an internal gravity wave and turbulence in the stably stratified nocturnal boundary layer. *J. Atmos. Sci., 41*, 2409–2436.

Harris, F.J., 1978: On the use of windows for harmonic analysis with the discrete Fourier transform. *Proc. IEEE, 66*, 51–83.

Haugen, D.A., 1978: Effects of sampling rates and averaging periods on meteorological measurements. In *Proc. Fourth Symposium on Meteorological Observations and Instrumentation*, April 1978, Denver, CO, American Meteorological Society, Boston, MA, 15–18.

Kaimal, J.C., and J.E. Gaynor, 1983: The Boulder Atmospheric Observatory. *J. Clim. Appl. Meteor., 22*, 863–880.

Kaimal, J.C., and L. Kristensen, 1991: Time series tapering for short data samples. *Bound.-Layer Meteor., 57*, 187–194.

Kaimal, J.C., S.F. Clifford, and R.J. Lataitis, 1989: Effect of finite sampling on atmospheric spectra. *Bound.-Layer Meteor., 47*, 337–347.

Kaimal, J.C., J.C. Wyngaard, Y. Izumi, and O.R. Coté, 1972: Spectral characteristics of surface layer turbulence. *Quart. J. Roy. Meteor. Soc., 98*, 563–589.

Lumley, J.L., and H.A. Panofsky, 1964: *The Structure of Atmospheric Turbulence.* Wiley-Interscience, New York, 239 pp.

McMillen, R.T., 1988: An eddy correlation technique with extended applicability to non-simple terrain. *Bound.-Layer Meteor., 43*, 231–245.

Pasquill, F., and F.B. Smith, 1983: *Atmospheric Diffusion* (3rd ed.), Ellis Horwood Ltd., Chichester, England, 437 pp.

Ramirez, R. W., 1985: *The FFT: Fundamentals and Concepts*. Prentice-Hall, Englewood Cliffs, NJ, 178 pp.

GENERAL LIST OF SYMBOLS

$\overline{(\qquad)}$ time average
$\underline{(\qquad)}$ running average
$(\qquad)'$ deviation from time average
$(\qquad)''$ deviation from spatial average
$\langle(\qquad)\rangle$ spatial average
$[(\qquad)]$ ensemble average in Lagrangian framework
A available thermal energy
c_p specific heat at constant pressure
C_n^2 structure parameter for refractive index
C_{TQ} structure parameter for temperature-humidity covariance
C_T^2 structure parameter for temperature
C_V^2 structure parameter for velocity
C_Q^2 structure parameter for humidity
$C_{uw}(f)$ stress cospectrum, function of f
$C_{w\theta}(f)$ heat flux cospectrum, function of f
$\hat{C}r_{\alpha\beta}(\kappa_1)$ two-sided cross spectrum of α and β, function of κ_1
$\hat{C}_{\alpha\beta}(r_1)$ two-sided cospectrum of α and β, function of r_1
$C_{\theta q}(\kappa_1)$ cospectrum of temperature and humidity, function of κ_1
C_d drag coefficient
$D_\alpha(r)$ structure function for variable α, function of r
e turbulent kinetic energy per unit mass, also base for natural logarithms $\simeq 2.72$ in Chapters 2 and 7
E vertical moisture flux
$E(\kappa)$ energy spectral density, function of κ
$E_{ij}(\mathbf{x}, \boldsymbol{\kappa})$ two-point spectrum tensor, generalized form
$\hat{E}_{\alpha\beta}(r_1)$ even part of $R_{\alpha\beta}(r_1)$
f cyclic frequency, also Coriolis parameter in Chapters 1 and 5
$F_\alpha(\kappa_1)$ one-sided spectrum of α, function of κ_1
$\hat{F}_\alpha(\kappa_1)$ two-sided spectrum of α, function of κ_1
g acceleration of gravity
G_s flux of heat into storage in the soil or vegetation
$G(z/L), H(z/L)$ cospectral coefficients for stress and heat flux cospectra
h depth of the stable boundary layer
H vertical heat flux

I	imbalance term in turbulent kinetic energy budget
k	von Karman constant
K_h, K_m, K_q	turbulent exchange coefficients for heat, momentum, and water vapor
L	Obukhov length
n	dimensionless surface layer frequency
n_i	dimensionless mixed layer frequency
N	Brunt-Väisälä frequency
N_θ	rate of dissipation of 1/2 $\times$ temperature variance
$\hat{O}_{\alpha\beta}(\kappa_1)$	odd part of $R_{\alpha\beta}(r_1)$
p	atmospheric pressure
q	specific humidity
q_*	surface layer humidity scale
Q_*	mixed layer humidity scale
$\hat{Q}_{\alpha\beta}(\kappa_1)$	two-sided quadrature spectrum of α and β, function of κ_1
r	distance between spatial correlation measurements
$\mathbf{r}$	separation vector
r_1	separation in the x direction
R_b	bulk Richardson number
R_f	flux Richardson number
Ri	gradient Richardson number
$R_{ij}(\mathbf{x}, \mathbf{r})$	two-point covariance tensor, generalized form
$R_{ij}(\mathbf{r})$	two-point covariance tensor in homogeneous turbulence
$R_L(\xi)$	Lagrangian correlation of $W(t)$, function of time lag ξ
R_n	net radiation flux
$R_{\alpha\beta}(r_1)$	cross covariance of α and β, function of r_1
$R_\alpha(\xi)$	autocovariance function, function of ξ
$S_\alpha(f)$	one-dimensional spectrum of α
$\hat{S}_\alpha(\omega)$	two-sided spectrum of α, function of ω
t	time
T	temperature
T_f	free convection layer scaling temperature
$T_r(z)$	hydrostatic adiabatic reference temperature, function of z
T_v	virtual temperature
T_*	surface layer scaling temperature
T_ϵ	eddy turnover time
u	longitudinal (streamwise) velocity component
u_f	free convection scaling velocity
u_*	surface layer scaling velocity (friction velocity)
v	lateral velocity component
w	vertical velocity component
w_*	mixed layer scaling velocity
$W(t)$	vertical velocity of Lagrangian particle, function of time
x	distance along the x direction
$\mathbf{x}$	position vector
x, y, z	longitudinal (streamwise), lateral, and vertical coordinate directions
z	height above ground
z_0	roughness length
z_h	boundary layer depth

z_i	height of capping inversion base taken as the height of the convective boundary layer
$Z(t)$	height of particle in the Lagrangian framework, function of time
α_1	Kolmogorov constant for velocity spectra
β	Bowen ratio
β_1	spectral constant for temperature (based on N_θ for 1/2 $\times$ temperature variance)
α, β	any two fluctuating variables
γ	dimensionless mixed layer temperature destruction rate
Γ	gamma function
ϵ	rate of dissipation of turbulent kinetic energy
η	Kolmogorov microscale
θ	potential temperature
θ_v	virtual potential temperature
θ_*	mixed layer scaling temperature
κ	wavenumber magnitude
$\boldsymbol{\kappa}$	wavenumber vector
κ_1	streamwise wavenumber component
λ	wavelength
λ_e	latent heat of vaporization
$(\lambda_m)_\alpha$	wavelength at spectral maximum for variable α
Λ_α	integral length scale for variable α
ν	kinematic viscosity
ν_e	extinction coefficient
ρ	density of air
σ	Stefan-Boltzmann constant
σ_α	standard deviation of variable α
σ_z	standard deviation of $Z(t)$
$\rho_\alpha(\xi)$	autocorrelation function for α, function of ξ
τ	shear stress
τ_0	surface shear stress
τ_x, τ_y, τ_z	shear stress components along x, y, and z directions
$\mathcal{T}_\alpha$	integral time scale for variable α
$\mathcal{T}_L$	Lagrangian time scale
ϕ	latitude
ϕ_h	dimensionless vertical temperature gradient
ϕ_m	dimensionless vertical wind shear
ϕ_q	dimensionless vertical humidity gradient
ϕ_t	dimensionless vertical turbulent transport of e
ϕ_N	dimensionless rate of destruction of 1/2 $\times$ temperature variance
ϕ_w	dimensionless form for σ_w
ϕ_ϵ	dimensionless rate of dissipation of turbulent kinetic energy
ϕ_θ	dimensionless form for σ_θ
ψ_ϵ	dimensionless mixed layer dissipation rate of turbulent kinetic energy
ψ_m	diabatic term in the wind profile equation
ω	circular frequency
Ω	earth's angular velocity

INDEX

www.ingramcontent.com/pod-product-compliance
Ingram Content Group UK Ltd.
Pitfield, Milton Keynes, MK11 3LW, UK
UKHW021308111225
9528UKWH00003B/32

9 780195 062397